D1292561

RECENT PROGRESS IN COMBINATORICS

CONTRIBUTORS

H. L. ABBOTT
RUTH BARI
CLAUDE BERGE
E. R. BERLEKAMP
N. G. DE BRUIJN
ROBERTO FRUCHT
J. M. GANDHI
B. GARDNER
ALLAN GEWIRTZ
JAY GOLDMAN
BRANKO GRÜNBAUM
RAM PRAKASH GUPTA
RICHARD K. GUY
R. HALIN
FRANK HARARY
D. A. HIGGS
ARTHUR M. HOBBS
A. J. HOFFMAN
MICHEL JEAN
J. G. KALBFLEISCH
LEROY M. KELLY
PATRICK A. KELLY
DAVID A. KLARNER
N. S. MENDELSOHN
U. S. R. MURTY
C. ST. J. A. NASH-WILLIAMS
OYSTEIN ORE
MICHAEL D. PLUMMER
LOUIS V. QUINTAS
J. M. S. SIMÕES PEREIRA
R. RADO
RONALD C. READ
FRED. S. ROBERTS
GIAN-CARLO ROTA
HORST SACHS
J. SCHÖNHEIM
J. J. SEIDEL
R. G. STANTON
DAVID. P. SUMNER
W. T. TUTTE
MARK E. WATKINS
P. H. WEILAND
D. H. YOUNGER
STEFAN ZNÁM

RECENT PROGRESS IN COMBINATORICS

Proceedings of the Third
Waterloo Conference on Combinatorics, May 1968

Edited by W. T. TUTTE

UNIVERSITY OF WATERLOO
WATERLOO, ONTARIO

QA
164
.W3
1968

1969

ACADEMIC PRESS New York and London

COPYRIGHT © 1969, BY ACADEMIC PRESS, INC.
ALL RIGHTS RESERVED
NO PART OF THIS BOOK MAY BE REPRODUCED IN ANY FORM,
BY PHOTOSTAT, MICROFILM, RETRIEVAL SYSTEM, OR ANY
OTHER MEANS, WITHOUT WRITTEN PERMISSION FROM
THE PUBLISHERS.

ACADEMIC PRESS, INC.
111 Fifth Avenue, New York, New York 10003

United Kingdom Edition published by
ACADEMIC PRESS, INC. (LONDON) LTD.
Berkeley Square House, London, W1X6BA

LIBRARY OF CONGRESS CATALOG CARD NUMBER: 69-18337
AMS 1968 SUBJECT CLASSIFICATION 0500

PRINTED IN THE UNITED STATES OF AMERICA

LIST OF CONTRIBUTORS

Numbers in parentheses indicate the pages on which the authors' contributions begin.

H. L. ABBOTT (211), University of Alberta, Edmonton, Alberta
RUTH BARI (217), George Washington University, Washington, D. C.
CLAUDE BERGE (49, 341), Institut Henri Poincaré, Paris, France
E. R. BERLEKAMP (3, 341), Bell Telephone Laboratories, Murray Hill, New Jersey
N. G. DE BRUIJN (59), Department of Mathematics, Technological University, Eindhoven, The Netherlands
ROBERTO FRUCHT (69), Universidad Técnica F. Santa Maria, Valparaiso, Chile
J. M. GANDHI (221), Department of Mathematics, University of Manitoba, Winnipeg, Manitoba
B. GARDNER (211), Memorial University of Newfoundland, St. John's, Newfoundland
ALLAN GEWIRTZ* (223), Mathematics Department, Pace College, New York, New York
JAY GOLDMAN (75), University of Colorado, Boulder, Colorado and Department of Statistics, Harvard University,Cambridge, Massachusetts
BRANKO GRÜNBAUM (85, 343), Department of Mathematics, University of Washington, Seattle, Washington
RAM PRAKASH GUPTA† (229) Department of Statistics, University of North Carolina, Chapel Hill, North Carolina
RICHARD K. GUY (237, 337), Department of Mathematics, University of Calgary, Calgary, Alberta
R. HALIN (91), University of Köln, Köln, Germany
FRANK HARARY (13), Department of Mathematics, University of Michigan, Ann Arbor, Michigan
D. A. HIGGS (245, 343), Department of Pure Mathematics, University of Waterloo, Waterloo, Ontario
ARTHUR M. HOBBS (255, 344), Faculty of Mathematics, University of Waterloo, Waterloo, Ontario

* Present address: Department of Mathematics, Brooklyn College of The City University of New York, Brooklyn, New York

† Present address: Department of Mathematics, Ohio State University, Columbus, Ohio

A. J. HOFFMAN (103), IBM Watson Research Center, Yorktown Heights, New York

MICHEL JEAN (265), Collège Militaire Royal, Saint-Jean, Québec

J. G. KALBFLEISCH (37, 273), University of Waterloo, Waterloo, Ontario

LEROY M. KELLY (111), Department of Mathematics, Michigan State University, East Lansing, Michigan

PATRICK A. KELLY (337), Department of Combinatorics and Optimization, University of Waterloo, Waterloo, Ontario

DAVID A. KLARNER* (337), McMaster University, Hamilton, Ontario

VICTOR KLEE (344), Department of Mathematics, University of Washington, Seattle, Washington

N. S. MENDELSOHN (123), University of Manitoba, Winnipeg, Manitoba

E. C. MILNER (345), Department of Mathematics, University of Calgary, Calgary, Alberta

U. S. R. MURTY (283), University of Waterloo, Waterloo, Ontario

C. ST. J. A. NASH-WILLIAMS (133, 346), Department of Combinatorics and Optimization, University of Waterloo, Waterloo, Ontario

OYSTEIN ORE† (287), Yale University, New Haven, Connecticut

MICHAEL D. PLUMMER‡ (287), Yale University, New Haven, Connecticut

LOUIS V. QUINTAS (223), Mathematics Department, Pace College, New York, New York

J. M. S. SIMÕES PEREIRA (295), University of Coimbra, Coimbra, Portugal

R. RADO (151), Department of Mathematics, The University, Reading, England

RONALD C. READ (161), University of the West Indies, Jamaica

FRED. S. ROBERTS (301), The RAND Corporation, Santa Monica, California

GIAN-CARLO ROTA (75), University of Colorado, Boulder, Colorado and Massachusetts Institute of Technology, Cambridge, Massachusetts

HORST SACHS (175), Mathematical Institute, Technical University, Ilmenau, German Democratic Republic

J. SCHÖNHEIM (311, 346), Department of Mathematics, University of Calgary, Calgary, Alberta and Tel Aviv University, Tel Aviv, Israel

J. J. SEIDEL (185), Department of Mathematics, Technological University, Eindhoven, The Netherlands

R. G. STANTON (21), Department of Mathematics, York University, Toronto, Ontario

* Present address: Mathematics Department, Technological University, Eindhoven, The Netherlands

† Deceased

‡ Present address: Department of Computer Science, City University of New York, New York

DAVID P. SUMNER (319), University of Massachusetts, Amherst, Massachusetts
W. T. TUTTE (199, 338), Department of Combinatorics and Optimization, University of Waterloo, Waterloo, Ontario
MARK E. WATKINS (323, 346), University of Waterloo, Waterloo, Ontario
P. H. WEILAND (37), York University, Toronto, Ontario
LOUIS WEINBERG (346), Department of Electrical Engineering, The City College of New York, New York, New York
D. H. YOUNGER (329), Department of Combinatorics and Optimization, University of Waterloo, Waterloo, Ontario
STEFAN ZNÁM (237), Slovenská Akadémie vied Kabinet Matematiky, Bratislava, Czechoslovakia

PREFACE

It is difficult to find a definition of combinatorics that is both concise and complete, unless we are satisfied with the statement "Combinatorics is what combinatorialists do." However, what combinatorialists usually do is study families of finite subsets of a given set or sets. Thus combinatorics includes graph theory, and a graph is often defined by a set of "vertices" and a family of pairs of vertices called "edges."

Some of us can remember a time when combinatorics was regarded as a deposit of a few amusing but trivial and many merely trivial problems, a refuge for a few eccentrics but not a field of activity for respectable mathematicians. We have watched with mingled awe and pleasure its development into a major field of mathematics, a mighty production line in which deep theorems, generated as the spin-off of the Four Color Problem, are applied to practical questions of transportation theory, communications theory, linear programming, and electrical engineering. And the field still has room for many workers and shows no signs of becoming worked out.

The Third Waterloo Conference on Combinatorics lasted from May 20 to May 31, 1968. It was supported by grants from N.A.T.O. and the National Research Council of Canada, and it was recognized as a N.A.T.O. Advanced Study Institute. Though most of the participants came from Canada and the United States we welcomed many distinguished visitors from outside North America.

No restriction was put upon the subject matter of the lectures, except that it should be "combinatorial." However, most of the lectures proved to be concerned with graph theory. Several lectures dealt with coloring problems, and in these and other lectures there were enough references to the Four Color Conjecture to justify the statement that this conjecture played a central though not dominating role.

Professor Oystein Ore told us something of his recent work in showing that a 5-chromatic planar map must have at least forty regions. His death a few months later was a heavy blow to the discipline of combinatorics, and is felt as a personal loss by most of its practitioners. It has prevented us from having an article in this book about the forty regions. Fortunately one such article exists, but it is being published elsewhere. We do have a joint paper by Ore and Plummer on another coloring problem.

Most of the lectures given at the conference are published in this book in revised, and sometimes in extended, form. Some, however, are to be published elsewhere, and these are represented here only by short abstracts. "Instructional courses" of five lectures each were given by F. Harary, E. R. Berlekamp, and R. G. Stanton. The other lectures were classified as "invited" (1 hour) and "contributed" (1/2 hour). Three lectures each were given by C. St. J. A. Nash-Williams and W. T. Tutte, of the University of Waterloo, and these lectures are counted here with the "invited" ones.

The conference took place in the new Mathematics and Computer Building of the University of Waterloo. By a happy coincidence participants in the conference were able to attend the official opening of this building on May 23, 1968.

W. T. TUTTE

June 1969

CONTENTS

CONTRIBUTED PAPERS

ABSTRACTS

UNSOLVED PROBLEMS

INSTRUCTIONAL COURSES AND RELATED PAPERS

A SURVEY OF CODING THEORY

E. R. Berlekamp

BELL TELEPHONE LABORATORIES
MURRAY HILL, NEW JERSEY

1. Nonconstructive Results

The subject of coding theory deals with the problem of transmitting information across noisy channels. The discrete, combinatorial case of this problem is formulated in terms of the following definitions.

A represents the input alphabet, which equals the symbol set or, equivalently, the set of letters.

$q = |A|$ is the alphabet size or, equivalently, the order of the alphabet.

B represents the output alphabet ($|B| \geq q$).

A "discrete memoryless channel" is a set of q probability distributions on B, one for each $a \in A$.

$n = N$ is the block length.

A codeword is an element of A^n.

A code is a set of codewords (typically $|\text{code}| \ll q^n$).

A list-L decoding algorithm is a function from B^n to $(\text{code} \cup \varnothing)^L$. ($\varnothing$ means "I haven't any idea," and L is the number of guesses decoder is allowed.)

$P_e(N, M, L)$ is the probability of error of the best code of length N with $|\text{code}| = M$ and best list-L decoding algorithm. ($\varnothing$ is counted as an error. If a transmitted word is on the decoder's list, then the decoder is considered correct. The dependence of P_e on the channel is implicit.)

The major results concerning the existence or nonexistence of good codes are as follows.

THEOREM [Shannon (1948).][1] *There exists a number C, called capacity, depending only on channel statistics, such that*

(a) *If* $R < C$, $\lim_{N\to\infty} P_e(N, [\exp RN], 1) = 0$.
(b) *If* $R > C$, $\lim_{N\to\infty} P_e(N, [\exp RN], 1) > 0$.

THEOREM [Wolfowitz (1961)].

(b′) *If* $R > C$, $\lim_{N\to\infty} P_e(N, [\exp RN], 1) = 1$.

THEOREM [Feinstein (1954, 1955)].

(*a*′) *There exists a function* $E(R, 1)$ *such that* $P_e(N, [\exp RN], 1) < \exp -NE(R, 1)$ *and* $E(R, 1) > 0$ *if* $R < C$.

Since Feinstein, more attention has been devoted to the question of how fast $P_e(N, M, L)$ approaches zero with increasing N and fixed $R = [\ln(M/L)]/N$. The following definitions have proved useful.

$$E(R, L) = \lim_{N\to\infty} -\frac{1}{N} \ln P_e(N, [L \exp RN], L).$$

[Note: The limit may not exist, but the bounds are valid anyway if properly interpreted as lim sup or lim inf.]

$$E(R, \infty) = \lim_{L\to\infty} E(R, L).$$

$$E(0^+, L) = \lim_{\varepsilon\downarrow 0} E(\varepsilon, L).$$

[Note: $E(0^+, L)$ is typically equal to or less than any of the many possible values of $E(0, L)$.]

$$E_M = \lim_{N\to\infty} -\frac{1}{N} \ln P_e(N, M, 1).$$

$$E_\infty = \lim_{M\to\infty} E_M.$$

The limit E_∞ is also known exactly, and it turns out that $E_\infty = E(0^+, 1)$. This point is called the "zero rate exponent." The function $E(R, \infty)$ is also known exactly at all rates between zero and capacity. This function is called the "sphere packing bound" or the "volume bound." In addition to giving the exact asymptotic form of the probability of decoding error with an infinite list, the volume bound serves as a convenient upper bound on $E(R, L)$ for

[1] References appearing after theorems are cited in full in the author's book "Algebraic Coding Theory," McGraw-Hill, 1968.

finite L. Surprisingly, it turns out that this bound is achieved exactly for sufficiently high rates. In fact, there exists an infinite sequence of rates, $0 \le R_\infty \le \cdots \le R_3 \le R_2 \le R_1 \le R_0 = C$ (capacity), such that if R is in the interval $R_L < R < C$, then $E(R, L) = E(R, \infty)$.

Several other bounds on $E(R, 1)$ are known. For most nonpathological channels, the general form of these bounds is shown in Fig. 1. For $0 < R < R_1$,

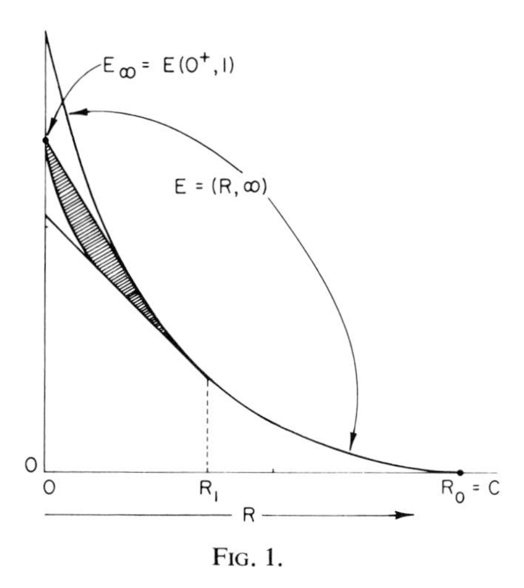

FIG. 1.

the function $E(R, 1)$ lies somewhere in the shaded region; for $R_1 < R < C$, $E(R, 1) = E(R, \infty)$.

Proofs of these results may be found in a number of books. The best are Jelinek's "Probabilistic Information Theory," McGraw-Hill, 1968, and Gallager's "Information Theory and Reliable Communications," Wiley, 1968.

Unfortunately, all of the known lower bounds on $E(R, 1)$ [i.e., upper bounds on $P_e(N, M, 1)$] have been derived by nonconstructive techniques. If we choose a long code at random, then there is a high probability that we will find a good code. However, it is very difficult to evaluate the quality of any particular code, except for a few poor codes which can easily be shown to be bad. Thus, very little is known about the "best code," although much is known about how good it is.

From a practical viewpoint, the greatest defect of the nonconstructive theory is that it fails to provide any feasible decoding algorithm. If we choose a code at random, we will probably obtain a good code. However, the decoding problem would probably require so much computation that we would still be unable to use the code for any practical purpose.

Because of these defects of the nonconstructive theory, considerable effort has been devoted to the construction of good, easily decodable codes. This effort has met with a certain degree of success, insofar as many classes of codes have been suggested, and simple decoding algorithms have been discovered for some of these codes. Also, some of the codes are now known to possess various properties of combinatorial interest. For short to moderate block lengths, the constructive codes are the best known, even though no one has yet been able to show that any of the known constructions succeed in attaining the nonconstructive bounds in the asymptotic limit of arbitrarily large block length. In fact, most of the constructions are known to fail this test, but there are a few constructions whose asymptotic properties are unknown.

2. Algebraic Codes

Algebraic codes comprise only a small subset of the codes defined in the beginning of Section 1. The additional structure imposes many restrictions, but it also provides additional tools by which we may attack the decoding problem.

Definitions. The symbol field is $GF(q)$, which equals the input alphabet.

$GF(q)$ or $GF(q) \cup \{?\}$ is equal to the output alphabet (" ?" is called "erasure" or "dit").

A linear code is a subspace of $(GF(q))^n$.

A codeword is a vector $\mathbf{C} = [C_0, C_1, C_2, \ldots, C_{n-1}]$, *or* a polynomial $C(x) = \sum_{i=0}^{n-1} C_i x^i$.

The Hamming weight of a codeword $\mathbf{C} = w_H(\mathbf{C}) = \sum_{i=0}^{n-1} |C_i|^0$, where $0^0 = 0$; and if $C_i \neq 0$, then $|C_i|^0 = 1$. The Hamming weight of a codeword is the number of its nonzero coordinates.

The Lee weight of $\mathbf{C} = w_L(\mathbf{C}) = \sum_{i=0}^{r-1} |C_i|$, where $|C_i| \equiv \pm C_i \mod q$ and $0 \leq |C_i| \leq q/2$.

k is the dimension of a linear code.

R is the rate of a linear code and equals k/n.

$$\left[\text{Note:} \qquad R = \frac{\log_q |\text{code}|}{n} \neq \frac{\ln |\text{code}|}{n} \quad \text{in nats.}\right]$$

$\mathscr{G}$ is the generator matrix of a linear code and is a matrix of k rows, and n columns, whose rows form a basis of the code.

$\mathscr{H}$ is the parity-check matrix of a code and equals $\mathscr{G}^{\perp}$, a matrix whose n–k linearly independent rows satisfy $\mathscr{H}\,\mathscr{G}^t = 0$.

The dual of the code generated by $\mathscr{G}$ is the code generated by $\mathscr{G}^{\perp}$.

Theorem. *For the binary symmetric channel of Fig. 2, linear codes are asymptotically as "good" as any known to exist.*

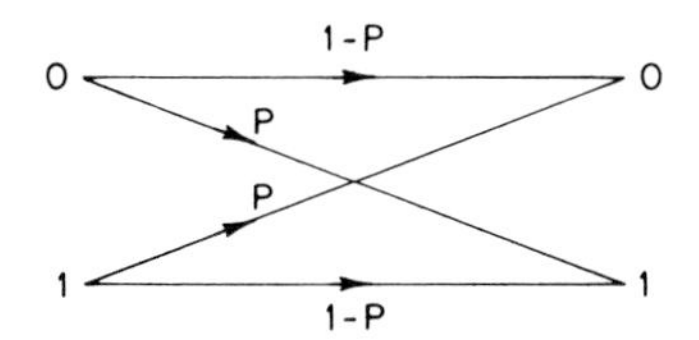

FIG. 2. The binary symmetric channel.

DECODING OF LINEAR CODES. Send **C**; channel adds error **E**, decoder receives $\mathbf{R} = \mathbf{C} + \mathbf{E}$ and computes *syndrome*

$$\mathbf{s}^t = \mathscr{H}\mathbf{R}^t = \mathscr{H}\mathbf{E}^t.$$

Given $\mathbf{s}^t$, find "most probable" **E** such that $\mathscr{H}\mathbf{E}^t = \mathbf{s}^t$ *or* "minimum weight" **E** such that $\mathscr{H}\mathbf{E}^t = \mathbf{s}^t$. (Note. $\dim \mathbf{s} = n\text{–}k < n$.)

Although the linear structure of the code allows some simplification of the decoding problem, additional structure is desirable. We assume that n and q are relatively prime. We may then consider codes with the following additional properties.

DEFINITIONS. A *cyclic code* is an ideal of polynomials mod $(x^n - 1)$.
A *negacyclic code* is an ideal of polynomials mod $(x^n + 1)$.
A *constacyclic code* is an ideal of polynomials mod $(x^n + \xi)$, $\xi \in GF(q)$.
α is a primitive root of $x^n - 1$ (or $x^n + 1$ or $x^n + \xi$).
m is the degree of the minimal polynomial in α over $GF(q)$.
$GF(q^m)$ is the locator field.
$g(x)$ is the generator polynomial and is equal to the monic polynomial of minimum degree in the code.
$h(x)$ is the check polynomial (*or* recursion polynomial) and equals $(x^n - 1)/g(x)$ [or $(x^n + \xi)/g(x)$].

THEOREM. *In* $GF(q^m)$, *for cyclic code,*

$$g(x) = \prod_{i \in K} (x - \alpha^i), \qquad h(x) = \prod_{i \in \bar{K}} (x - \alpha^i)$$

where K is a subset of integers mod n, *closed under multiplication by q.*

DEFINITIONS OF IMPORTANT CYCLIC CODES

DEFINITION. The set of powers of α which are roots of the generator polynomial of the BCH code (Bose–Chaudhuri–Hocquenghem) of designed distance d is

$$K = \{jq^i | 1 \leq j < d;\ i \text{ arbitrary}\}.$$

THEOREM. *The actual distance is larger than or equal to the designed distance of the* BCH *code.*

THEOREM. *If the designed distance of an asymptotically long* BCH *code is chosen as a fixed fraction of the block length, $d = un$, then the number of information symbols in the code, k, is approximately given by*

$$k \approx n^{s(u)}$$

where s is a known continuous (singular) function of u.

DEFINITION. An RS code (Reed–Solomon) is a BCH code with $n = q - 1$.

DEFINITION. The set of powers of α which are roots of the generator polynomial of the shortened GRM code (generalized Reed–Muller) of order r and length $q^m - 1$ is

$$K = \{j \mid 0 \leq w(j) < (q-1)m - r\}$$

where $w(j)$ is the real sum of digits of q-ary expansion of j.

DEFINITION. A GRM code is the direct sum of a shortened GRM code and an "all ones" vector of length q^m.

DEFINITION. An RM code is equal to a GRM code over $GF(2)$.

DEFINITION. The set of powers of α which are roots of the generator polynomial of the QR code (quadratic residue) is represented by

$$K = \{r \mid r \text{ is a quadratic residue mod } n\}.$$

(Defined only if n is prime and q is the quadratic residue mod n.)

VIEWPOINTS OF CYCLIC CODES

Viewpoint 1:

DEFINITIONS. $C(x)$ is the codeword polynomial,

$$\sum_{i=0}^{n-1} C_i x^i, \qquad C_i \in GF(q).$$

The nonzero codeword locations are equal to $\{\alpha^i \mid C_i \neq 0\}$.
The *codeword locator polynomial*

$$\sigma(z) = \prod_{X_i = \text{codeword location}} (1 - X_i z) = \sum_{j=0}^{\deg \sigma} \sigma_j z^j.$$

The σ_j are elementary symmetric functions of locations, $\sigma_j \in GF(q^m)$. The *codeword evaluator polynomial*

$$\omega(z) = \sigma(z) + \sum_{\substack{X_i = \text{codeword} \\ \text{location}}} zX_i C_i \prod_{j \neq i} (1 - X_j z).$$

The *weighted power-sum symmetric functions*

$$S_j = \sum_{x_i = \text{locations}} C_i X_i^{\,j}; \qquad S(z) = \sum_{j=1}^{\infty} S_j z^j.$$

THEOREM. $S_j = C(\alpha^j)$.

THEOREM. $(1 + S(z))\sigma(z) = \omega(z)$.

DECODING BCH CODES. Compute $E(\alpha^j) = R(\alpha^j)$ for $j = 1, 2, \ldots, 2t$. Find the error-locator $\sigma(z)$ and error-evaluator $\omega(z)$ by solving $(1 + S(z))\sigma(z) \equiv \omega(z) \bmod z^{2t}$. Find the reciprocal roots of $\sigma(z)$ (error locations). Find the error values from $\omega(z)$ and reciprocal roots of $\sigma(z)$.

Viewpoint 2:

Instead of $C(x)$, we may consider the Mattson–Solomon polynomial,

$$MS(x) = \frac{1}{n} \sum_{j=0}^{n-1} S_j x^{n-j}.$$

THEOREM. $C_j = MS(\alpha^j)$.

THEOREM. *For a* BCH *codeword*, $\deg MS(x) \leq n\text{–}d$.

Viewpoint 3:

Instead of $C(x)$, for the RM code $n = 2^m$, let $C_j = F(X_1, X_2, \ldots, X_m)$, where F is an m-variate binary and $\alpha^j = \sum_{i=1}^{m} X_{i,\,j} \alpha^i$.

THEOREM. *If $C(x)$ is an element of the* rth *order* RM *code, then F is a polynomial of degree less than or equal to r in $X_1, X_2, \ldots, X_m$. Locations of any minimum weight codeword form an affine subspace over $GF(2)$.*

Weight Distribution Theorems

Theorem (MacWilliams). *Let A_i be equal to the number of codewords of weight i in linear code and B_i be the number of codewords of weight i in dual code,*

$$A(z) = \sum_i A_i z^i, \qquad B(z) = \sum_i B_i z^i,$$

then

$$A(z) = q^k \left(q^{-1} + \frac{q-1}{q} z \right)^n B\left(\frac{1-z}{1+(q-1)z} \right).$$

Theorem (Pless).

$$\sum_{j=0}^{n} (2w - 2j)^r A_j = q^k \sum_{j=0}^{r} B_j F_r^{(j)}(n)$$

where w may be any constant, and

$$F_r^{(j)}(n) = \frac{d^r}{dx^r} \left[\left(q^{-1} + \frac{q-1}{q} \exp -2x \right)^{n-j} \times (q^{-1} - q^{-1} \exp -2x)^j \exp 2wx \right]_{x=0}.$$

Weight Restriction Theorems

Theorem (McEliece). *If there does not exist any $\beta_1, \beta_2, \ldots, \beta_j$ such that $h(\beta_i) = 0$, $i = 1, 2, \ldots, j$, and $\prod_{i=1}^{j} \beta_i = 1$, then the weight of every codeword in the binary code with check polynomial $h(x)$ is a multiple of 2^j.*

Theorem (Kasami). *In second order* RM *code, all weights are of the form $w = 2^{m-1} \pm \varepsilon 2^{(m+i)/2-1}$ where $i \equiv m \bmod 2$ and $\varepsilon = 0 \;\; +1$, or -1.*

Theorem (Carlitz). *If w is a weight of a codeword in the dual code of an extended binary* BCH *code of length 2^m and designed distance d, then*

$$|2^{m-1} - w| \le (d-1)2^{m/2}.$$

Kasami first suggested combining best results known on minimum weights and weight restrictions (on both code and dual) with the Pless theorem. This enables us to obtain complete weight enumerations of many binary codes, including all of the following:

BCH codes of length $2^m - 1$, $d = 1, 3, 5, 7$, m arbitrary;
First and second order RM codes;
various low-rate BCH codes which are subcodes of second order RM code;
QR codes of length less than or equal to 71.

Proofs, references, and a detailed discussion of algebraic coding theory may be found in my book, "Algebraic Coding Theory," McGraw-Hill, 1968. Many unsolved problems may be found on pages listed in the index under " unsolved problems."

THE GREEK ALPHABET OF "GRAPH THEORY"

*Frank Harary**

DEPARTMENT OF MATHEMATICS
UNIVERSITY OF MICHIGAN
ANN ARBOR, MICHIGAN

The purpose of this series of lectures was to develop a coherent consistent system of notation accompanied by a natural intuitive terminology for the invariants and other properties of graph theory This was accomplished by following the system used in *Graph Theory*, [11]. Although both the Greek and Roman alphabets were exploited in these talks (the efforts of the 100 students present to add the Cyrillic alphabet and also that of a dialect of Chinese were largely ignored), only the Greek alphabet and its use in *Graph Theory* is reported in this article. As every mathematics student must know, the Greek alphabet is

alpha	α	A	iota	ι	I	rho	ρ	P
beta	β	B	kappa	κ	K	sigma	σ	Σ
gamma	γ	Γ	lambda	λ	Λ	tau	τ	T
delta	δ	Δ	mu	μ	M	upsilon	υ	Υ
epsilon	ε	E	nu	ν	N	phi	φ	Φ
zeta	ζ	Z	xi	ξ	Ξ	chi	χ	X
eta	η	H	omicron	o	O	psi	ψ	Ψ
theta	θ	Θ	pi	π	Π	omega	ω	Ω

By definition,[1] a *graph* G consists of a *finite* set V of *points* (or *vertices* or *nodes* or 0-*simplexes*) together with a prescribed subset X of $V^{(2)}$, the collection of 2-subsets of V. The elements of X are called the *lines* (or *edges* or *arcs* or 1-*simplexes*) of G. Therefore it is not necessary to include here the

*The preparation of this article was supported in part by a grant from the United States Air Force Office of Scientific Research.

[1] This definition was accepted by almost all the other speakers, who began, "In this talk I will use the Michigan definition of a graph."

traditional sentence of the form, "The only graphs which we consider have a finite number of vertices and edges and contain no loops or multiple edges."

If $\{u, v\} = x \in X$, then u and v are *adjacent points*, and point u and line x are *incident*. Two graphs G_1 and G_2 are *isomorphic*, written $G_1 \cong G_2$, if there exists a one-to-one correspondence between their point sets V_1 and V_2 which preserves adjacency. Of course isomorphism is an equivalent relation. An *invariant* of a graph G is a number (or sequence of numbers) associated with some property of G which is constant over the isomorphism class of G.

Because we are developing the properties of G expressed by the Greek alphabet, we begin with *alpha* and *beta*. To do this, we need to say that (1) if point u and line x are incident, then they *cover* each other, (2) a set $U \subset V$ of *points* is *independent* if no two are adjacent in G, (3) two distinct lines x, y are *adjacent lines* if $x \cap y$ is a singleton, (4) a set of *lines* is *independent* if no two are adjacent, (5) two adjacent points *cover each other* and so do two adjacent lines *cover each other*, and for completeness, (6) every point covers itself and every line does likewise.

Then the alphabetically first invariant of a graph G is its *point covering number* $\alpha_0 = \alpha_0(G)$, which is the minimum number of points needed to cover all the lines Analogously, $\alpha_1 = \alpha_1(G)$ is the *line covering number* which is the minimum number of lines needed to cover all the points. Continuing, $\beta_0 = \beta_0(G)$ is the *point independence number*, the maximum number of independent points of G; similarly $\beta_1 = \beta_1(G)$ is the *line independence number*. Throughout this article $p = p(G)$ will denote $|V|$, the number of points of G. The fundamental theorem linking these five invariants was discovered by Gallai [6], who proved that for every graph G,

$$\alpha_0 + \beta_0 = p = \alpha_1 + \beta_1. \tag{1}$$

These equations were independently but subsequently discovered in a joint effort by Harary and Norman who were fortunate that P. Halmos sent their manuscript to a referee acquainted with Gallai's paper.

The alpha-invariants can be refined further by introducing α_{00} as the minimum number of points of G needed to cover V, and similarly α_{11} as the number of lines needed to cover X. In addition one can define α'_{00} as the minimum number of *independent* points needed to cover V. (It is easy to verify that α'_{00} is a meaningful invariant for any graph.) Similarly one can denote by α_{11} and α'_{11} the corresponding invariants for lines covering the lines of G. In these terms one could write $\alpha_0 = \alpha_{01}$ since it involves points which cover lines, and similarly write $\alpha_1 = \alpha_{10}$. Then the four invariants α_{ij} with $i, j = 0$ or 1 can be called the *i, j covering numbers* of G. On the surface, it appears that there exist two additional covering numbers for a graph, namely α'_{00} and α'_{11}. The graph G of Fig. 1 shows that in the obvious inequality $\alpha_{00} \leq \alpha'_{00}$, strict inequality can occur. On the other hand, Gupta [7] found

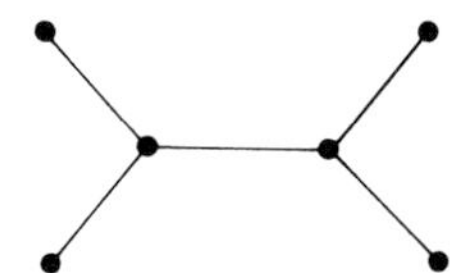

FIG. 1. A graph to illustrate α and β numbers.

that $\alpha'_{11} = \alpha_{11}$ for all graphs Thus we have so far the five covering invariants α_0, α_1, α_{00}, α'_{00}, and α_{11}.

$$\begin{array}{llll} \alpha_0 = 2 & \alpha_1 = 4 & \alpha_{00} = 2 & \alpha_{11} = 1 \\ \beta_0 = 4 & \beta_1 = 2 & \alpha'_{00} = 3 & \end{array} \tag{2}$$

Even this list does not include all covering numbers which have been used in the literature The *complete graph* K_p has every pair of points adjacent (see Fig. 2). The *complete covering number* α_K of G is the smallest number of

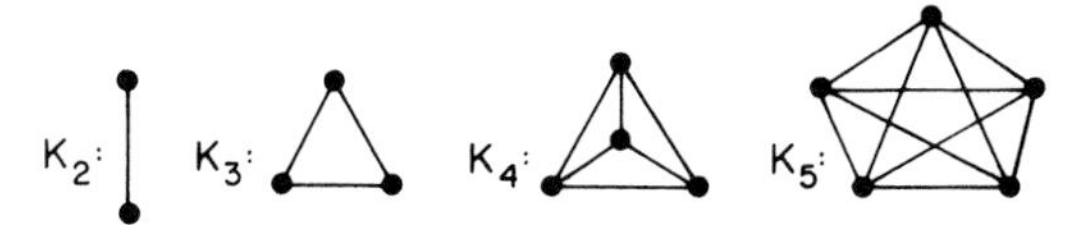

FIG. 2. Some complete graphs.

complete subgraphs whose union contains all the points and lines of G. This particular invariant has appeared in the literature under another name is a slightly different setting. A *clique* of G is a maximal complete subgraph. It is easy to verify that α_K is just the number of cliques contained in G.

One final covering number is suggested by α_K. We can define α_K', the *line-disjoint complete covering number* of G, to be the smallest number of line-disjoint complete subgraphs whose union is G. Taking a graph at random in Fig. 3, we see that $\alpha_K = 2$ by taking the two triangles as the complete subgraphs. However $\alpha_K' = 3$, demonstrating that the obvious inequality $\alpha_K \leq \alpha_K'$ is not an equality.

FIG. 3. The random graph.

As a semantic aid to the reader, we note that Berge [4] calls α_{00} the "coefficient of external stability," and β_0 the "coefficient of internal stability."

The next letter γ places us quickly in the midst of topological graph theory because of the first letter of the word, genus. The *genus* $\gamma(G)$ of a given graph G is the minimum genus n of an orientable surface S_n (sphere with n handles) on which G can be embedded with no pair of its edges intersecting (except at a vertex of course).

The genus of several families of graphs has only recently been determined. The *complete bipartite graph* $K_{m,n}$ (or more briefly the *complete bigraph*) has $p = m + n$ points partitioned into two sets V_1 and V_2 of m and n points such that two points are adjacent if and only if one is in V_1 and the other in V_2 (see Fig. 4).

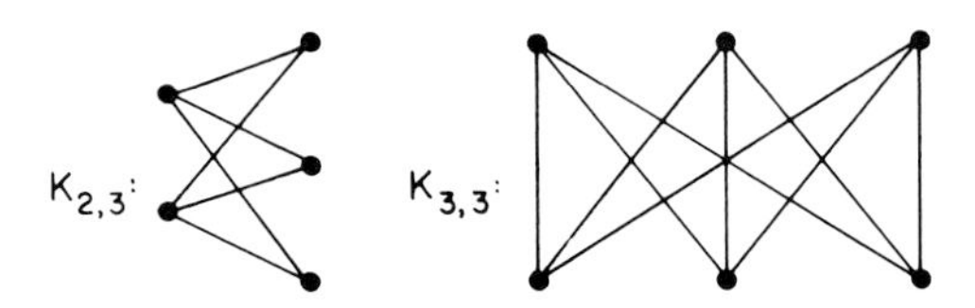

FIG. 4. Two complete bigraphs.

The *n-cube* Q_n is that graph with $p = 2^n$ points which are all the binary sequences (of zeros and ones) of length n, with two points adjacent whenever their sequences differ in exactly one place. The genus of the complete graph K_p has been a long unsolved problem finally settled in late 1967 by the efforts of Ringel and Youngs [19]. In this formula, $\{r\}$ is the least integer not less than the real number r,

$$\gamma(K_p) = \left\{\frac{(p-3)(p-4)}{12}\right\}. \tag{3}$$

The genus of the complete bigraph $K_{m,n}$ was determined by Ringel [18] and that of the n-cube by Beineke and Harary [5],

$$\gamma(K_{m,n}) = \left\{\frac{(m-2)(n-2)}{4}\right\}, \tag{4}$$

$$\gamma(Q_n) = 1 + (n-4)2^{n-3}. \tag{5}$$

Although they are not in alphabetical order, it is logically convenient to handle the other topological invariants of a graph next. The reader will recall Kuratowski's theorem [14] that a graph is planar if and only if it has no subgraph homomorphic to K_5 or $K_{3,3}$ (see Figs. 2 and 4).

The *thickness* of a graph G, denoted by $\theta(G)$, is the smallest number of planar subgraphs whose union is G. The thickness $\theta(K_p)$ of the complete graph was investigated by Beineke [1]. For $p \neq 9$ and $p \not\equiv 4 \bmod (6)$, we have

$$\theta(K_p) = \left[\frac{p+7}{6}\right]. \tag{6}$$

Some of the exceptional values are $\theta(K_4) = 1$, $\theta(K_9) = \theta(K_{10}) = 3$, and $\theta(K_{22}) = 4$. The value of $\theta(K_{16})$ is still not known, but is conjectured to be 4.

It has been shown by Beineke that $\theta(K_{28}) = 6$, and by Jean Mayer (unpublished) that $\theta(K_{34}) = 7$ and $\theta(K_{40}) = 8$.

The thickness $\theta(K_{m,n})$ of the complete bigraph was also studied by Beineke [2], who found that

$$\theta(K_{m,n}) = \left\{\frac{mn}{2(m+n-2)}\right\}, \tag{7}$$

except possibly when $m < n$, mn is odd, and there exists an integer k such that $n = [2k(m-2)/(m-2k)]$.

The corresponding problem for the cube was settled by Kleinert [13]:

$$\theta(Q_n) = \left\{\frac{n+1}{4}\right\}. \tag{8}$$

The *coarseness* $\xi(G)$ is the maximum number of nonplanar subgraphs whose union is G. The formulas for $\xi(K_p)$ and $\xi(K_{m,n})$ which have been proved for coarseness are rather complicated and not quite complete. They appear in two papers by Beineke and Guy [3, 9].

The *crossing number* $\nu(G)$ is the minimum number of pairs of edges which must intersect each other when G is drawn in the plane. Only inequalities are known for the graphs K_p and $K_{m,n}$ as reported by Guy [8]:

$$\nu(K_p) \leq \frac{1}{4}\left[\frac{p}{2}\right]\left[\frac{p-1}{2}\right]\left[\frac{p-2}{2}\right]\left[\frac{p-3}{2}\right], \tag{9}$$

$$\nu(K_{m,n}) \leq \frac{1}{2}\left[\frac{m}{2}\right]\left[\frac{m-1}{2}\right]\left[\frac{n}{2}\right]\left[\frac{n-1}{2}\right]. \tag{10}$$

The next letter not yet discussed is Γ. An *automorphism* of a graph G is an isomorphism of G onto itself. It is well-known that the set of all automorphisms of G form a permutation group which acts on the set of points $V(G)$. We write $\Gamma(G)$ for the (automorphism) *group of* G. The group $\Gamma(G)$ is sometimes written $\Gamma_0(G)$ to emphasize that it acts on the points of G. Two related groups are $\Gamma_1(G)$, the group induced by $\Gamma(G)$ acting on $X(G)$, and $\Gamma_{0,1}(G)$, the group induced by $\Gamma(G)$ and acting on $V(G) \cup X(G)$. These three different groups are useful in graphical enumeration problems, as can be seen, for example, in the work of Robinson [20].

The next letter in the alphabet, *delta*, is most easily discussed in connection with *pi*. The *degree* d_i of the point v_i is the number of lines incident with it. Then $\pi(G) = (d_1, d_2, \ldots, d_p)$ is defined to be the *degree sequence* of G. It is customary to label the points of G so that $d_1 \geq d_2 \geq \cdots \geq d_p$; with this convention, we define $\Delta(G) = d_1$, the *maximum degree* of G, and $\delta(G) = d_p$, the *minimum degree* of G.

The first theorem of graph theory, due to Euler [5], states that

$$2q = \sum_{i=1}^{p} d_i . \tag{11}$$

In this situation we say that the sequence of numbers $(d_1, \ldots, d_p)$ forms a *partition* of the number $2q$. This explains the use of the letter π for the degree sequence.

Closely related to δ are the invariants κ, λ, μ. Let u and v be two nonadjacent points of G. The *local point connectivity* $\kappa(u, v)$ is defined as the smallest number of points whose removal separates u and v. The *point connectivity* $\kappa(G)$ of a graph G is then defined to be the minimum of $\kappa(u, v)$ over all nonadjacent points u and v if G is not complete; we define $\kappa(K_p) = p - 1$. In an analogous manner, the *local line connectivity* $\lambda(u, v)$ is the smallest number of lines whose removal separates u and v. The *line connectivity* $\lambda(G)$ of a graph G is the minimum over all pairs of points u and v of $\lambda(u, v)$. For every connected, nontrivial graph G, we have the relationship

$$0 < \kappa \leq \lambda \leq \delta. \tag{12}$$

Moreover, for any three integers a, b, and c satisfying $0 < a \leq b \leq c$, there is a graph G with $\kappa(G) = a$, $\lambda(G) = b$, and $\delta(G) = c$.

Two paths from u to v are called *point disjoint* if u and v are the only points they have in common. We define $\mu_0(u, v)$ to be the maximum number of point-disjoint u–v paths. Likewise $\mu_1(u, v)$ is the maximum number of line-disjoint u–v paths. The far-reaching theorem due to Menger [15] is that

$$\kappa(u, v) = \mu_0(u, v). \tag{13}$$

The analogous result for lines is also true,

$$\lambda(u, v) = \mu_1(u, v). \tag{14}$$

A survey of related results may be found in [10].

After these latest detours, we continue alphabetically with *epsilon*. An *elementary homomorphism* ε of a graph G is an identification of two nonadjacent points. A *homomorphism* of G, denoted by φ, is a sequence of elementary homomorphisms. The path P_4 with four points has just four homomorphic images, as seen in Fig. 5.

Homomorphisms are closely related to colorings. A *coloring* of a graph is an assignment of colors to its points so that no two adjacent points have the

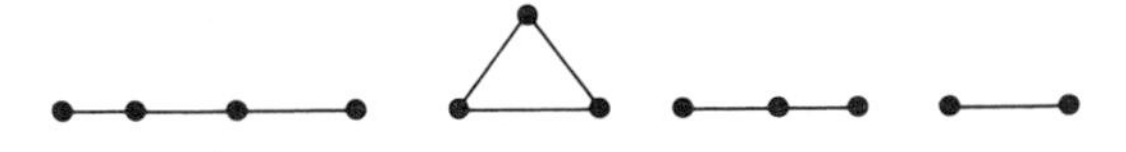

FIG. 5. The homomorphic images of P_4.

same color; an *n-coloring* uses n colors. The *chromatic number* $\chi(G)$ is defined as the minimum n for which G has an n-coloring. A homomorphism φ of G is *complete of order n* if $\varphi(G) = K_n$. Thus any complete homomorphism φ of order n corresponds to an n-coloring of G since the points of K_n can be regarded as colors, and by definition of homomorphism no two points of G with the same color are adjacent. It is natural to consider the maximal order of all complete homomorphisms of G. This new invariant, $\psi(G)$, is called the *achromatic number* of G. It is clear that

$$\chi(G) \leq \psi(G) \leq p. \tag{15}$$

A more important result [12], called the "homomorphism interpolation theorem," states that for each integer n with $\chi(G) \leq n \leq \psi(G)$, there exists a complete homomorphism of G of order n.

We can also define a *line-coloring* of a graph G as an assignment of colors to its lines so that no two adjacent lines have the same color; an *n-line-coloring* uses n colors. The *line-chromatic number* $\chi'(G)$ is the minimum n for which G has an n-line-coloring. The following result, due to Vizing, is presented in Ore [17, p. 248],

$$\Delta \leq \chi' \leq \Delta + 1. \tag{16}$$

The next Greek letter was chosen for its suggestive shape rather than for any letter in the name of its invariant, the reason being that capital *upsilon* has a tree-like appearance. A subgraph of G is called a *forest* if it contains no cycles. It is *spanning* if it contains all the points of G. Then clearly any graph can be expressed as a union of spanning forests by letting each forest contain only one line of G. The *arboricity* $\Upsilon(G)$ is the minimum number of line-disjoint spanning forests into which G can be decomposed. A formula for the arboricity of any graph was discovered by Nash-Williams [16],

$$\Upsilon(G) = \max_n \left\{ \frac{q_n}{(n-1)} \right\}, \tag{17}$$

where q_n is the maximum number of lines in any subgraph of G having n points.

We naturally conclude with *omega*. Let F be a family of distinct, non-empty sets S_i whose union is S. The *intersection graph* of F, denoted $\Omega(F)$, has the sets in F as its points, with S_i and S_j adjacent whenever $i \neq j$ and $S_i \cap S_j \neq \varnothing$. (A capital *omega* looks like an intersection sign.) Clearly every graph is isomorphic to an intersection graph. Thus it makes sense to define the *intersection number* $\omega(G)$ as the minimum number of elements in a set S such that G is an intersection graph on S. One easily obtainable result is that

$$\omega(G) \leq q + p_0 \tag{18}$$

where q is the number of lines and p_0 the number of isolated points. Consequently, every connected nontrivial graph G has $\omega(G) \leq q$.

It is to be hoped that this summary of the use of Greek letters in *Graph Theory* will constitute a modest contribution toward the uniformization of notation in graph theory, rather than an increase in confusion.

REFERENCES

1. L. W. BEINEKE, The Decomposition of Complete Graphs into Planar Subgraphs, in in *Graph Theory and Theoretical Physics* (F. Harary, ed.), Chapt. 4, pp. 139–153, Academic Press, New York, 1967.
2. L. W. BEINEKE, Complete Bipartite Graphs: Decomposition into Planar Subgraphs, in *A Seminar in Graph Theory* (F. Harary, ed.), Chapt. 7, pp. 42–53, Holt, New York, 1967.
3. L. W. BEINEKE and R. K. GUY, The coarseness of $K_{m,n}$. *Canad. J. Math.* (to appear).
4. C. BERGE, *The Theory of Graphs and its Applications*, Methuen, London, 1962.
5. L. EULER, Solutio Problematis ad Geometriam Situs Pertinentis, *Comment. Academaie Sci I. Petropolitanae* **8** (1736), 128–140.
6. T. GALLAI, Über extreme Punkt und Kantenmengen, *Ann. Univ. Sci. Budapest, Eötvös Sect. Math.* **2** (1959), 133–138.
7. R. P. GUPTA, Independence and Covering Numbers of Line Graphs and Total Graphs, in *Proof Techniques in Graph Theory* (F. Harary, ed.), pp. 61–62. Academic Press, New York, 1969.
8. R. K. GUY, The Decline and Fall of Zarankiewicz' Theorem, in *Proof Techniques in Graph Theory* (F. Harary, ed.), pp. 63–69. Academic Press, New York, 1969.
9. R. K. GUY and L. W. BEINEKE, The Coarseness of the Complete Graph, *Canad. J. Math.* **20** (1968), 888–894.
10. F. HARARY, Variations on a Theorem by Menger, *J. SIAM Appl. Math.* (to appear).
11. F. HARARY, *Graph Theory*, Addison-Wesley, Reading, Massachusetts, 1969.
12. F. HARARY, S. Hedetniemi, and G. Prins, An Interpolation Theorem for Graphical Homomorphisms, *Portugal. Math.* (to appear).
13. M. KLEINERT, Die Dicke des n-dimensionalen Würfel-Graphen, *J. Combinatorial Theory* **3** (1967), 10–15.
14. K. KURATOWSKI, Sur le Problème des Courbes Gauches en Topologie, *Fund. Math.* **15** (1930), 271–283.
15. K. MENGER, Zur allgemeinen Kurventheorie. *Fund. Math.* **10** (1927), 96–115.
16. C. ST. J. A. NASH-WILLIAMS, Edge-Disjoint Spanning Trees of Finite Graphs, *J. London Math. Soc.* **12** (1960), 555–567.
17. O. ORE, *The Four-Color Problem*, Academic Press, New York, 1967.
18. G. RINGEL, Das Geschlecht des vollständiger paaren Graphen, *Abh. Math. Sem. Univ. Hamburg* **28** (1965), 139–150.
19. G. RINGEL, and J. W. T. YOUNGS, Solution of the Heawood Map-Coloring Problem, *Proc. Nat. Acad. Sci. U.S.A.* **60** (1968), 438–445.
20. R. W. ROBINSON, Enumeration of Colored Graphs, *J. Combinatorial Theory* **4** (1968), 181–190.

COVERING THEOREMS IN GROUPS (or: How to Win at Football Pools)

R. G. Stanton

DEPARTMENT OF MATHEMATICS
YORK UNIVERSITY, TORONTO, ONTARIO

1. Introduction-The Group Formulation

Taussky and Todd [7] introduced the problem of group coverings. One considers an Abelian group with n base elements of order p (not necessarily a prime). Call these (independent) elements $g_1, g_2, \ldots, g_n$; a typical element in G then has the form

$$g_1^{\alpha_1} g_2^{\alpha_2} \cdots g_n^{\alpha_n}.$$

Now let S be the set of all powers of the base elements (counting 1 only once). S contains 1 and all elements

$$g_i^{\alpha_i} \qquad (\alpha_i \neq 0;\ i = 1, 2, \ldots, n).$$

Clearly there are

$$v = 1 + n(p - 1)$$

elements in the set S.

The problem posed by Taussky and Todd is to find the integer $\sigma(n, p)$, defined as the minimal number of elements in a subset H of G with the property that every element $g \in G$ can be written in the form $g = hs$, where $h \in H$, $s \in S$. Thus $G = HS$.

Such a set H is called a *minimal covering set*; when the meaning is clear from the context, we simply use the term *covering set*.

If we add the restriction that H be a subgroup, we denote the cardinality of the minimal covering set by $\sigma^*(n, p)$. Clearly

$$\sigma(n, p) \leq \sigma^*(n, p).$$

2. The Combinatorial Formulation

Taussky and Todd [8] reformulated the problem in combinatorial terms (actually, they originally encountered the problem in this context, in connection with a copyright problem that arose in connection with football pools).

Suppose that we represent the elements of G by vectors $(x_1, x_2, \ldots, x_n)$. We let x_i range from 0 to $p - 1$, and add vectors in the usual way. Obviously, we are really just looking at the exponents in the representation using the g_i's of Section 1.

We say that a vector $X = (x_1, \ldots, x_n)$ *covers* a vector $Y = (y_1, \ldots, y_n)$ if $x_i = y_i$ for at least $n - 1$ values of the subscript i (thus X and Y match in all components, or in all components but one). A minimal covering set is then defined to be a set of $\sigma(n, p)$ vectors, call it H, such that the vectors of H cover all vectors of G, and also $\sigma(n, p)$ is as small as possible. Again, we refer to $\sigma^*(n, p)$ as the cardinality of the minimal subset H which is a group (under vector addition of components—modulo p).

It is easy to note that any specific vector can cover itself and also $n(p - 1)$ other vectors obtained from it by permitting any one of the $p - 1$ possible alterations in each of its n components. The total number of vectors coverable by a specific vector is thus

$$v = 1 + n(p - 1),$$

and so we have the basic inequality

$$\sigma(n, p) \geq p^n/v. \tag{1}$$

Furthermore, if we have any covering set for vectors with $n - 1$ components, we can derive from it a covering set for vectors with n components by using all p possible values of the nth component along with each vector of the covering set for $n - 1$. Thus

$$\sigma(n, p) \leq p\sigma(n - 1, p). \tag{2}$$

For instance, $\sigma(3, 2) = 2$, since the vectors (0, 0, 0) and (1, 1, 1) certainly cover the set (0, 0, 0); (0, 0, 1), (0, 1, 0), (1, 0, 0); (0, 1, 1), (1, 0, 1), (1, 1, 0); (1, 1, 1). Consequently, $\sigma(4, 2) \leq 4$, since we can derive a covering set (possibly not minimal) by taking

$$\left(0, 0, 0, {0 \atop 1}\right) \quad \text{and} \quad \left(1, 1, 1, {0 \atop 1}\right).$$

Actually, since $\sigma(4, 2) \geq 2^4/[1 + 4] > 3$, we see that we can not get by with fewer vectors; so $\sigma(4, 2) = 4$.

3. The Mauldon-Zaremba-Mattioli Theorem

In 1950, papers [5], [6], and [10] discussed independently the special case when p is a prime; we follow the presentation given in [10]. Our hypothesis is that p is a prime, and that $v = 1 + n(p - 1) = p^r$, that is, v divides p^n.

Theorem. If p is a prime and $v = p^r$, then $\sigma(n, p) = \sigma^*(n, p) = p^{n-r}$.

Proof: Define an abstract Abelian group A generated by r independent generators c_i of order p; then A is of order p^r, and contains elements

$$\prod c_i^{\alpha_i}, \qquad 0 \leq \alpha_i \leq p - 1.$$

A contains $p^r - 1$ elements of order p (all but the identity). Pick an element a_1, and let A_1 denote its powers $a_1, a_1^2, \ldots, a_1^{p-1}$. Then pick a_2 not in A_1, and define A_2 as the set of all powers of a_2 other than 1. In this way, A can be stratified into the identity and a number of disjoint sets A_i; the number of such A_i is

$$(p^r - 1)/(p - 1) = n.$$

Thus we have n elements $a_1, \ldots, a_n$.

Now let θ be a mapping of G onto A by

$$\prod_{i=1}^{n} g_i^{\alpha_i} \xrightarrow{\theta} \prod_{i=1}^{n} a_i^{\alpha_i}.$$

It is easily verified that θ is a homomorphism. We define $H = \ker(\theta)$, i.e., H is made up of all elements of G mapped on 1 in A.

To show that $HS = G$, note that

$$g = \prod_{i=1}^{n} g_i^{\alpha_i} \xrightarrow{\theta} \prod_{i=1}^{n} a_i^{\alpha_i} = a_k^{\,e}.$$

Hence

$$\prod_{i \neq k} g_i^{\alpha_i} \to h \in H, \qquad \text{and} \qquad g = hg_k^{\,e}(g_k^{\,e} \in S).$$

Indeed, if we now write $g = h_1 s_1$, we see that

$$g \xrightarrow{\theta} s_1^{\,\theta}.$$

If $s_1^{\,\theta} = a_k^{\,e}$, then $s_1 = g_k^{\,e}$; hence s_1 and h_1 are uniquely determined, and g can be factored uniquely into an element in this set H and an element in S.

As an example, consider $\sigma(4, 3)$; G has base elements g_1, g_2, g_3, g_4, and $p^n/v = 9$. We define A as made up of

$$1, c_1, c_2, c_1c_2, c_1^{\,2}, c_2^{\,2}, c_1^{\,2}c_2, c_1c_2^{\,2}, c_1^{\,2}c_2^{\,2}.$$

Stratify A as follows:

$$\begin{array}{l} 1 \\ c_1, c_1^2 \\ c_2, c_2^2 \\ c_1^2c_2^2, c_1c_2 \\ c_1^2c_2, c_1c_2^2 \end{array}$$

Then take

$$a_1 = c_1, \quad a_2 = c_2, \quad a_3 = c_1^2c_2^2, \quad a_4 = c_1^2c_2;$$

$$\ker(\theta) = H = \{g_1^{\alpha_1}g_2^{\alpha_2}g_3^{\alpha_3}g_4^{\alpha_4} : \prod_i a_i^{\alpha_i} = 1\}.$$

Then

$$(c_1)^{\alpha_1}(c_2)^{\alpha_2}(c_1^2c_2^2)^{\alpha_3}(c_1^2c_2)^{\alpha_4} = 1,$$

$$\alpha_1 + 2a_3 + 2\alpha_3 \equiv 0 \quad \mod 3,$$

$$\alpha_2 + 2\alpha_3 + \alpha_4 \equiv 0 \quad \mod 3.$$

We find: $\alpha_1 - \alpha_2 + \alpha_4 \equiv 0$. It seems easiest to let α_1 and α_2 assume all 9 pairs of values made up of 0's and 1's; then $\alpha_4 \equiv \alpha_2 - \alpha_1$, $\alpha_3 \equiv \alpha_1 - \alpha_4$. We thus get the covering set:

$$\begin{array}{lllll} 1, & g_1g_2g_3, & g_2g_3^2g_4, & g_1^2g_2^2g_3^2, & g_2^2g_3g_4^2, \\ & g_1g_2^2g_4, & g_1g_3^2g_4^2, & g_1^2g_2g_4^2, & g_1^2g_3g_4. \end{array}$$

We might remark that Mauldon [6], among other results, showed that for p a prime, v not a divisor of p^n, one can write

$$p^r < 1 + n(p-1) < p^{r+1},$$

and establish that $\sigma^*(n, p) = p^{n-r}$.

4. Discussion of the Case $q = p^k$, $v \mid q^n$

Once the result

$$\sigma(n, p) = p^n/v$$

has been proved for p a prime, it would seem natural to investigate the behaviour of $\sigma(n, q)$, where $q = p^k$ is a prime power. This was done by Zaremba [10]. His result is the following:

Theorem. *If G is an Abelian group with n base elements $g_1, \ldots, g_n$, each of order $q = p^k$ (p a prime), and if S is the set of $v = 1 + n(q-1)$ elements*

$$1, g_i^{\alpha_i} \qquad (i = 1, \ldots, n; \quad 1 \leq \alpha_i \leq q),$$

and if v is a power of q, then there exists a covering subset H of G containing q^n/v elements (and $G = HS$).

We indicate the proof of the theorem, since there is such a close analogy to the case where $q = p$. First we obtain a lemma.

LEMMA. *Let Γ be an Abelian group with kn base elements g_{ij} ($i = 1, \ldots, n$; $j = 1, \ldots, k$) each of prime order p. Let Σ be the set of $v = 1 + n(p^k - 1)$ elements of the form*

$$\prod_{j-1}^{k} g_{ij}^{\gamma_i} \qquad (0 \leq \gamma_i \leq p - 1; \quad j = 1, \ldots, k; \quad i = 1, \ldots, n).$$

If v is a power of q, then there exists a subgroup X of Γ such that $\Gamma = X\Sigma$, where the order of $X = q^n/v$.

PROOF: We set $v = 1 + n(q - 1) = q^r$.

Let A be an abstract Abelian group with kr base elements of order p. Then we can stratify A into subgroups forming a *geometric set*.[1] The features of this set $A_1, \ldots, A_n$ are:

(1) each A_i has order p^k,
(2) $A_i \cap A_j = 1$ $(i \neq j)$, and
(3) the number of elements in the geometric set is $(p^{kr} - 1)/(p^k - 1) = n$.

Now pick $a_{i1}, a_{i2}, \ldots, a_{im}$ as a basis for A_i; then any element of Γ can be represented as

$$\prod_{i,j} g_{ij}^{\alpha_{ij}}$$

and we can define a mapping θ by

$$\prod_{i,j} g_{ij}^{\alpha_{ij}} \xrightarrow{\theta} \prod_{i,j} a_{ij}^{\alpha_{ij}}.$$

It can easily be checked that θ is a homomorphism; the set $X = \ker(\theta)$ then has the required properties.

Note that the construction in the lemma specializes to that of the previous section in the case when $k = 1$.

PROOF OF THEOREM: Now let us apply the lemma to proving the theorem. We consider G, and have $v = q^r = p^{kr}$. Any α_i in the range

$$0 \leq \alpha_i \leq p^k - 1 = q - 1$$

[1] For properties of a geometric set, see Carmichael [1].

can be represented uniquely as a series in powers of p, say

$$\alpha_i = \alpha_{i1} p^0 + \alpha_{i2} p + \alpha_{i3} p^2 + \cdots + \alpha_{ik} p^{k-1},$$

where $0 \le \alpha_{ij} \le p - 1$ ($j = 1, \ldots, k$). We now introduce a mapping ψ from G to Γ (the group defined in the lemma).

$$\psi: \quad g_1^{\alpha_1} g_2^{\alpha_2} \cdots g_n^{\alpha_n} \xrightarrow{\psi} \prod_{i,\, j} g_{ij}^{\alpha_{ij}}.$$

This is a one-to-one mapping of G onto Γ. We define H to be the inverse image of X. Hence the set H contains p^{kn}/v elements.

It is now easily checked that $G = HS$, and so

$$\sigma(n, q) = \frac{q^n}{1 + n(q-1)}, \qquad \text{for} \qquad 1 + n(q-1) = q^r.$$

5. Some Results on $\sigma(n, 2)$

If we now restrict ourselves to the case $p = 2$, we have immediately

$$2^n/(n+1) \le \sigma(n, 2) \le 2\sigma(n-1, 2).$$

We define the length l of any vector to be the number of 1's appearing in the vector; with this convention, there are

$$\binom{n}{i}$$

vectors of length i. We now strengthen the inequality

$$(n+1)\sigma(n, 2) \ge 2^n$$

by dividing the total set of 2^n vectors into subsets according to length. Suppose that H is a covering set with m elements, and that H contains exactly y_i elements of length i. We at once note:

(a) A vector of length i can be covered by itself.

(b) Any vector of length $i - 1$ will cover $n - i + 1$ vectors of length i (since any zero can be altered to a 1).

(c) Any vector of length $i + 1$ will cover $i + 1$ vectors of length i (since any 1 can be altered to a zero).

The vectors just noted are the only vectors available to cover the

$$\binom{n}{i}$$

vectors of length i; hence

$$(n - i + 1)y_{i-1} + y_i + (i + 1)y_{i+1} \geq \binom{n}{i}$$

for all i. Writing these equations down in a triangular array, we obtain:

$$\begin{aligned}
y_0 + y_1 &\geq \binom{n}{0} \\
ny_0 + y_1 + 2y_2 &\geq \binom{n}{1} \\
(n-1)y_1 + y_2 + 3y_3 &\geq \binom{n}{2} \\
(n-2)y_2 + y_3 + 4y_4 &\geq \binom{n}{3} \\
(n-3)y_3 + y_4 + 5y_5 &\geq \binom{n}{4} \\
&\cdots \\
y_{n-1} + y_n &\geq \binom{n}{n}
\end{aligned}$$

This set of inequalities is stronger than our earlier set, since it places limits on the number of vectors of a covering set which appear in each subset of vectors of length i ($i = 1, 2, \ldots, n$). Indeed, if we add these inequalities, we obtain the earlier result

$$(n + 1)\sum_{i=0}^{n} y_i = (n + 1)m \geq \sum_{i=0}^{n} \binom{n}{i} = 2^n.$$

For small values of n, the preceding inequalities are easily discussed by hand; for n of any appreciable size, a computer algorithm is desirable. We will summarize some relevant considerations from [9].

6. Consideration of the Positive-Class Inequalities

We define two covering sets H_1 and H_2 as *equivalent* if, for some $g \in G$, H_1 and $g\,H_2$ contain the same elements (up to a permutation of $g_1, g_2, \ldots, g_n$). It is clear that, if $H_2\,S = G$, then $g\,H_2\,S = H_1\,S$ and $g\,H_2\,S = g\,G = G$. Hence $H_1\,S = G$, and we should not regard H_1 and H_2 as being essentially different solutions of the covering problem. For example, $\sigma(5, 2) = 7$, and $H_1 = 1, g_1, g_2, g_3g_4g_5, g_1g_2g_3g_4, g_1g_2g_3g_5, g_1g_2g_4g_5$, is a covering set. If we form

$$H_2 = g_1g_2 H_1 = g_1g_2, g_1, g_2, g_1g_2g_3g_4g_5, g_3g_4, g_3g_5, g_4g_5,$$

we see that it is an equivalent covering set.

The above example illustrates the fact that if $h \in H$, then hH is a covering set which contains $hh = 1$. So, in our algorithm for solving the inequalities involving the y_i's, we may always take $y_0 = 1$. Also, given a covering set H, we can always form the complementary covering set H^* defined by

$$H^* = g_1g_2 \cdots g_n H.$$

If $y_0^*, y_1^*, \ldots, y_n^*$ are the parameters of H^*, we see that y_i^* vectors of length i in H^* produce vectors of length $n - i$ in H; thus

$$y_i^* = y_{n-i}.$$

In short, H^* and H have the same parameter sets $(y_0, \ldots, y_n)$, but the values occur in reverse order; it is often convenient to be able to use either set.

The actual solution of the inequalities given in the last section is best accomplished by introducing "slack variables" z_i so that we obtain equations

$$(n - i + 1)y_i = y_i + (i + 1)y_{i+1} = \binom{n}{i} + z_i.$$

The z_i are nonnegative variables, and by addition,

$$(n + 1)\sum_{i=0}^{n} y_i = 2^n + \sum_{i=0}^{n} z_i.$$

This set of equations has a tridiagonal matrix whose inverse is easily computed. Since only integral values are permitted for y_i and z_i, only a finite number of parameter sets are possible.

For example, $\sigma(5, 2) \geq 2^5/6 > 5$. Hence

$$(n + 1)\sum y_i \geq 32 + \sum z_i \qquad (n = 6)$$

tells us that $\sum y_i = 36$ or 42 or $48 \cdots$, and $\sum z_i = 4$ or 10 or $16 \cdots$. Actually $\sum z_i = 10$ yields a covering set, as exemplified in [9]. The main difficulty with the algorithm for finding parameter sets $(y_0, \ldots, y_n)$ lies in the fact that many parameter sets do not yield coverings. Furthermore, if there is no solution for the equation

$$\sigma(n, 2) = \sum y_i = a,$$

there may be a solution for

$$\sigma(n, 2) = \sum y_i = a + n + 1.$$

Indeed, any parameter set for a (even if it yields no covering) will produce several parameter sets for $a + n + 1$. It is this fact that too many parameter sets do not correspond to coverings that restricts the use of the y_i-method. However, it will show (see [9]) that $\sigma(2, 2) = \sigma(3, 2) = 2$, $\sigma(4, 2) = 4$, $\sigma(5, 2) = 7$, $\sigma(6, 2) = 12$, $\sigma(7, 2) = 16$, $\sigma(8, 2) \geq 31$, $\sigma(9, 2) \geq 54$, $\sigma(10, 2) \geq 96$. The value for 6, as well as the three inequalities, stems from this method. In a later section, we shall note that $\sigma(8, 2) = 32$.

Perhaps we should record the (6, 2) covering set here; it is

$$H = \{1,\ g_2 g_4,\ g_2 g_5,\ g_1 g_3 g_5 g_6,\ g_1 g_3 g_4 g_6,\ g_1 g_2 g_3 g_4 g_5 g_6,$$
$$g_1 g_2 g_3,\ g_1 g_4 g_5,\ g_1 g_2 g_6,\ g_4 g_5 g_6,\ g_2 g_3 g_6,\ g_3 g_4 g_5\}.$$

7. Upper Bounds on $\sigma(3, p)$

In this section, we sketch some of the results given in [2]. Clearly $\sigma(1, p) = 1$, $\sigma(2, p) = p$; so the first interesting case, for fixed n, is $\sigma(3, p)$. Our method is somewhat analogous to that of the last sections, except that our stratification of the vectors of the covering sets into subsets proceeds by columns (positions) rather than by rows (vectors).

We consider the p^3 vectors (a, b, c) where a, b, c, range from 1 to p. Pick any number r less than p, and form r^2 vectors

$$(a_i, b_i, c_i).$$

Here (a_i, b_i) runs through all possible r^2 ordered pairs, and c_i is obtained from the congruence

$$c_i \equiv a_i + b_i \pmod{r}.$$

We readily establish a fundamental lemma.

Lemma. *In this set of vectors, every ordered pair from* 1, 2, ..., r, *occurs once in each of the possible positions* (1, 2), (1, 3), (2, 3). *Consequently, the vectors of this set cover all vectors with* 2 *or more components from* 1, 2, ..., r.

The same construction, defining c_i modulo $p - r$, will produce a set of $(p - r)^2$ vectors with components from $r + 1$, $r + 2$, ..., p, with the property that each ordered pair from $r + 1$, $r + 2$, ..., p, occurs once in each possible position. Hence, the vectors of this set cover all vectors with 2 or more components from $r + 1$, $r + 2$, ..., p.

Putting these 2 sets together produces a covering set containing $r^2 + (p - r)^2$ vectors. For $p = 2t$, we choose $r = t$; for $p = 2t + 1$, we choose $r = t$, and obtain the following theorem.

THEOREM. $\sigma(3, 2t) \leq 2t^2$, $\sigma(3, 2t+1) \leq 2t^2 + 2t + 1$.

Thus, the theorem shows that $\sigma(3, 5) \leq 13$, and we can produce a (possibly nonminimal) covering set as

112, 121, 211, 222,
333, 344, 355, 434, 445, 453, 535, 543, 554.

8. VALUE OF $\sigma(3, p)$

In this section, we establish that the bounds just obtained are actually the required values. We let H be a covering set of triples (a_i, b_i, c_i) with $i = 1, 2, \ldots, m$, and partition H into 4 sets H_1, H_2, H_3, H_4.

H_1 contains α vectors (a, b_1, c_1), (a, b_2, c_2), $\ldots$, (a, b_α, c_α), where all the components in one position are identical, and α is chosen as small as possible. Then $\alpha \leq [m/p]$.

Let r be the number of distinct b_i in H_i; clearly, $r \leq \alpha$. We define H_2 as the set of vectors (a_i, b_i, c_i) such that $a_i \neq a$, but the b_i's are the same as those appearing in H_1. Since α is minimal, each one of these r b's occurs at least α times; hence, the cardinality of H_2 is at least $r\alpha - \alpha$

H_3 is the set of x vectors which differ from vectors in H_1 in both the first 2 positions, but whose third components are selected from among $c_1, \ldots, c_\alpha$.

Finally, H_4 is the set of vectors having $a_i \neq a$; $b_i \neq b_1, \ldots, b_\alpha$; $c_i \neq c_1, \ldots, c_\alpha$. The number of vectors in H_4 is at most $m - x - r\alpha$. Now look at vectors (a, b, c) where a is as in H_1, but $b \neq b_i$ $(i = 1, \ldots, \alpha)$, $c \neq c_i$ $(i = 1, \ldots, \alpha)$. These vectors differ from those in H_1, H_2, H_3, in 2 components, and so can only be covered by vectors from H_4. Also, they can only be covered one at a time (by changing the first component to a). Since the number of such vectors is at least

$$(p - r)(p - \alpha),$$

and since they can only be covered one at a time by vectors from H_4, we find

$$(p - r)(p - \alpha) \leq m - r\alpha - x.$$

CASE 1. $p = 2t$, $m \leq 2t^2$.

Suppose $m = 2t^2 - \beta$ $(\beta \geq 0)$. Then

$$(2t - r)(2t - \alpha) \leq 2t^2 - r\alpha - \beta - x.$$

$$2(t - r)(t - \alpha) \leq -\beta - x. \tag{3}$$

But

$$\alpha \leq \left[\frac{2t^2 - 1}{2t}\right] < t, \qquad r \leq \alpha < t.$$

The LHS of (3) is positive, and this is impossible. Hence, $\beta = 0$, $x = 0$; $r = \alpha = t$; and we have $m = 2t^2$.

CASE 2. $p = 2t + 1$, $m \leq 2t^2 + 2t + 1$.

Again, let $m = 2t^2 + 2t + 1 - \beta$, and we find

$$2\{t + \tfrac{1}{2} - r\}\{t + \tfrac{1}{2} - \alpha\} - \tfrac{1}{2} \leq -\beta - x.$$

$$(2t + 1 - 2r)(2t + 1 - 2\alpha) \leq -\beta - x + 1. \tag{4}$$

Since

$$\alpha \leq \left[\frac{2t^2 + 2t}{2t + 1}\right] = \left[t + \frac{t}{2t + 1}\right],$$

we see that the LHS of (4) is positive. Thus, the only possibility is to have $\beta = x = 0$, $r = \alpha = t$.

We can unify Cases 1 and 2 into one expression in the following theorem.

THEOREM. $\sigma(3, p) = \left[\dfrac{p^2 + 1}{2}\right]$.

9. THE INTERSECTION INEQUALITIES

We pursue the idea of stratifying the p^n vectors of our problem into subsets by columns. We again let H be a covering set of m vectors, but we use A, B, C, to denote first, second, third, etc., components. We define the following numbers, as in [3].

$(A)_i$ is the number of times i appears as first component,
$(B)_i$ is the number of times i appears as second component, etc.
Clearly $\sum_i (A)_i = \sum_i (B)_i = \cdots = m$.

Similarly, $(AB)_{ij}$ = number of times i appears as a first component, along with j as a second component. We find

$$\sum_i (XY)_{ij} = Y_j, \qquad \sum_j (XY)_{ij} = X_i,$$

for all X, Y selected from A, B, C, ...

Finally, $(ABC)_{ijk}$ is the number of times that i appears as first component along with j as second component and k as third component. Thus

$$\sum_i (XYZ)_{ijk} = (YZ)_{jk}, \quad \sum_j (XYZ)_{ijk} = (XZ)_{ik}, \quad \sum_k (XYZ)_{ijk} = (XY)_{ij},$$

for all X, Y, Z, selected from A, B, C, etc. This notation can be extended further in the obvious manner.

We now deduce a series of inequalities presented in [3]. First, the number of elements in G with first component i is p^{n-1}. By our usual argument, the number of these vectors covered by a vector of H with first component i is simply $1 + (n-1)(p-1)$. The remaining vectors of G with first component i must be covered by vectors of H whose first component is not i, that is, they are covered one at a time by other members of H. We thus have

$$(A)_i\{(n-1)(p-1)+1\} + \sum_{x \neq i} (A)_x \geq p^{n-1},$$

$$(A)_i\{(n-1)(p-1)\} + \sum_{x} (A)_x \geq p^{n-1}.$$

Since $\sum_x (A)_x = m$, we have the following theorem.

THEOREM. $(A)_i\{(n-1)(p-1)\} \geq p^{n-1} - m.$

This sort of argument can be extended; if we present the next stage, the generalization should become obvious.

The number of elements of G with i in the first component, j in the second component, is p^{n-2}; of these, some will be covered by a vector from H that begins $(ij \cdots)$. The number of these is $(n-2)(p-1)+1$.

The other vectors of the form $(ij \cdots)$ are covered one at a time by vectors of H beginning $(xy \cdots)$, where either $x = i$, $y \neq j$; or $x \neq i$, $y = j$. Thus we have

$$(AB)_{ij}\{(n-2)(p-1)+1\} + \sum_{x \neq i} (AB)_{xj} + \sum_{y \neq j} (AB)_{iy} \geq p^{n-2}.$$

Hence

$$(AB)_{ij}\{(n-2)(p-1)-1\} + \sum_{x} (AB)_{xj} + \sum_{y} (AB)_{iy} \geq p^{n-2},$$

$$\boxed{(AB)_{ij}\{(n-2)(p-1)-1\} \geq p^{n-2} - (A)_i - (B)_j .}$$

This result gives us our pattern; a similar argument will give the next inequality of this type as

$$(ABC)_{ijk}\{(n-3)(p-1)-2\} \geq p^{n-3} - (AB)_{ij} - (AC)_{ik} - (BC)_{jk} .$$

Clearly, the sequence of inequalities is extensible.

10. APPLICATION TO $\sigma(5, 3)$ AND $\sigma(8, 2)$

The basic inequalities of Section 2 give us, for $\sigma(5, 3)$, the result

$$\frac{243}{11} \leq \sigma(5, 3) \leq 3\sigma(4, 3) = 27.$$

Thus $\sigma(5, 3) = 23$, 24, 25, 26, or 27. We will show how the inequalities of Section 9 easily rule out the first three possibilities.

If $m = 23$, we find $8(A)_i \geq 3^4 - 23 = 58$. Thus $(A)_i \geq 58/8 > 7$; hence $(A)_i \geq 8$.

Since $m = (A)_1 + (A)_2 + (A)_3$, we have $m \geq 24$; this is a contradiction, since $m = 23$.

For $m = 24$, $8(A)_i \geq 57$; again $(A)_i \geq 8$. Since $m = 24$ and $m = \sum(A)_i \geq 24$, we see that $(A)_i = 8$ for all i (and all positions A). It then follows from the second-order inequalities that

$$5(AB)_{ij} \geq 27 - 16 = 11;$$

hence

$$(AB)_{ij} \geq 3.$$

Thus $(A)_i = \sum_j (AB)_{ij} \geq 9$; this is a contradiction.

It remains to discuss the case $m = 25$. Here we have $8(A)_i \geq 56$; $(A)_i \geq 7$. We suppose if possible that some $(A)_i$, say $(A)_1$, is equal to 7. Then

$$5(AB)_{ij} \geq 27 - (A)_1 - (B)_j = 20 - (B)_j.$$

Add over j; then

$$35 = 5(A)_1 \geq 60 - m = 35.$$

Hence the inequality is an equality, and

$$5(AB)_{ij} = 20 - (B)_j.$$

Since $(B)_j \geq 7$, we have (by divisibility) $(B)_j \geq 10$, and this is impossible since $\sum_j (B)_j = 25$. It follows that, for $X = A, B, C, D, E$, we must have $X_1 = X_2 = 8$, $X_3 = 9$.

Now $5(XY)_{\alpha\beta} \geq 27 - (X)_\alpha - (Y)_\beta = 11$, for $\alpha, \beta = 1$ or 2. Thus $(XY)_{\alpha\beta} \geq 3$ for $\alpha, \beta = 1$ or 2. If exactly one of α, β, is 3, we get

$$5(XY)_{\alpha\beta} \geq 10; \qquad (XY)_{\alpha\beta} \geq 2.$$

But

$$8 = (X)_1 = \sum_j (XY)_{1j} \geq 3 + 3 + 2.$$

We thus have

$$\begin{aligned} &(XY)_{\alpha\beta} = 3 \quad && \text{for neither} \quad \alpha, \beta = 3; \\ &(XY)_{\alpha\beta} = 2 \quad && \text{for exactly one of} \quad \alpha, \beta = 3; \\ &(XY)_{33} = (X)_3 - (XY)_{31} - (XY)_{32} = 5. \end{aligned}$$

Now continue to

$$2(XYZ)_{ijk} \geq 9 - (XY)_{ij} - (XZ)_{ik} - (YZ)_{jk}.$$

If just 1 of i, j, k, is 3, we find $(XYZ)_{ijk} \geq 1$. But $(XY)_{\alpha 3} = (XY)_{3\alpha} = 2 = \sum_k (XYZ)_{\alpha 3k}$ where $\alpha \neq 3$. It follows that $(XYZ)_{ijk} = 1$ if one subscript is 3, and 0 if two subscripts are 3's. Finally,

$$(XY)_{33} = 5 = \sum_k (XYZ)_{33k} = 0 + 0 + (XYZ)_{333}.$$

Thus $(ABC)_{333} = (ABD)_{333} = (ABE)_{333} = 5$. We thus see that, since $(XYZ)_{ijk} = 0$ for two subscripts equal to 3, the set H must contain five repeated vectors (3, 3, 3, 3, 3). Delete four of these to leave a covering set of 21 vectors; this is impossible.

We thus have $\sigma(5, 3) = 26$ or 27; this is much sharper than the result of [8], but the discussion of 26 is more strenuous. Recently, Kamps and Van Lint [4], using different methods, have established that $\sigma(5, 3) = 27$.

We now give a brief indication of the beginning of the discussion in [3] concerning $\sigma(8, 2)$. We assume $m = 31$, and rule out this case by considering the result

$$7(A)_i \geq 97;$$

thus $(A)_i \geq 14$.

It is easy to prove the lemma that, by considering an equivalent covering set, we always have

$$(X)_1 \leq (X)_2.$$

Furthermore, we have the following lemma.

LEMMA. *There is at most one column with* $(X)_1 = 14$, $(X)_2 = 17$.

PROOF: If there are two, we let them be the first and second; then

$$5(AB)_{11} \geq 64 - 14 - 14 = 36; \qquad \text{thus} \qquad (AB)_{11} \geq 8$$
$$5(AB)_{12} \geq 64 - 14 - 17 = 33; \qquad \text{thus} \qquad (AB)_{12} \geq 7.$$

Add, to obtain $(A)_1 = (AB)_{11} = (AB)_{12} \geq 15$, a contradiction. Our covering set H thus contains at least seven columns with $(X)_1 = 15$, $(X)_2 = 16$. A straightforward discussion (see [3]) rules out the various possibilities, and we end up ruling out the possibility $m = 31$. Thus, we obtain the following theorem.

THEOREM. $\sigma(8, 2) = 32$.

11. CONCLUDING REMARKS

Table I gives known values of $\sigma(n, p)$, for n and p not exceeding 8. We omit a number of inequalities known concerning certain blank tabular entries.

TABLE I

n \ p	2	3	4	5	6	7	8
2	2	3	4	5	6	7	8
3	2	5	8	13	18	25	32
4	4	9	24		72		
5	7	27	4^3				
6	12			5^4			
7	16						
8	32					7^6	

Recently, Kalbfleisch and Weiland have had good results concerning some values of $\sigma(4, p)$, and their methods seem capable of further development. Basically, they note that, in the case of $\sigma(3, p)$, the covering sets may be derived from two Latin Squares (of unequal size if p is odd). For instance, let $p = 5$. Then $t = 2$, $t + 1 = 3$, and we write:

	1	2
1	1	2
2	2	1

	3	4	5
3	3	4	5
4	4	5	3
5	5	3	4

The covering set is then obtained by using triplets made up of row, column, entry; the above two squares give the set

111, 122, 212, 221, 333, 344, 355, 434, 445, 453, 535, 543, 554.

In the chapter which follows this, Kalbfleisch and Weiland show, among other results, that

$$\sigma(4, 3t) \leq p^3/3 = 9t^3.$$

We might point out that a perfect covering, where no redundancies occur, is equivalent to a perfect code. With redundancy, the connection with coding theory is less direct, but it can be hoped that the two areas will interact fruitfully.

Note added, July 1968: Subsequent to the writing of this survey, several results have been obtained at the Canadian Mathematical Congress Summer Research Institute, 1968. G. O. Losey has obtained a very brief and usable proof of the Zaremba theorem of Section 4, and has pointed out that the fact that $\sigma(n, p) = \sigma^*(n, p)$ occurs only for p a prime follows trivially from a theorem of Kantorowicz. T. J. Dickson has a number of results on $\sigma^*(n, p)$; among them are complete determinations of $\sigma^*(3, p)$ and $\sigma^*(4, p)$. R. G. Stanton, J. D. Horton, and J. G. Kalbfleisch have obtained a number of general results which will be reported elsewhere. The most easily summarized are: $\sigma(n, rs) \leq s^{n-1}\sigma(n, r)$; $\sigma(4, 4) = 24$; and much better upper bounds on $\sigma(4, p)$ and $\sigma(5, p)$.

References

1. R. D. Carmichael, *Introduction to the Theory of Groups of Finite Order*, Ginn, New York, 1937.
2. J. G. Kalbfleisch and R. G. Stanton, A Combinatorial Problem in Matching, *J. London Math. Soc.* **44** (1969), 60–64.
3. J. G. Kalbfleisch and R. G. Stanton, Intersection Inequalities for the Covering Problem, *Stan . J. Appl. Math.* (to appear).
4. H. J. L. Kamps and J. H. Van Lint, The Football Pool Problem for 5 Matches, *J. Combinatorial Theory* **3** (1967), 315–325.
5. E. Mattioli, Sopra una Particolare Proprieta dei Gruppi Abeliani Finiti, *Ann. Scuola Norm. Sup. Pisa* (3) **3** (1950), 59–65.
6. J. G. Mauldon, Covering Theorems for Groups, *Quart. J. Math. Oxford Ser.* (2) **1** (1950), 284–287.
7. O. Taussky and J. Todd, Covering Theorems for Groups, *Ann. Soc. Polon. Math.* **21** (1948), 303–305.
8. O. Taussky and J. Todd, Some Discrete Variable Computations, *Proc. Symp. Appl. Math.* **10** (1960), 201–209.
9. R. G. Stanton and J. G. Kalbfleisch, Covering Problems for Dichotomized Matchings, *Aequat. Math.* **1** (1968), 94–103.
10. S. K. Zaremba, A Covering Theorem for Abelian Groups, *J. London Math. Soc.* **26** (1950), 71–72.
11. S. K. Zaremba, Covering Problems Concerning Abelian Groups, *J. London Math. Soc.* **27** (1952), 242–246.

SOME NEW RESULTS FOR THE COVERING PROBLEM

J. G. Kalbfleisch

UNIVERSITY OF WATERLOO
WATERLOO, ONTARIO

and

P. H. Weiland

YORK UNIVERSITY
TORONTO, ONTARIO

1. Introduction

In this paper we shall present some new results for the covering problem discussed by R. G. Stanton in the immediately preceding chapter of this volume. We shall use his notation, and refer the reader to his paper for background material and references.[1]

We first give a method of representing a covering set which is convenient for checking its covering properties. We then show how a perfect $(p + 1, p)$ covering set can be expanded to give a $(p + 1, pk)$ covering set. Thus for p a prime or prime power we obtain

$$\sigma(p + 1, pk) \leq k^p p^{p-1}.$$

We then improve the lower bound for $\sigma(n, p)$ with $n \leq p$ by means of a simple counting argument. In particular, we show that

$$\sigma(k + 1, 2k) \geq 2^k k^{k-1},$$

with equality if k is a prime or prime power. For example,

$$\sigma(4, 6) = 72, \sigma(5, 8) = 1024.$$

Finally, we consider covering groups, and prove that

$$\sigma^*(n, p) \geq p^{n-1}/(n - 1).$$

[1] Numbered references appearing in the text of this chapter refer to the correspondingly numbered references given at the end of the chapter by Stanton.

2. Upper Bounds

We recall that the set G consists of the p^n vectors $(x_1, x_2, \ldots, x_n)$ with components from $\{1, 2, \ldots, p\}$. We define a *block* to be a set of p vectors in G whose first $n - 1$ components are all equal. A *row* is a set of p blocks (p^2 vectors) for which $n - 2$ of the first $n - 1$ components are equal. Clearly there are p^{n-1} blocks and $(n - 1)p^{n-1}$ rows.

Table I illustrates the representation of a (3, 6) covering set H. Here $n = 3$, and we have a 2-dimensional array of blocks. If (x_1, x_2, x_3) is a vector of H, we enter x_3 in block (x_1, x_2). Now if block (x_1, x_2) contains an entry, all vectors of G in that block are covered. We say that they are *directly covered.*

TABLE I

(3, 6) Covering Set H

Vectors of H

1 1 1	4 4 4
1 2 2	4 5 5
1 3 3	4 6 6
2 1 2	5 4 5
2 2 3	5 5 6
2 3 1	5 6 4
3 1 3	6 4 6
3 2 1	6 5 4
3 3 2	6 6 5

Representation (x_1 down, x_2 across; Entry is x_3)

1	2	3			
2	3	1			
3	1	2			
			4	5	6
			5	6	4
			6	4	5

For example, in Table 1 all vectors $(1, 1, x_3)$ are directly covered. If a block is not directly covered, its vectors must be *indirectly covered* from other blocks in the same row (vertical or horizontal). Each of $1, 2, \ldots, p$, must appear as an entry in one of the rows containing such a block. In Table I, the rows containing a block such as (1, 4) which is not directly covered, contain entries 1, 2, 3, 4, 5, and 6. We have thus verified that H is a (3, 6) covering set.

In constructing a $(3, 2s)$ covering set, we begin with the (3, 2) covering set in Table II. Each block is expanded into an s by s set of blocks. The s^2 blocks derived from an empty block are empty; those from a block with entry 1 will contain a Latin square on elements $1, 2, \ldots, s$; and the s^2 blocks

TABLE II

(3, 2) COVERING SET

Vectors	Representation
1 1 1 2 2 2	—x_1→ ; x_2 ↓ 1 | – – | 2 Entry is x_3

obtained from the block with entry 2 will contain a Latin square on elements $s + 1, \ldots, 2s$. We thus have a $(3, 2s)$ covering set with $2s^2$ members, and $\sigma(3, 2s) \leq 2s^2$. This is shown to be the exact value in [2].

For $n = 4$, the array of blocks is three-dimensional, and a (4, 3) covering set may be represented as in Table III. Vectors in blocks like (2, 2, 1) which

TABLE III

REPRESENTATION OF A (4, 3) COVERING SET

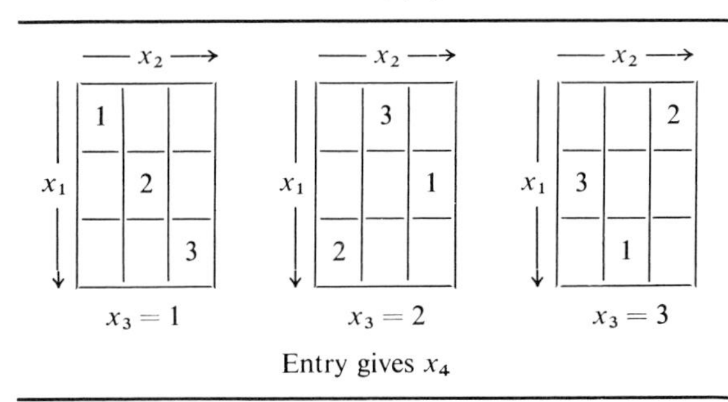

Entry gives x_4

contain an entry are directly covered. Vectors in blocks like (1, 2, 1) must be indirectly covered from other blocks in the same row and plane, the same column and plane, or the same position in another plane. The vectors in block (1, 2, 1) are all covered from blocks (1, 1, 1), (2, 2, 1) and (1, 2, 2) which contain entries 1, 2, and 3. Similar comments apply to the other blocks which are not directly covered.

We now construct a (4, 6) covering set from this (4, 3) covering set by expanding each block into a 2 by 2 by 2 set of blocks. The eight blocks obtained from any empty block will be empty. The eight blocks obtained from a block containing entry i will form a Latin cube on elements $2i - 1, 2i$ $(i = 1, 2, 3)$. Table IV gives the resulting configuration. Consider a block such as

TABLE IV

REPRESENTATION OF A (4, 6) COVERING SET

$x_3 = 1$

	x_2					
	1	2				
	2	1				
x_1			3	4		
			4	3		
					5	6
					6	5

$x_3 = 4$

	x_2					
			6	5		
			5	6		
x_1					2	1
					1	2
	4	3				
	3	4				

$x_3 = 2$

	x_2					
	2	1				
	1	2				
x_1			4	3		
			3	4		
					6	5
					5	6

$x_3 = 5$

	x_2					
					3	4
					4	3
x_1	5	6				
	6	5				
			1	2		
			2	1		

$x_3 = 3$

	x_2					
			5	6		
			6	5		
x_1					1	2
					2	1
	3	4				
	4	3				

$x_3 = 6$

	x_2					
					4	3
					3	4
x_1	6	5				
	5	6				
			2	1		
			1	2		

Entry gives x_4.

(1, 3, 1) which is not directly covered. Its six vectors are all covered indirectly, two from the same row in the same plane, two from the same column in the same plane, and two from the same position in other planes. Thus we have a (4, 6) covering set with $9 \cdot 2^3 = 72$ vectors, and $\sigma(4, 6) \leq 72$. We shall later prove that $\sigma(4, 6) = 72$.

If Latin cubes of side 2 are replaced by Latin cubes of side k in the above construction, one obtains a $(4, 3k)$ covering set with $9k^3$ members. Thus $\sigma(4, 3k) \leq 9k^3$. The construction is extended to higher dimensions by replacing Latin cubes by Latin "hypercubes." The starting point is a $(p+1, p)$ covering set with p^{p-1} members, and Zaremba [11] proved that these exist if p is a prime or prime power. We thus obtain the following theorem.

THEOREM 1. $\sigma(p+1, sp) \leq s^p p^{p-1}$.

Further extensions of the construction due to R. G. Stanton, J. D. Horton, and J. G. Kalbfleisch, will be reported elsewhere.

3. LOWER BOUNDS

In Section 2 we defined the blocks and rows of set G. Now suppose that H is an (n, p) covering set with m vectors in $m - r$ blocks $(r \geq 0)$. Suppose that the ith row of G contains x_i vectors of H in t_i different blocks. We may assume that H contains no repeated vectors and then

$$0 \leq t_i \leq x_i \leq pt_i \qquad (1 \leq i \leq (n-1)p^{n-2}). \tag{1}$$

Now adding over all rows of G, each vector of H is counted $n - 1$ times, once for each row containing it. Thus

$$\sum x_i = (n-1)m. \tag{2}$$

Similarly, by counting occupied blocks one obtains

$$\sum t_i = (n-1)(m-r). \tag{3}$$

Since $m - r$ blocks contain vectors of H, exactly $p(m - r)$ different vectors of G are covered directly, and thus at least $p^n - p(m - r)$ vectors of G must be covered indirectly. But if a row contains x_i vectors in t_i blocks, there are at most $x_i(p - t_i)$ different indirect covers along that row (at most x_i in each of the $p - t_i$ unoccupied blocks). Adding over all rows counts each indirect cover once, and thus

$$\sum x_i(p - t_i) \geq p^n - p(m-r). \tag{4}$$

From (2), (3), and (4) one readily obtains the following lemma.

LEMMA. *For any* a, $\sum \left(x_i - \dfrac{p}{n-1}\right)(t_i - a) \leq -(p-a)[p^{n-1} - m(n-1)]$.

As a first example of the use of the Lemma, we prove a theorem.

THEOREM 2. $\sigma(k+1, 2k) \geq 2^k k^{k-1}$ *for all* $k \geq 1$.

PROOF: In the lemma, put $n = k+1$, $p = 2k$, and $a = 1$. Then

$$\sum (x_i - 2)(t_i - 1) \leq -(2k-1)[(2k)^k - km]$$

where m is the number of vectors in a $(k+1, 2k)$ covering set. Consider the terms in the sum on the left. If $x_i = 0$, then $t_i = 0$ and $(x_i - 2)(t_i - 1) > 0$; if $x_i = 1$, then $t_i = 1$ and $(x_i - 2)(t_i - 1) = 0$; finally, if $x_i \geq 2$, then $t_i \geq 1$, and $(x_i - 2)(t_i - 1) \geq 0$. Thus the sum on the left is nonnegative, and $(2k)^k - km \leq 0$. This implies $m \geq 2^k k^{k-1}$, and Theorem 2 follows.

The results of Theorems 1 and 2 combine to give a third theorem.

THEOREM 3. $\sigma(k+1, 2k) = 2^k k^{k-1}$ *for k a prime or prime power.* (Values of particular interest are $\sigma(4, 6) = 72$, and $\sigma(5, 8) = 1024$.)

Theorem 2 is a special case of the following theorem.

THEOREM 4. *Let n and p be such that* $n - 1 < p \leq 2(n-1)$. *Then* $\sigma(n, p) \geq p^{n-1}/(n-1)$.

PROOF: With $a = 1$ the lemma gives

$$\sum \left(x_i - \frac{p}{n-1}\right)(t_i - 1) \leq -(p-1)[p^{n-1} - m(n-1)].$$

Noting that $p/(n-1) \leq 2$, we employ the argument in the proof of Theorem 2 to show that each term in the sum on the right is nonnegative. Thus $p^{n-1} - m(n-1) \leq 0$ and the theorem follows.

Examples of bounds from Theorem 4 are

$$\sigma(4, 4) \geq 22, \qquad \sigma(4, 5) \geq 42, \qquad \sigma\sigma(5, 5) \geq 157,$$
$$\sigma(5, 6) \geq 324, \qquad \sigma(5, 7) \geq 601.$$

It is likely that all of these can be improved. In [2] it is proved that $\sigma(4, 4) \geq 24$, and we shall show in Section 4 that $\sigma(4, 5) \geq 45$. However, the bounds of Theorem 4 represent a substantial improvement over the only general lower bound previously given: namely, $\sigma(n, p) \geq p^n/v$ where $v = n(p-1) + 1$.

Some additional lower bounds are provided by a fifth theorem.

THEOREM 5. *Let* $p = 2(n-1) + c$ *where* $0 < c \leq n - 1$. *Let s be the greatest number for which* $p^{n-1}/(n-1) - s$ *is an integer and*

$$s\left[\frac{(p-2)(n-1)}{p(n-2)} + \frac{(p-1)(n-1)}{c}\right] \leq p^{n-2}. \tag{5}$$

Then $\sigma(n, p) \geq p^{n-1}/(n-1) - s$.

PROOF: Suppose there exists an (n, p) covering set with $m = p^{n-1}/(n-1) - s$ members. To prove the theorem it is sufficient to show that s must satisfy (5). Applying the lemma with $a = 1$ we obtain

$$\sum \left(x_i - \frac{p}{n-1}\right)(t_i - 1) \leq -(p-1)(n-1)s.$$

Since $2 < p/(n-1) \leq 3$, a term on the left is negative only if $x_i = t_i = 2$, and then its value is $-c/(n-1)$. Thus $x_i = t_i = 2$ for at least $(p-1)(n-1)^2 s/c$ values of i. Also, for $a = 2$ the lemma implies

$$\sum \left(x_i - \frac{p}{n-1}\right)(t_i - 2) \leq -(p-2)(n-1)s.$$

A term on the left is negative only if $t_i = 1$ and $x_i \geq 3$, and since then by (1) $x_i \leq p$, its value is at least $-[p - p/(n-1)] = -p(n-2)/(n-1)$.
Thus $t_i = 1$ and $x_i \geq 3$ for at least $(p-2)(n-1)^2 s/[p(n-2)]$ values of i. But $1 \leq i \leq (n-1)p^{n-2}$, and thus

$$\frac{(p-2)(n-1)^2 s}{p(n-2)} + \frac{(p-1)(n-1)^2 s}{c} \leq (n-1)p^{n-2}.$$

Division by $n - 1$ gives (5).

Examples of bounds yielded by Theorem 5 are

$$\sigma(4, 7) \geq 112, \qquad \sigma(4, 8) \geq 166, \qquad \sigma(4, 9) \geq 235.$$

For $p = 3k$ and $n = k + 1$, the coefficient of s in (5) is greater than $3k$. Thus $s < (3k)^{k-2}$, and we obtain the following corollary.

COROLLARY. $\sigma(k + 1, 3k) \geq (9k - 1)(3k)^{k-2}$.

For k a prime or prime power, this may be compared with the upper bound of $9k(3k)^{k-2}$ provided by Theorem 1, and the previous lower bound, $p^n/v = (3k)^{k+1}/(3k^2 + 2k)$.

One may apply the lemma for other values of n and p, with resulting improvements in the lower bounds when $p \geq n$. However, the general expressions are quite complicated, and will not be given here.

4. AN IMPROVED BOUND FOR $\sigma(4, 5)$

The smallest unknown covering number is $\sigma(4, 5)$, and it is of some interest to improve the lower bound of 42 obtained in the preceding section. We accomplish this by means of the intersection inequalities discussed in Section

5 of the preceding paper by R. G. Stanton. The three inequalities we require are

$$12(A)_i \geq 125 - m, \tag{6}$$

$$7(\mathrm{AB})_{ij} \geq 25 - (A)_i - (B)_j, \tag{7}$$

$$2(ABC)_{ijk} \geq 5 - (AB)_{ij} - (AC)_{ik} - (BC)_{jk} \tag{8}$$

We first note that if $m \leq 52$, then (1) implies that $(X)_i \geq 7$ for all X and i. Secondly, if $m \leq 46$, then $(X)_i \geq 8$ for all X and i; for if $(A)_1 = 7$, then (2) gives $7(AB)_{1j} \geq 18 - (B)_j$ where $18 - (B)_j \leq 11$ and $\sum (18 - (B)_j) = 18.5 - \sum (B)_j \geq 90 - 46 = 44$. It is easily verified that the minimum possible value of $\sum (AB)_{1j}$ for integers $(AB)_{1j}$ satisfying these inequalities is 8, contradicting $\sum (AB)_{1j} = (A)_1 = 7$.

If $m = 44$, $(X)_i \geq 8$, and we may take $(A)_1 = 8$. Now (2) gives $7(AB)_{1j} \geq 17 - (B)_j$ where $17 - (B)_j \leq 9$ and $\sum (17 - (B)_j) = 41$. The only solution to the inequalities is

$$(B)_j = 8, 8, 8, 10, 10; \qquad (AB)_{lj} = 2, 2, 2, 1, 1 \qquad (j = 1, 2, 3, 4, 5).$$

These results hold for any pair of columns X, Y, with components suitably named, and we have

$$(X)_i = 8, 8, 8, 10, 10 \qquad (1 \leq i \leq 5)$$

$$(XY)_{ij} = 2 \qquad (1 \leq i, j \leq 3)$$

$$(XY)_{ij} = 1 \qquad (1 \leq i \leq 3 \text{ and } j \geq 4, \text{ or } i \geq 4 \text{ and } 1 \leq j \leq 3).$$

Now (3) gives $(ABC)_{114} \geq 1$ and $(ABC)_{124} \geq 1$. Adding gives $(AC)_{14} \geq (ABC)_{114} + (ABC)_{124} \geq 2$, contradicting $(AC)_{14} = 1$. Thus $\sigma(4, 5) \geq 45$.

We remark that, with very minor modifications, the arguments of the last paragraph indicate that if $m = 45$, then $(X)_i = 9$ for all X and i. We summarize our results in a sixth theorem.

THEOREM 6. $\sigma(4, 5) \geq 45$. *If a* (4, 5) *covering set with* 45 *members exists, then* $(X)_i = 9$ *for all* X *and* i.

The best available upper bound for $\sigma(4, 5)$ seems to be 51, which comes from the general results of Stanton, Horton, and Kalbfleisch which were mentioned at the end of Section 2.

5. A Result for Covering Groups

It is not difficult to show that if one requires that the (n, p) covering set H be a group, then each occupied bock of G must contain the same number of elements of H. Thus in the lemma of Section 3 we have $x_i = qt_i$ for all i, where q is a positive integer, and we obtain

$$\sum \left(qt_i - \frac{p}{n-1}\right)(t_i - a) \leq -(p-a)[p^{n-1} - m(n-1)]$$

With $qa = p/(n-1)$, this becomes

$$q \sum (t_i - a)^2 \leq -(p-a)[p^{n-1} - m(n-1)],$$

and thus $p^{n-1} - m(n-1) \leq 0$. Furthermore, if $m = p^{n-1}/(n-1)$ then $t_i = a$ for all i, and a must be an integer. Thus $n-1$ divides p, and we have a seventh theorem.

THEOREM 7. $\sigma^*(n, p) \geq p^{n-1}/(n-1)$, *with strict inequality unless* p *is s multiple of* $n-1$.

INVITED PAPERS

THE RANK OF A FAMILY OF SETS AND SOME APPLICATIONS TO GRAPH THEORY

Claude Berge

INSTITUT HENRI POINCARÉ
PARIS, FRANCE

1. Introduction

Given a family of sets $U_1, U_2, \ldots, U_m \subset X$, we define the *rank* $r(A)$ of a set $A \subset X$ to be integer $r(A) = \max |U_i \cap A|$. This concept appears in the theory of matroids, and a fruitful idea developed by Nash-Williams [6] was to express some coefficients of a matroid using only its rank function. The purpose of the present paper is to show that some of the statements for matroids hold for more general families of sets, called here "graphoids."

In Sections 2 and 3, we show that in considering a graphoid as a direct generalization of a graph, we can define for a graphoid the same concepts as for a graph, and that this yields some simplifications. (For instance, the relationship graph-line-graph becomes a duality in the graphoids.)

In Section 4, we study a special class of graphoid, called γ-perfect, which has the property that, for any subset $A \subset X$, the minimum number of colors needed to color its vertices (such that two adjacent vertices have different colors) is equal to the rank $r(A)$. For these graphoids it is possible to express the stability number $\alpha(A)$ by using only the rank (see Theorem 1). In addition, the conditions of Theorem 2, which are known for matroids, appear to be true also for perfect graphoids.

2. Rank of a General Graphoid

In this section we give the basic definitions and general background which are necessary for an understanding of the theory of graphoids.

A *graphoid* $G = (X, \mathscr{U})$ is the pair constituted by a set $X = \{x_1, x_2, \ldots, x_n\}$ ("the points" or "elements"), and a family of subsets $\mathscr{U} = \{U_1, U_2, \ldots, U_m\}$ ("the edges"), with the two conditions:

(1) $U_i \neq \varnothing$ $(i = 1, 2, \ldots, m)$.
(2) $\bigcup U_i = X$.

If $|U_i| \leq 2$ for all i, G is called a graph (undirected), and an edge U with $|U| = 1$ is called a loop.

A *cycle of length* k is a sequence $(x_1, U_1, x_2, U_2, x_3, \ldots, x_{k+1})$ with

(1) $x_{k+1} = x_1$,
(2) the U_i all different,
(3) the x_i all different,
(4) $x_i \in U_i$ $(i = 1, 2, \ldots, k)$, and
(5) $x_{i+1} \in U_i$ $(i = 1, 2, \ldots, k)$.

We shall say for short that two points x and x' are *adjacent* if there exists an edge U which contains both of them; similarly, two edges U and U' are said to be *adjacent* if they intersect. The *incidence matrix* of a graphoid $G = (X, \mathscr{U})$ is the $m \times n$ matrix $A = ((a_j^{\,i}))$, with

$$a_j^{\,i} = \begin{cases} 1 & \text{if} \quad x_i \in U_j, \\ 0 & \text{if} \quad x_i \notin U_j. \end{cases}$$

For such a matrix, there exists at least one 1 in each column, and one 1 in each row.

The dual G^* of graphoid G is defined by the set of points

$$M = \{1, 2, \ldots, m\},$$

and by the edges

$$M_i = \{j / j \in M, \quad U_j \ni x_i\}.$$

We have $M_i \neq \varnothing$, $\bigcup_{i=1}^{m} M_i = M$; therefore G^* is a graphoid, and its incidence matrix is the transpose A^*. Note that $(G^*)^* = G$.

Given a graphoid G, its *rank-function* is a function r, defined for all subsets $A \subset X$ by

$$r(A) = \max_i |A \cap U_i|.$$

Note that:

$$r(A) = 0 \qquad \text{iff} \qquad A = \varnothing.$$

$r(X)$ is the *rank* of G. These definitions are the same as in matroid theory.

PROPOSITION 1. *Consider a set* $A \subset X$, *and the family*

$$\mathscr{U}_A = \{U \cap A \mid U \in \mathscr{U}, U \cap A \neq \varnothing\}.$$

The pair $G_A = (A, \mathscr{U}_A)$ *is also a graphoid, and its rank-function* r_A *is given by*

$$r_A(B) = r(B) \qquad (B \subset A).$$

This graphoid G_A is called the *subgraphoid of G spanned by A*.

3. Truncated Graphoids

Consider an integer $k > 0$, and the family

$$\mathscr{U}_{(k)} = \{V \mid V \subset X; \quad 1 \leq |V| \leq k; \quad V \subset U \quad \text{for some} \quad U \in \mathscr{U}\}.$$

We have:

Proposition 2. *The pair* $G_{(k)} = (X, \mathscr{U}_{(k)})$ *is also a graphoid, and its rank-function* $r_{(k)}$ *is given by*

$$r_{(k)}(A) = \min \{k, r(A)\}.$$

$G_{(k)}$ is called the *truncature by k* of the graphoid G.

From the above definitions, it follows that $G_{(2)}$ is a graph, obtained by joining all the adjacent pairs of points; the graph $G^*_{(2)}$ is called the *representing graph* for graphoid G, i.e., each point represents a set $U \in \mathscr{U}$, and two points are joined if and only if the two corresponding sets intersect. A graph can always be considered as a representing graph for a family of sets, since we have:

Proposition 3. *If* G_0 *is a graph, let* $G = (X, \mathscr{U})$ *be the graphoid where* X *is the set of the vertices of* G_0, *and where* $\mathscr{U}$ *is a family of cliques of* G_0 *such that each edge of* G_0 *is contained in at least one* $U \in \mathscr{U}$; *then* G_0 *is the representing graph of* G^*.

Proof: This follows from the equalities:

$$G_0 = G_{(2)} = G^{**}_{(2)}.$$

In Fig. 1, for instance, G_0 is the representing graph of $G_1{}^*$, $G_2{}^*$, and $G_3{}^*$. A graphoid $G = (X, \mathscr{U})$ is said to have a *faithful graph representation* if the maximal sets of family $\mathscr{U}$ are the maximal cliques of $G_{(2)}$.

Example: Unimodular graphoids.

A matrix A is said to be *totally unimodular* if every subdeterminant is equal to 0, +1, or −1. If A contains only zeros and ones, with at least one 1 in each row and in each column, then it is the incidence matrix of a graphoid $(X, \mathscr{U})$, and we shall call it a *unimodular graphoid*. A unimodular graphoid has a faithful graph representation [1].

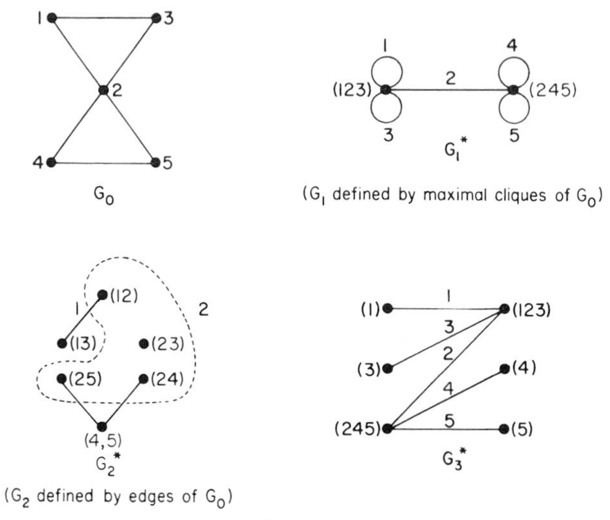

FIG. 1.

A graphoid $(X, \mathscr{U})$ is said to be *balanced* if every odd cycle (i.e., cycle with an odd number of edges) is such that one of its edges contains at least three of its points (or "balanced cycle"). For instance, the graphoid of Fig. 2 is balanced.

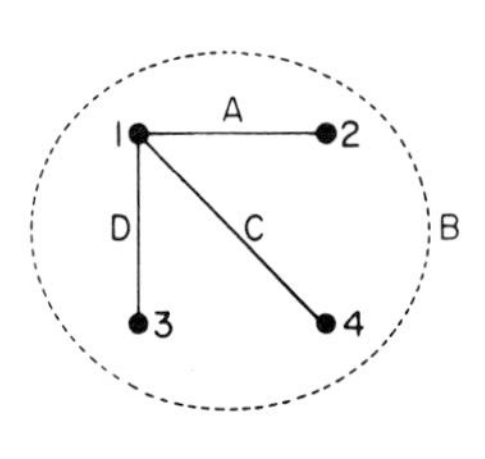

FIG. 2.

EXAMPLE 1. A graph is balanced if and only if it is a bipartite graph.

EXAMPLE 2. A unimodular graphoid is balanced.

If G be a unimodular graphoid, and if there existed an odd cycle $(x_1, U_1, x_2, \ldots, U_p, x_1)$ where each U_i contained only two points in $S = \{x_1, x_2, \ldots, x_p\}$, then the subgraphoid generated by S and by $\mathscr{U}' = \{U_1, U_2, \ldots, U_p\}$ would be an odd cycle, which would contradict the fact that G is unimodular.

EXAMPLE 3. If G is a balanced graphoid, its dual G^* is also a balanced graphoid. (If G^* contained an unbalanced cycle, the corresponding cycle in G would also be unbalanced.)

EXAMPLE 4. If $G = (X, \mathscr{U})$ is a balanced graphoid, every subgraphoid G_A is also balanced; and every graphoid $(X, \mathscr{V})$, where $\mathscr{V} \subset \mathscr{U}$, is also balanced, by the same argument.

PROPOSITION 4. *If G is a balanced graphoid, it has a faithful graph representation.*

PROOF: (1) Let us show that *each clique C of $G_{(2)}$ is contained in an edge of G*. If $|C| = 2$, it is obvious.

Assume that this statement is true for each clique with less than k points, and let us prove it for a clique C with $|C| = k \geq 3$. Let c_1, c_2, c_3 be three distinct points of C; for $i = 1, 2, 3$, there exists an edge U_i of G containing $C - \{c_i\}$ (by the hypothesis of induction). Therefore $(c_1, U_3, c_2, U_1, c_3, U_2, c_1)$ is an odd cycle. As G is balanced, U_1 (or U_3 or U_2) contains $\{c_1, c_2, c_3\}$, and therefore contains C.

(2) Let us denote by $\mathscr{V}$ the family of the maximal sets of $\mathscr{U}$. The graphoid $H = (X, \mathscr{V})$ is balanced (Example 4) and $G_{(2)} = H_{(2)}$.

If C is a maximal clique of $H_{(2)}$, it is contained in a $V \in \mathscr{V}$, and by (1), there exists a clique C_0 with

$$C \subset V \subset C_0.$$

Since C is maximal, we have $C = V$, that is each maximal clique of $G_{(2)}$ belongs to $\mathscr{V}$.

Conversely, if $V \in \mathscr{V}$, it is contained in a maximal clique C of $G_{(2)}$, and there exist a $V_0 \in \mathscr{V}$ with

$$V \subset C \subset V_0.$$

Again, the maximality of the sets of $\mathscr{V}$ implies $V = C$, and this proves that V is a maximal clique of $G_{(2)}$.[1]

4. γ-PERFECT GRAPHOIDS

For a graphoid $G = (X, \mathscr{U})$, a set $S \subset X$ is said to be *stable* if $r(S) = 1$; or in other words, if

$$|S \cap U_j| \leq 1 \qquad (\mathrm{j} = 1, 2, \ldots, \mathrm{m}).$$

The *stability number* $\alpha(G)$ of G is the maximum number of points in a stable set of G. The *chromatic number* $\gamma(G)$ of G is the maximum number of

[1] The concept of families of maximal sets with faithful graph representations has also been considered by Paul Gilmore [5] who has proved that a necessary and sufficient condition for G to have a faithful graph representation is that, for $U_1, U_2, U_3 \in \mathscr{U}$, there exists a $U \in \mathscr{U}$ containing $(U_1 \cap U_2) \cup (U_1 \cap U_3) \cup (U_2 \cap U_3)$.

colors needed to color the points of G such that two points in the same edge have different colors.

We shall denote by $\theta(G)$ the minimum number of edges needed to cover all the points of G.

Graphoid G is said to be *γ-perfect* if

$$\gamma(G_A) = r(A) \qquad (A \subset X).$$

Graphoid G is said to be *α-perfect* if

$$\alpha(G_A) = \theta(G_A) \qquad (A \subset X).$$

EXAMPLE 1 (Bipartite Graphs). If G is a bipartite graph, we have $r(A) = 2$ if A does contain an edge, or $r(A) = 1$, otherwise. Therefore, $\gamma(G_A) = r(A)$ and the bipartite graph is γ-perfect. (Note that Königs' theorem states that a bipartite graph is α-perfect.)

EXAMPLE 2 (The Family of the Cliques in a Triangulated Graph). A graph G_0 is said to be triangulated if every cycle of length greater than 3 posseses a chord. It is known [1] that its cliques constitute a γ-perfect and α-perfect graphoid.

The same results hold for comparability graphs [1], for unimodular graphoids [1], and also for some classes of graphs recently studied by T. Gallai.

For a γ-perfect graphoid, we have the following properties.

PROPERTY 1. *If G is a γ-perfect graphoid, and if $r(X) = k$, then X is the union of k stable sets $S_1, S_2, \ldots, S_k$.*

PROOF: If $r(X) = k$, set X can be colored with k colors, each set S of vertices with the same color being a stable set, since

$$r(S) = \gamma(G_S) = 1.$$

PROPERTY 2. *If $G = (X, \mathscr{U})$ is a γ-perfect graphoid, it has a faithful graph representation.*

PROOF: Let $\mathscr{V}$ be the family of all the maximal sets of $\mathscr{U}$, and consider the graphoid $H = (X, \mathscr{V})$. We have

$$r_H(A) = r_G(A) = \gamma(G_A) = \gamma(H_A)$$

Therefore, the graphoid H is also γ-perfect.

We want to prove that $\mathscr{V}$ is the family of all the maximal cliques of $G_{(2)} = H_{(2)}$.

(1) If C is maximal clique of $H_{(2)}$, we have $\gamma(H_c) = |C|$, therefore $r_H(C) = |C|$, and C is contained in an edge $V \in \mathscr{V}$. There exists a clique C_0 such that $C \subset V \subset C_0$, so that $C = V$: each maximal clique of $H_{(2)}$ belongs to $\mathscr{V}$.

(2) Conversely, if $V \in \mathscr{V}$, it is contained in a maximal clique C of $H_{(2)}$ and by (1) there exist, an edge $V_0 \in \mathscr{V}$ such that $V \subset C \subset V_0$. Again, from the maximality of the sets of $\mathscr{V}$, we have $V = C$: each edge of $\mathscr{V}$ is a maximal clique of $H_{(2)}$.

PROPERTY 3. *If G is a γ-perfect graphoid, for every odd cycle $(x_1, U_1, x_2, U_2, \ldots, U_{2k+1}, x_{2k+2} = x_1)$, there exists in $\mathscr{U}$ an edge which contains at least three points of the cycle.*

PROOF: Assume that there exists an odd cycle for which no three points belong to the same edge, and consider the shortest cycle of that kind:

$$(x_1, U_1, x_2, U_2, \ldots, x_{2k+1}, U_{2k+1}, x_1).$$

Two points x_i, x_j of the cycle, with $j \neq i-1, i+1$, and $i \neq j-1, j+1$, are nonadjacent; because if they were adjacent, assuming $j > i$, one of the cycles $[x_1, x_2, \ldots, x_i, x_j, x_{j+1}, \ldots, x_1]$ or $[x_i, x_{i+1}, \ldots, x_j, x_i]$ would be an odd cycle shorter than the initial one, with no edge U_0 containing three of its points; this would contradict the minimality of the length of our initial cycle.

Let us put $A = \{x_1, x_2, \ldots, x_{2k+1}\}$; we have $\gamma(G_A) = 3$, and $r(A) = 2$, which contradicts the fact that graphoid G is γ-perfect.

THEOREM 1. *If $G = (X, \mathscr{U})$ is a γ-perfect graphoid, then its stability number $\alpha(G)$ is equal to*

$$k_0 = \min\left\{k \;/\; k \text{ integer} \geq 0; \quad \min_{A \subset X}(kr(A) - |A|) = 0\right\}.$$

LEMMA 1. *For every graphoid G, we have $\alpha(G) \leq k_0$.*

PROOF: If k is an integer less than $\alpha(G)$, there exists a stable set S with $|S| > k$, therefore:

$$\min_{A \subset X}(kr(A) - |A|) \leq kr(S) - |S| = k - |S| < 0.$$

Since $\min\,(k_0\, r(A) - |A|) = 0$, we have necessarily $k < k_0$. As this is true for every integer $k < \alpha(G)$, we have $\alpha(G) \leq k_0$.

Note that this lemma is, in some sense, best possible; for instance, if G is a polygon with seven vertices, we have $\alpha = 3$, $k_0 = 4$.

Let us now turn to the proof of Theorem 1.

PROOF: If G is γ-perfect, consider a set $A \subset X$, colored with $t = r(A)$ colors. Let A_i be the set of all the vertices in A having color (i).

$$|A| = |A_1| + |A_2| + \cdots + |A_t| \leq t\alpha(G_A) \leq r(A)\alpha(G)$$

Therefore

$$\min_{A \subset X}(\alpha(G)r(A) - |A|) = 0;$$

hence $\alpha(G) \geq k_0$. But by Lemma 1, we have $\alpha(G) \leq k_0$, and therefore:

$$\alpha(G) = k_0 .$$

THEOREM 2. *For a graphoid G, both α-perfect and γ-perfect, k edges are enough to cover all the points if and only if*

$$kr(A) - |A| \geq 0 \qquad (A \subset X).$$

PROOF: If G is α-perfect and γ-perfect, by Theorem 1 we have

$$\theta(G) = \alpha(G) = k_0 .$$

It is possible to cover X with k edges if and only if $k \geq k_0$, that is,

$$\min\,(kr(A) - |A|) = 0$$

or:

$$kr(A) - |A| \geq 0 \qquad (A \subset X).$$

The same condition is known for matroids (see Edmonds [3] and Nash-Williams [6]).

5. CONJECTURES

In some previous papers (see Berge [2]), we stated the following conjecture:

CONJECTURE 1 (the "strong" coloring conjecture). *For a graphoid $G = (X, \mathcal{U})$ with a faithful graph representation, the condition*

$$\gamma(G_A) = r(A) \qquad (A \subset X)$$

is equivalent to

$$\alpha(G_A) = \theta(G_A) \qquad (A \subset X).$$

In other words, the "α-perfect" property is equivalent to the "γ-perfect" property. (It is obvious that if one yields the other, they are equivalent.)

The concepts developed in the present paper bring two new conjectures, closely related to the preceding one.

CONJECTURE 2 (the "weak" covering conjecture). *For a γ-perfect graphoid $G = (X, \mathcal{U})$, k edges are enough to cover all the points if and only if*

$$kr(A) - |A| \geq 0 \qquad (A \subset X)$$

(the "only if" part is obvious).

CONJECTURE 3. *For a graphoid $G \neq (X, \mathcal{U})$, we have*

$$\gamma(A, \mathcal{V}_A) = r(A, \mathcal{V}_A) \qquad (A \subset X, \mathcal{V} \subset \mathcal{U})$$

if and only if G is a balanced graphoid (the "only if" part is obvious).

By Theorem 1, we see that Conjecture 3 implies Conjecture 1.

REFERENCES

1. C. BERGE, Some Classes of Perfect Graphs, in *Graph Theory and Theoretical Physics* (F. Harary, ed.), Academic Press, New York, 1967.
2. C. BERGE, Färbung von Graphen, deren sämtliche bzw deren ungerade Kreise starr sind (Zusammenfassung), *Wiss. Z. Martin-Luther-Univ. Halle-Wittenberg Math.-Natur. Reihe* (1961), 114.
3. J. EDMONDS, Partition of a Matroid into Independent Subsets, *J. Res. Nat. Bur. Standards Sect. B* (1965), 67–72.
4. J. EDMONDS, A Minimum Element-Covering from a Finite Class of Weighted Sets, April meeting of Amer. Math. Soc., 1961, pp. 578–592.
5. P. GILMORE, Families of Sets with Faithful Graph Representations, Res. Note N.C. 184, IBM, December 1962.
6. C. ST. J. A. NASH-WILLIAMS, Decomposition of Finite Graphs into Forests, *J. London Math. Soc.* **39** (1964), 12.
7. D. K. RAY-CHAUDHURI, An Algorithm for a Maximum Cover of an Abstract Complex, *Canad. J. Math.* **15** (1963), 11–24.

ENUMERATION OF TREE-SHAPED MOLECULES

N. G. de Bruijn

DEPARTMENT OF MATHEMATICS
TECHNOLOGICAL UNIVERSITY
EINDHOVEN, THE NETHERLANDS

1. Introduction

In 1937 Pólya published his now famous paper "Kombinatorische Anzahlbestimmungen für Gruppen, Graphen und Chemische Verbindungen" [3], which was devoted mainly to the enumeration of several kinds of molecules which we shall refer to as tree-shaped. For rooted trees and rooted planar trees, which are special cases of tree-shaped molecules, some results had been obtained already by Cayley (see [3] for references).

In Pólya's paper the exposition depended on geometric intuition without proper formal definition of the objects that were enumerated. For part of the contents of the paper, i.e., what we now call Pólya's Fundamental Theorem, it is comparatively easy to translate the original exposition into the modern terminology of mappings and equivalences. (We refer to [1] for the terminology to be used in the present paper.) The part about trees and molecules, however, is much harder to translate in terms of formal mathematics. It is the purpose of this paper to begin this translation.

As is often the case with intuitive notions, there are several ways to put them into a formal framework, and it is by no means easy to express the equivalence between the different ways. We shall take only one point of view and disregard the others. In particular, we disregard the possibility of defining trees and molecules as figures in a Euclidean space of dimension 2 or 3 and of defining equivalences on the basis of topological transformations. We shall take a purely combinatorial point of view, which we can even maintain for things like planar trees. (For planar graphs this is much harder: we do have Kuratowski's combinatorial characterization for embeddability of a graph into a plane, but we do not seem to have a definition of the notion of a planar graph as a combinatorial object in such a way that there is a one-to-one correspondence between these objects and topological equivalence classes of graphs in a plane.)

An atom can be considered as a "kernel", i.e. a set whose elements are called "places," and a group of permutations of the places. Molecules can be built by taking a number of atoms and connecting the places pairwise by "bonds" (a bond has to link two places belonging to distinct atoms). For our purpose we shall consider a type of atom where one of the places is specially designated as a "root place," and this root place is to be invariant under the permutations. Therefore we shall discuss the set E of all remaining places, and a group H of permutations of E. We want to consider tree-shaped molecules built from such atoms. These trees should be orientable such that all bonds are directed towards some root place. If we use this terminology, a single extra place has to be created in order to serve as root of the whole tree: it has to be linked by a bond directed towards the root place of the atom at the root of the tree. However, we can modify the description slightly to the effect that the root places are left out entirely (see Section 4).

A large part of Pólya's paper deals with formulas for generating functions of the set of all tree-shaped molecules produced from a given store of atoms. This includes cases like planar trees and topological trees.

In this paper we shall go a bit further along this line. Instead of counting the number of molecules of a given type we shall evaluate the sum of the cycle indices of the automorphism groups of such molecules. Such sums can be used for various combinatorial problems concerning the structures under consideration (cf. [2]).

As a preliminary notion we introduce the word *crown*. For this kind of object we shall prove an enumeration theorem that generalizes both Pólya's Fundamental Theorem and Pólya's theorem about the wreath product ("Kranzgruppe").

2. Notations

$\mathscr{R}$ is a commutative ring containing the rationals as a subring.

If S is a finite set, then $|S|$ denotes the number of its elements.

If A and B are sets, then A^B is the set of mappings of B into A.

If D is a finite set, and g is a permutation of D, then g splits into cycles. The number of cycles of length j is denoted by $b_j(g)$.

If G is a group of permutations of D, then its *cycle index* $P_G(x_1, x_2, \ldots)$ is defined by

$$P_G(x_1, x_2, \ldots) = |G|^{-1} \sum_{g \in G} x_1^{b_1(g)} x_2^{b_2(g)} \cdots .$$

3. Crowns and Their Groups

An *atomic store* is a finite set I with a triple (E_i, H_i, v_i) attached to every $i \in I$, where E_i is a finite set, H_i a group of permutations of E_i, and v_i an element of $\mathscr{R}$. P_{H_i} is the cycle index of H_i. If E_i is empty, then H_i consists of the identity permutation only, and we define $P_{H_i} = 1$.

The v_i's are called *weights*, $\sum_{i \in I} v_i$ is called the *inventory* of the store, and

$$\sum_{i \in I} v_i P_{H_i}(x_1, x_2, \ldots)$$

is called the *cycle index inventory* of the store.

Let D be a finite set, and G be a group of permutations of D. We shall use the word "crown" if an atom from the store is attached to every $d \in D$. That is, a *crown* is simply a mapping of D into I.

We want to consider crowns as new atoms, and to build up a new atomic store. It will consist of an index set F with a triple (B_f, K_f, w_f) for every $f \in F$.

We have an equivalence relation in the set of crowns: $f_1 \in I^D, f_2 \in I^D$ are called equivalent iff $f_1 \in f_2 G$. Accordingly, I^D splits into equivalence classes, to be called *(D, G)-crown patterns*. Let F be some complete set of representatives. That is, F has just one element in every equivalence class.

If $f \in I^D$, we define the set

$$B_f = \{(d, s) \mid d \in D, s \in E_{f(d)}\}.$$

Moreover we define

$$G_f = \{g \in G \mid fg = f\},$$

i.e., G_f is the group of all $g \in G$ leaving f invariant. And Ω_f will denote the set of all choice functions ω that attach to every $d \in D$ an $\omega(d) \in H_{f(d)}$. Hence

$$|\Omega_f| = \prod_{d \in D} |H_{f(d)}|. \tag{1}$$

For every $g \in G_f$, $\omega \in \Omega_f$ we define $\kappa^{(f)}(g, \omega)$ as follows. $\kappa^{(f)}(g, \omega)$ is a permutation of B_f, defined by

$$(\kappa^{(f)}(g, \omega))(d, s) = (g(d), \omega(d)s) \tag{2}$$

for all $(d, s) \in B_f$. It is not difficult to show that the permutations $\kappa^{(f)}(g, \omega)$ and $\kappa^{(f)}(g', \omega')$ are distinct for distinct pairs (g, ω) and (g', ω'), and that all such permutations form a group of permutations of B_f. Multiplication is effected as follows:

$$(\kappa^{(f)}(g', \omega'))(\kappa^{(f)}(g, \omega)) = \kappa^{(f)}(g'', \omega'')$$

with

$$g'' = g'g, \qquad \omega''(d) = \omega'(g(d))\omega(d).$$

Note that $(\kappa^{(f)}(g, \omega))^{-1} = \kappa^{(f)}(g''', \omega''')$ where $g''' = g^{-1}$, and ω''' is defined by $\omega'''(d) = (\omega(g^{-1}(d)))^{-1}$ for all $d \in D$.

The set $\{\kappa^{(f)}(g, \omega) \mid g \in G, \omega \in \Omega_f\}$ with the multiplication just defined is *crown group* K_f which is a group of permutations of B_f. Obviously

$$|K_f| = |G_f| \cdot |\Omega_f|. \tag{3}$$

Finally we define the weight w_f:

$$w_f = \prod_{d \in D} v_{f(d)}. \tag{4}$$

We want to know what happens to (B_f, K_f, w_f) if we replace the function f by an equivalent one, viz., fp, with some $p \in G$. Let τ_p denote the one-to-one mapping of B_f onto B_{fp} given by

$$\tau_p(d, s) = (p^{-1}(d), s) \qquad (d \in D, s \in E_{f(d)}).$$

We can check that

$$\kappa^{(f)}(g, \omega) = \tau_p^{-1}\kappa^{(fp)}(g^*, \omega^*)\tau_p$$

for all $g \in G_f$, $\omega \in \Omega_f$, where

$$g^* = p^{-1}gp \in G_{fp}, \qquad \text{and} \qquad \omega^* = \omega p \in \Omega_{fp}.$$

If $\kappa^{(f)}_{g, \omega}$ runs through K_f then $\kappa^{(fp)}_{g^*, \omega^*}$ runs through K_{fp}. So K_f and K_{fp} are conjugate permutation groups:

$$K_f = \tau_p^{-1} K_{fp} \tau_p. \tag{5}$$

The special (B_f, K_f, w_f)'s with f's taken from our complete set of representatives F, form a new atomic store, to be called the *crown store*, described by the index set F with triples (B_f, K_f, w_f). Its *cycle index inventory* is

$$U(x_1, x_2, \ldots) = \sum_{f \in F} w_f P_{K_f}(x_1, x_2, \ldots). \tag{6}$$

Although the crown store depends on the special selection of the complete set of representatives, the cycle index inventory does not. This is true because in the first place, equivalent f's have the same weight (since the factors in (4) are permuted if f is replaced by fg). Secondly, equivalent f's produce conjugate K_f's [see (5)], and conjugate permutation groups have the same cycle index. Thus $U(x_1, x_2, \ldots)$ depends only on D, G, and on the original atomic store I. The following theorem expresses U in terms of the cycle indices of G and of the H_i's.

THEOREM 1. *If* V_m, $m = 1, 2, \ldots$, *is defined by*

$$V_m = \sum_{i \in I} (v_i)^m P_{H_i}(x_m, x_{2m}, x_{3m}, \ldots), \tag{7}$$

then

$$U(x_1, x_2, x_3, \ldots) = P_G(V_1, V_2, V_3, \ldots). \tag{8}$$

PROOF: If $f \in I^D$ then the orbit of f (i.e., the equivalence class to which f belongs) contains $|G|/|G_f|$ elements. Since w_f and P_{K_f} are constant on each class, we have

$$\begin{aligned} U(x_1, x_2, \ldots) &= \sum_{f \in I^D} w_f(|G|/|G_f|)^{-1} P_{K_f}(x_1, x_2, \ldots) \\ &= |G|^{-1} \sum_{i \in I^D} w_f(|G_f|/|K_f|) \sum_{\gamma \in K_f} x_1^{b_1(\gamma)} x_2^{b_2(\gamma)} \cdots \\ &= |G|^{-1} \sum_{f \in I^D} w_f(|G_f|/|K_f|) \sum_{g \in G_f} \sum_{\omega \in \Omega_f} x_1^{b_1(f, g, \omega)} x_2^{b_2(f, g, \omega)} \cdots. \end{aligned}$$

Here $b_j(f, g, \omega)$ denotes the number of cycles of length j in the permutation $\kappa^{(f)}(g, \omega)$ [see (2)]. Using (3) we obtain

$$\begin{aligned} &U(x_1, x_2, \ldots) \\ &\quad = |G|^{-1} \sum_{f \in I^D} w_f \sum_{g \in G, fg = f} \mathscr{M}_{\omega \in \Omega_f}\{x_1^{b_1(f, g, \omega)} x_2^{b_2(f, g, \omega)} \cdots\} \\ &\quad = |G|^{-1} \sum_{g \in G} \sum_{f \in I^D, fg = f} w_f \mathscr{M}_{\omega \in \Omega_f}\{x_1^{b_1(f, g, \omega)} x_2^{b_2(f, g, \omega)} \cdots\}. \end{aligned}$$

Here $\mathscr{M}_{\omega \in \Omega}$ denotes the mean taken over Ω_f, i.e.,

$$\mathscr{M}_{\omega \in \Omega_f} = |\Omega_f|^{-1} \sum_{\omega \in \Omega_f}.$$

In order to prove the theorem, it suffices to show that for every $g \in G$

$$\sum_{f \in I^D, fg = f} w_f \mathscr{M}_{\omega \in \Omega_f}\{x_1^{b_1(f, g, \omega)} x_2^{b_2(f, g, \omega)} \cdots\} = V_1^{b_1(g)} V_2^{b_2(g)} \cdots, \tag{9}$$

since (8) follows from (9) by taking $\mathscr{M}_{g \in G}$ on both sides.

From now on we keep $g \in G$ fixed.

D splits into cycles $Z_1, \ldots, Z_t$. Each Z_ν is permuted cyclically by g. Let $l(\nu)$ denote the length of the νth cycle: $l(\nu) = |Z_\nu|$. The $f \in I^D$ with $fg = f$ are constant on each Z_ν, whence they can be described uniquely by mappings φ of $\{1, \ldots, t\}$ into I: $f(d) = \varphi(\nu)$ if $d \in Z_\nu$. We can describe the weight w_f in terms of φ:

$$w_f = z_\varphi, \qquad \text{where} \quad z_\varphi = \prod_{\nu=1}^{t} (v_{\varphi(\nu)})^{l(\nu)}. \tag{10}$$

The left-hand side of (9) can now be written as

$$\sum_{\varphi} z_{\varphi} \mathcal{M}_{\omega \in \Omega_f} \{x_1^{b_1(f, g, \omega)} x_2^{b_2(f, g, \omega)} \cdots\}; \tag{11}$$

φ runs through $I^{\{1, \ldots, t\}}$.

The set B_f, permuted by $\kappa^{(f)}(g, \omega)$, can be written as the disjoint union $\bigcup_{v=1}^{t} (Z_v \times E_{\varphi(v)})$, and $\kappa^{(f)}(g, \omega)$ permutes each $Z_v \times E_{\varphi(v)}$ separately. And we can describe ω by means of its restrictions $\lambda_1, \ldots, \lambda_t$ to $Z_1, \ldots, Z_t$, respectively. Accordingly, the restriction of $\kappa^{(f)}(g, \omega)$ to $Z_v \times E_{\varphi(v)}$ can be expressed as $\sigma(g, \lambda_v)$, where

$$\sigma(g, \lambda_v)(d, s) = (g(d), (\lambda_v(d))s) \qquad (d \in Z_v, s \in E_{\varphi(v)}).$$

If $b_j(v, \lambda_v)$ is the number of cycles of length j into which $\sigma(g, \lambda_v)$ splits Z_v, then we have

$$b_j(f, g, \omega) = \sum_{v=1}^{t} b_j(v, \lambda_v).$$

It is now easy to see that the mean in (11) equals $M_1 \cdots M_t$, where

$$M_v = \mathcal{M}_{\lambda \in \Lambda_v} \{x_1^{b_1(v, \lambda)} x_2^{b_2(v, \lambda)} \cdots\},$$

with $\Lambda_v = (H_{\varphi(v)})^{Z_v}$.

De Bruijn [1 (proof of Pólya's Kranz theorem, p. 177)] obtains a survey of all $b_j(v, \lambda)$,with the result

$$M_v = P_{H_{\varphi(v)}}(x_{l(v)}, x_{2l(v)}, \ldots).$$

So finally we obtain for (11)

$$\sum_{\varphi} \prod_{v=1}^{t} \{(v_{\varphi(v)})^{l(v)} P_{H_{\varphi(v)}}(x_{l(v)}, x_{2l(v)}, \ldots)\},$$

where the summation is over all $\varphi \in I^{\{1, \ldots, t\}}$. The latter sum is obviously

$$\prod_{v=1}^{t} \sum_{i \in I} (v_i)^{l(v)} P_{H_i}(x_{l(v)}, x_{2l(v)}, \ldots) = \prod_{v=1}^{t} V_{l(v)},$$

and this is the same thing as

$$V_1^{b_1(g)} V_2^{b_2(g)} \cdots.$$

This completes the proof.

Special Cases: 1. If we take $E_i = \varnothing$ for all i, then we have

$$P_{K_f}(x_1, x_2, \ldots) = 1, \qquad P_{H_i}(x_1, x_2, \ldots) = 1,$$

and (7) becomes

$$\sum_{f \in F} w_f = P_G\left(\sum_i v_i, \sum_i v_i^2, \ldots\right);$$

this is Pólya's fundamental theorem.

2. If we have only one element 0, say, in I, then we have only one f, and in this case P_{K_f} is Pólya's Kranz group (see [1] or [3]). Indeed, if we take $v_i = 1$ we have $w_f = 1$, and (7) gives

$$P_K(x_1, x_2, \ldots) = P_G(P_H(x_1, x_2, \ldots), P_H(x_2, x_4, \ldots), P_H(x_3, x_6, \ldots), \ldots).$$

4. Tree-Shaped Molecules Generated by an Atomic Store

Let Λ be an atomic store. It consists of an index set I and a triple (E_i, H_i, v_i) for each $i \in I$ (see Section 3). We shall assume that I has an element 0 for which E_0 is the empty set.

A Λ-*molecule* is a quadruple (S, θ, σ, π) with the following specifications.

(i) S is a finite nonempty set (its elements are called points).

(ii) θ is a mapping of S into I. For all $P \in S$ we have $|E_{\theta(P)}| = |\sigma(P)|$.

(iii) σ attaches a subset $\sigma(P)$ of S to every $P \in S$. If $P \in S$, $Q \in S$, $P \neq Q$, then $\sigma(P) \cap \sigma(Q) = \varnothing$. There is a point $P_0 \in S$ (called the *root*) such that $P_0 \in \sigma(P)$ for no $P \in S$, and for every $P \in S$, $P \neq P_0$ there exists a finite sequence $P_0, P_1, \ldots, P_m = P$ such that $P_j \in \sigma(P_{j-1})$ $(j = 1, \ldots, m)$.

(iv) π attaches an element $\pi(P)$ of Γ_P to every $P \in S$. This Γ_P is defined as follows. In the set of all one-to-one mappings of $E_{\theta(P)}$ onto $\sigma(P)$ we can consider equivalence by defining $f_1 \sim f_2$ iff $f_1 \in f_2 H_{\theta(P)}$. The set of equivalence classes is called Γ_P. So for every P there is an f that maps $E_{\theta(P)}$ one-to-one onto $\sigma(P)$, such that $\pi(P) = f H_{\theta(P)}$.

The *weight* of the molecule (S, θ, σ, π) is defined as $\Pi_{P \in S}\, v_{\theta(P)}$. (It is not to be confused with the notion of molecular weight in chemistry, which is a sum instead of a product of atomic weights.)

Let (S, θ, σ, π) be a molecule, and let λ be a one-to-one mapping of S onto a set S_1. We shall define a new Λ-molecule $(S_1, \theta_1, \sigma_1, \pi_1)$. The operation depends on λ, and will be denoted by Ξ_λ:

$$\Xi_\lambda(S, \theta, \sigma, \pi) = (S_1, \theta_1, \sigma_1, \pi_1),$$

where

$$S_1 = \lambda S, \qquad \theta_1 = \theta\lambda^{-1},$$

$$\sigma_1 = \lambda\sigma\lambda^{-1} \qquad (\text{i.e., } \sigma_1(Q) = \{\lambda(P) \mid P \in \sigma(\lambda^{-1}(Q))\} \quad \text{for all} \quad Q \in S_1),$$

$$\pi_1 = \lambda\pi \qquad (\text{i.e., } \pi_1(Q) = \{\lambda f \mid f \in \pi(\lambda^{-1}(Q))\} \quad \text{for all} \quad Q \in S_1).$$

Notice that $H_{\theta(P)} = H_{\theta_1(P_1)}$. It is not difficult to show that $(S_1, \theta_1, \sigma_1, \pi_1)$ is again a Λ-molecule, and that $\Xi_\lambda \Xi_\mu = \Xi_{\lambda\mu}$ whenever $\lambda\mu$ makes sense.

Two Λ-molecules are called *equivalent* iff there is a Ξ_λ transforming one into the other.

If λ is such that

$$\Xi_\lambda(S, \theta, \sigma, \pi) = (S, \theta, \sigma, \pi),$$

then λ is called an automorphism of (S, θ, σ, π). These λ's form a group of permutations of S, the *automorphism group* of (S, θ, σ, π).

Equivalent Λ-molecules have the same weight. Their automorphism groups are conjugate, whence they have the same cycle index.

We form a collection T by selecting a molecule from each equivalence class of Λ-molecules, and for every n ($n = 1, 2, 3, \ldots$) we construct a new atomic store T_n, consisting of the triples $(S_t, M_t, (W_t)^n)$. Here t runs through T, S_t is the point set of t, M_t the automorphism group of t, and W_t the weight of t.

The cycle index inventory of T_n is

$$U_n(x_1, x_2, \ldots) = \sum_{t \in T} (W_t)^n P_{M_t}(x_1, x_2, x_3, \ldots). \tag{12}$$

Note that U_n is a formal power series in $x_1, x_2, x_3, \ldots$ with infinitely many terms, although the atomic store is finite.

The coefficients of the U_n's can be evaluated by means of the following theorem.

THEOREM 2. *For* $n = 1, 2, 3, \ldots$ *we have*

$$\begin{aligned} &U_n(x_1, x_2, x_3, \ldots) \\ &\quad = x_1 \sum_{i \in I} v_i^n P_{H_i}(U_n(x_1, x_2, x_3, \ldots), U_{2n}(x_2, x_4, x_6, \ldots), \\ &\qquad U_{3n}(x_3, x_6, x_9, \ldots), \ldots). \end{aligned}$$

PROOF: Every Λ-molecule has a root. If P_0 is the root of (S, θ, σ, π), then $\theta(P_0)$ indicates the atom placed at that root. This $\theta(P_0)$ is simply an element of I. If $\theta(P_0) = i$, then we say that the molecule is rooted by atom i. We can split U_n [defined by (12)] according to the value of this i. Therefore it suffices to show that the sum

$$\sum_{t \in T,\, t \text{ rooted by } i} (W_t)^n P_{M_t}(x_1, x_2, x_3, \ldots) \tag{13}$$

is equal to

$$x_1 v_i^n P_{H_i}(U_n(x_1, x_2, x_3, \ldots), U_{2n}(x_2, x_4, \ldots), \ldots). \tag{14}$$

The Λ-molecules rooted by i (with weights $(W_t)^n$) can be brought into one-to-one correspondence with the (E_i, H_i)-crown patterns formed with the atomic store T_n. It can be shown that the automorphism group of such a molecule is the same as the crown group apart from the fact that the automorphism group operates on one extra element, the root, which it leaves invariant. Moreover taking the weight of the root into account we conclude that (13) equals $x_1 v_i^n$ times the cycle index inventory of the (E_i, H_i)-crown patterns. By Theorem 1 the latter inventory equals

$$P_{H_i}(V_1, V_2, V_3, \ldots),$$

where, according to (7)

$$V_m = \sum_{t \in T} (W_t)^{mn} P_{M_t}(x_m, x_{2m}, \ldots) = U_{nm}(x_m, x_{2m}, \ldots).$$

5. EXAMPLES

A. If we take $v_i = 1$ for all i, then Theorem 2 gives a functional equation for a single function $U = U_1 = U_2 = \cdots$.

$$U(x_1, x_2, \ldots) = x_1 \sum_{i \in I} P_{H_i}(U(x_1, x_2, \ldots), U(x_2, x_4, \ldots), \ldots). \tag{15}$$

B. If $v_i = 1$ for all i, and if we put $x_k = x^k$ for all k, then

$$U(x_1, x_2, \ldots) = \sum_{t \in T} x^{\nu(t)} = \sum_{n=1}^{\infty} a_n x^n, \tag{16}$$

where $\nu(t)$ is the number of points of the molecule t, and a_n is the number of $t \in T$ with $\nu(t) = n$. Putting $\sum_1^\infty a_n x^n = f(x)$, we have by (15):

$$f(x) = x \sum_{i \in I} P_{H_i}(f(x), f(x^2), f(x^3), \ldots) \tag{17}$$

C. If again $v_i = 1$ for all i, we have [see (12)]

$$U_n(x, 0, 0, \ldots) = \sum_{t \in T} P_{M_t}(x, 0, 0, \ldots) = \sum_{t \in T} x^{\nu(t)}/|M_t|,$$

which does not depend on n. Note that $(\nu(t))!/|M_t|$ is the number of essentially different (i.e., different up to automorphism) ways to label the points of the molecule t with labels $1, \ldots, \nu(t)$. So if b_n is the number of essentially different labelled molecules with n points, then

$$g(x) = \sum_{n=1}^{\infty} b_n x^n/n!. \tag{18}$$

From Theorem 2 we obtain the equation

$$g(x) = x\psi(g(x)) \tag{19}$$

where

$$\psi(z) = \sum_{i \in I} P_{H_i}(z, 0, 0, \ldots) = \sum_{i \in I} z^{|E_i|}/|H_i|.$$

Now $g(x)$ can be obtained from (19) by Lagrange's inversion formula.

D. We can obtain the so-called planted planar trees as special tree-shaped molecules by taking the following atomic store: $I = \{0, 1, 2, \ldots\}$; E_i is a set with i elements; H_i consists of the unit permutation only. Thus we obtain for the generating function, as a special case of (17),

$$f(x) = x \sum_{i \in I} (f(x))^i$$

which leads to the known result $a_n = (2n-2)!/(n!(n-1)!)$.

E. If we modify the previous example by making H_i equal to the full symmetric group, we get the case of rooted topological trees. We now have

$$\sum_{i \in I} x^i P_{H_i}(x_1, x_2, \ldots) = \exp\left(\frac{xx_1}{1} + \frac{x^2 x_2}{2} + \frac{x^3 x_3}{3} + \cdots\right)$$

(see [1] or [3]), and accordingly, if

$$U(x_1, x_2, \ldots) = \sum_{t \in T} P_{M_t}(x_1, x_2, \ldots),$$

we obtain by Theorem 2 (all $v_i = 1$):

$$U(x_1, x_2, \ldots) = x_1 \exp(U(x_1, x_2, \ldots) + \tfrac{1}{2} U(x_2, x_4, \ldots) + \tfrac{1}{3} U(x_3, x_6, \ldots) + \cdots).$$

The special case $x_j = x^j$ gives for $f(x) = \sum_{t \in T} x^{v(t)}$ the equation

$$f(x) = x \exp\left(f(x) + \frac{f(x^2)}{2} + \frac{f(x^3)}{3} + \cdots\right). \tag{20}$$

Similarly, the case $x_1 = x$, $x_2 = x_3 = \cdots = 0$ gives for the generating function of the different labeled topological trees [see (18)]

$$g(x) = x \exp(g(x)). \tag{21}$$

Formulas (20) and (21) are due to Cayley (see [3]).

References

1. N. G. de Bruijn, Pólya's Theory of Counting, in *Applied Combinatorial Mathematics* (E. F. Beckenbach, ed.), Chapt. 5, Wiley, New York, 1964.
2. N. G. de Bruijn, Enumerative Combinatorial Problems Concerning Structures, *Nieuw Arch. Wisk.* (3) **17** (1963), 142–161.
3. G. Pólya, Kombinatorische Anzahlbestimmungen für Gruppen, Graphen und Chemische Verbindungen, *Acta Math.* **68** (1937), 145–254.

A COMBINATORIAL APPROACH TO THE BELL POLYNOMIALS AND THEIR GENERALIZATIONS

Roberto Frucht

UNIVERSIDAD TÉCNICA F. SANTA MARIA
VALPARAISO, CHILE

The Bell polynomials were known long before Eric Temple Bell studied them. Except for a slight change in notation, they are nothing more than the well-known formulas for the higher-order derivatives of a function of a function.

Let $y = f(g(x))$; then by repeated application of the chain rule of ordinary differential calculus:

$$\begin{aligned} y' &= f'(g(x)) \cdot g'(x), \\ y'' &= f''(g(x)) \cdot [g'(x)]^2 + f'(g(x)) \cdot g''(x), \\ y''' &= f'''(g(x)) \cdot [g'(x)]^3 + 3f''(g(x)) \cdot g'(x) \cdot g''(x) + f'(g(x))g'''(x) \\ &\cdots. \end{aligned} \tag{1}$$

The Bell polynomials are now obtained by suppressing the arguments x and $g(x)$, and replacing primes by subindexes:

$$\begin{aligned} A_1(f_1; g_1) &= f_1 g_1, \\ A_2(f_1, f_2; g_1, g_2) &= f_1 g_2 + f_2 g_1^2, \\ A_3(f_1, f_2, f_3; g_1, g_2, g_3) &= f_1 g_3 + 3 f_2 g_1 g_2 + f_3 g_1^3, \\ A_4(f_1, \ldots, f_4; g_1, \ldots, g_4) &= f_1 g_4 + f_2(4 g_1 g_3 + 3 g_2^2) + 6 f_3 g_1^2 g_2 + f_4 g_1^4, \\ &\cdots. \end{aligned} \tag{2}$$

It is obvious that the nth Bell polynomial is a function of $2n$ independent variables $f_1, f_2, \ldots, f_n; g_1, g_2, \ldots, g_n$; it is linear in f_i and (for $n > 1$) not linear in g_i. A table up to $n = 7$ is to be found in [7, p. 49]. In the same book, explicit and recurrence formulas, generating functions, etc., are also given for these polynomials. [In the case of the recursion formulas, it is convenient to add to (2) a polynomial of order zero:

$$A_0 = 1.] \tag{2'}$$

Let us now consider another problem which, at first glance, has nothing to do with Bell polynomials: that of the partially ordered set of the partitions of a set containing n elements (where the partial order is introduced in a natural way via refinements). For example, for $n = 3$ we have the "trivial" partition

abc

where the three elements form one block; there exist three cases where one element has been separated from the other two; and, finally, the "discrete" partition comprised only of "atoms" (i.e., blocks of one element each). (See Fig. 1.)

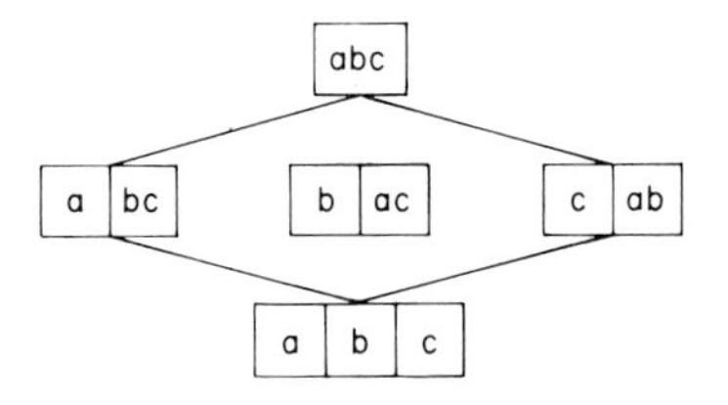

FIG. 1.

In general, to any partition can be assigned a type $(k_1, k_2, \ldots, k_n)$, where k_1 is the number of atoms, k_2 the number of couples, k_3 the number of blocks with three elements, etc. For example, in the case $n = 3$ just considered, we have the type (0, 0, 1) for the trivial partition, three partitions of type (1, 1, 0), and (3, 0, 0) as the type of the discrete partition.

It is, of course, obvious (and well known) that

$$k_1 + 2k_2 + 3k_3 + \cdots + nk_n = n, \tag{3}$$

while

$$i = k_1 + k_2 + k_3 + \cdots + k_n \tag{4}$$

gives us the number of blocks (or "parts") present in the partition under consideration.

It seems, however, that until the publication of [1] in 1965 it had not yet been realized[1] that the nth Bell polynomial could be obtained easily in the following manner:

[1] It seems that only the related case of the cycle indicator of the symmetric group was known; see [7].

Given a partition of the type $(k_1, k_2, \ldots, k_n)$ with the corresponding power product $f_i g_1^{k_1} g_2^{k_2} \cdots g_n^{k_n}$ where i is defined by (4), then it can be easily proved that the sum of these products over all the partitions of a set of n elements results in the nth Bell polynomial!

In other words:

$$A_n(f_1, \ldots, f_n; g_1, \ldots, g_n) = \sum_{\pi_n} f_i g_1^{k_1} g_2^{k_2} \cdots g_n^{k_n}, \tag{5}$$

where π_n stands for any partition of a set of n elements, and i is the number of blocks present in π_n, according to (4).

For example, in the case $n = 3$, we have $f_1 g_3$ for the trivial partition, $f_3 g_1{}^3$ for the discrete partition, and three products $f_2 g_1 g_2$ for the partitions with two blocks. The sum of these five products yields the Bell polynomial $A_3(f_1, f_2, f_3; g_1, g_2, g_3)$.

Thus far we have considered only the "complete" Bell polynomials of $2n$ variables. For many purposes it is, however, sufficient to consider the "special" Bell polynomials $Y_n(g_1, g_2, \ldots, g_n)$ of only n variables, derived by assigning to f_i $(i = 1, 2, \ldots, n)$ the value one:

$$Y_n(g_1, g_2, \ldots, g_n) = A_n(1, 1, \ldots, 1; g_1, \ldots, g_n) = \sum_{\pi_n} g_1^{k_1} g_2^{k_2} \cdots g_n{}^{k_n}. \tag{6}$$

[Actually, Y_n is no more "special" than A_n, since the latter can be reconstructed from Y_n by readding the missing factors f_i, using (4) for the value of i.]

The special Bell polynomials lend themselves to an even simpler combinatorial interpretation: I recall many years ago being in a Chinese restaurant which offered, let us say, g_1 dinner choices for one person eating alone, g_2 different choices for two people eating together, g_3 still different choices for a group of three persons, etc. Considering the event that n persons patronize this restaurant, in how many different ways may they choose their dinner? The answer is, of course, that $Y_n(g_1, g_2, \ldots, g_n)$ is the required number of possible choices!

This approach to the Bell polynomials (in both the "special" and "complete" cases) possesses the following advantages.

1. It allows us to obtain, by purely combinatorial arguments, the principal identities holding for Bell polynomials.
2. It enables us to write down almost immediately, the solutions of some combinatorial problems related to partitions of finite sets.
3. Interesting generalizations of the Bell polynomials can be obtained by replacing in (5) the sum over the set of partitions of a finite set by a sum over other sets of the partition type.

I can give here only some typical examples. To illustrate the first advantage of our approach, let us consider the principal recurrence formula for the special Bell polynomials

$$Y_{n+1}(g_1, g_2, \ldots, g_{n+1}) = \sum_{k=0}^{n} \binom{n}{k} g_{k+1} Y_{n-k}(g_1, g_2, \ldots, g_{n-k}) \tag{7}$$

where $Y_0 = 1$ in accordance with (2′). This formula can be easily obtained by the following purely combinatorial argument.

Consider again the restaurant with g_j choices of dinners for a group of j persons ($j = 1, 2, 3, \ldots$). Suppose that one more person enters the restaurant to have dinner together with the n persons that have just arrived. The number of possible choices for the whole group of $n + 1$ persons is now $Y_{n+1}(g_1, g_2, \ldots, g_{n+1})$, but the same number can also be expressed by the right-hand side of (7), since the $(n + 1)$st person can join any of the

$$\binom{n}{k}$$

possible groups of k persons, thus forming a block of $k + 1$ persons with g_{k+1} choices of dinners, while the remaining $n - k$ persons have still $Y_{n-k}(g_1, g_2, \ldots, g_{n-k})$ choices [$k = 0, 1, 2, \ldots, n$; the case $k = 0$ corresponds, of course, to the possibility that the $(n + 1)$st person prefers to have his own dinner].

I wish to mention here the possibility of obtaining, in an analogous fashion, new identities for the Bell polynomials. Consider, e.g., the case that our n persons, after having chosen their dinners, begin to quarrel, and some of them want to form smaller groups. This consideration (for details [1] should be consulted) furnishes the interesting fact that in the partially ordered set of partitions already considered the power product $Y_1^{k_1} Y_2^{k_2} \cdots Y_n^{k_n}$ (where, for clearer notation, the arguments of the special Bell polynomials have been omitted) is the *summatory* function of the power product $g_1^{k_1} g_2^{k_2} \cdots g_n^{k_n}$. The formula (6) is, of course, the special case that results when this fact is applied to the trivial partition of the type $(0, 0, \ldots, 0, 1)$.

Coming now to the second advantage of our combinatorial approach, I wish only to mention the following example. Let us ask for the number of forests with n labeled vertices. Since a forest is by definition either a tree or a disjoint union of two or more trees (where a singleton or isolated vertex is also to be considered as a tree), each forest corresponds to a partition of the set of n labeled vertices. Hence the number asked for is a special Bell polynomial $Y_n(g_1, g_2, \ldots, g_n)$; here g_j is now the number of trees with j labeled vertices, i.e.,

$$g_j = j^{j-2} \qquad (j = 1, 2, 3, \ldots), \tag{8}$$

by Cayley's famous formula for counting trees. (For different proofs of (8) see Chapter 11 in [6] written by J. W. Moon.) Thus, the number of forests with n labeled vertices is equal to

$$Y_n(1, 1, 3, 4^2, 5^3, \ldots, n^{n-2}).$$

For another combinatorial problem that can be solved easily by our approach to Bell polynomials, namely the number of permutations with limited repetitions, see [4].

Coming finally to the third advantage of our approach, let us remark that our basic definition (5) furnishes immediately generalizations of the "classical" Bell polynomials by postulating that π_n on the right-hand side of (5) should apply to the elements of some other set of the "partition type," i.e., any finite set to whose elements are associated nonnegative integers $k_1, k_2, \ldots, k_n$, where k_j represents the number of blocks consisting of j elements in some prescribed kind of partition ($j = 1, 2, \ldots, n$).

The partitions of the *number n* might be an instance (e.g., 3, 2 + 1, and 1 + 1 + 1 for $n = 3$); but this example is a rather trivial one as it turns out that the generalized Bell polynomials differ from the classical ones only by the fact that all the coefficients are now equal to 1.

More interesting is the case of the *compositions* of an integer n, i.e., partitions of the number n supposing that the commutative law of addition does not hold. For example, for $n = 3$ we have the four partitions of Fig. 2,

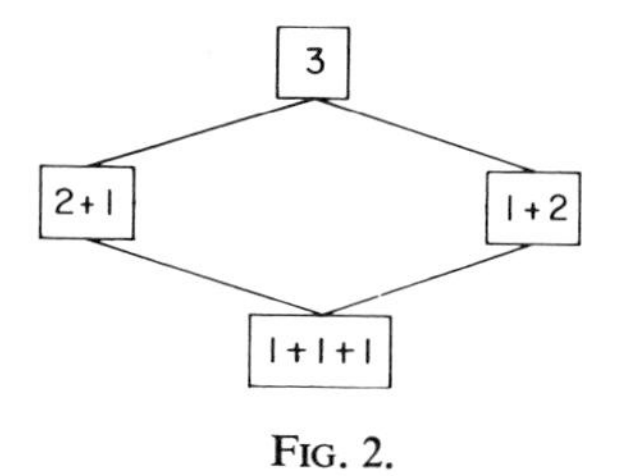

FIG. 2.

giving rise to the generalized polynomial

$$f_1 g_3 + 2f_2 g_1 g_2 + f_3 g_1{}^3$$

with a coefficient 2 in the middle term whereas the classical one has a coefficient 3. There is also another interpretation of these compositions as given by partitions of an interval of length n into subintervals whose lengths are integers. See [2] for more details; see also [5] where the same generalized Bell polynomials are used to determine the number of combinations with limited repetition.

Another generalization of the Bell polynomials, proposed by the author in [3], refers to the set of partitions of a rectangular matrix into rectangular submatrices. A recurrence formula for these polynomials is, however, given only for matrices with (at most) two rows. For a general $m \times n$ matrix it is likely that not even the number of its partitions into submatrices is known.

(In general, of course, the number of partitions of a certain kind can be obtained from the corresponding generalized Bell polynomial by giving to all its variables the value 1.)

REFERENCES

1. R. FRUCHT and G.-C. ROTA, Polinomios de Bell y Particiones de Conjuntos Finitos, *Scientia* (*Valparaiso, Chile*) No. 126 (1965), 5–10.
2. R. FRUCHT, Polinomios para Composiciones de Números, una Generalización de los Polinomios de Bell, *Scientia* (*Valparaiso, Chile*) No. 128 (1965), 49–54.
3. R. FRUCHT, Polinomios Análogos a los de Bell para Particiones de Matrices, *Scientia* (*Valparaiso, Chile*) No. 130 (1966), 67–74.
4. R. FRUCHT, Permutations with Limited Repetitions, *J. Combinatorial Theory* **1** (1966), 195–201.
5. R. FRUCHT, Combinaciones con Repetición Limitada, *Scientia* (*Valparaiso, Chile*) No. 132 (1967), 57–63.
6. *A Seminar on Graph Theory.* (F. Harary, ed.), Holt, New York, 1967.
7. JOHN RIORDAN, *An Introduction to Combinatorial Analysis*, Wiley, New York, 1958.

THE NUMBER OF SUBSPACES OF A VECTOR SPACE

*Jay Goldman**

UNIVERSITY OF COLORADO
BOULDER, COLORADO
AND
DEPARTMENT OF STATISTICS
HARVARD UNIVERSITY
CAMBRIDGE, MASSACHUSETTS

and

*Gian-Carlo Rota**

UNIVERSITY OF COLORADO
BOULDER, COLORADO
AND
MASSACHUSETTS INSTITUTE OF TECHNOLOGY
CAMBRIDGE, MASSACHUSETTS

1. Introduction

The number G_n of subspaces of a finite-dimensional vector space over a finite field $GF[q]$ has received very little attention. We are guided by the analogy between the numbers G_n and the numbers B_n of partitions of a set of n elements, studied in a previous paper [1]. We establish some of the fundamental enumerative properties of the sequence of numbers G_n, which we call the *Galois numbers*, using the symbolic method as our main device. Our purpose is, in fact, to display the suppleness of the symbolic method, already used in [1], and to show its adaptability to various enumerative situations; in addition to the Galois numbers, we use it to derive a number of identities for the Gaussian coefficients.

We have omitted throughout any discussion of convergence of our infinite series. The simplest way to establish that all infinite series and products are in fact convergent is not to assume suddenly $|q| < 1$, as is usually done in the literature, but rather, to use convergence in the q-adic norm (since q is the power of a prime). In this way, convergence questions simplify to the point of triviality.

* This work is the result of research supported by the Office of Naval Research under contracts No. N00014-67A-2098-0017 at Harvard University and No. N00014-67A-0204-0016 at the Massachusetts Institute of Technology.

This is the first of a projected series of papers dealing with the combinatorial theory of finite vector space.

2. A q-Binomial Theorem

Let V_n be a vector space of dimension n over the finite field $GF(q)$ (field with q elements); thus V_n has altogether q^n distinct points. Let X be a vector space over the same field having x vectors. We shall count in two ways the set of all linear transformations of V_n into X, thereby obtaining an identity.

Let $T: V_n \to X$ be such a linear transformation and let $v_1, v_2, \ldots, v_n$ be a basis for the vector space V_n. The linear transformation T is uniquely determined once the images of the v_i are assigned. The image of each v_i can be any one of x vectors, hence there are x^n choices for T.

Now we count the set of all linear transformations according to the dimension of their null spaces. Having chosen a subspace N of dimension k in V_n, the set of linear transformations into X whose null space is N is counted as follows. Let $v_1, v_2, \ldots, v_n$ be a basis for V_n, of which $v_1, v_2, \ldots, v_k$ is a basis for the subspace N. A linear transformation T has N as its null space if and only if the vectors $v_1, v_2, \ldots, v_k$ are mapped into zero and no other nontrivial linear combination of the v_i is mapped into zero. This gives for the image of v_{k+1} the choice of $x - 1$ vectors, all vectors of X but the zero vector: for v_{k+2} the choice of $x - q$ vectors, all vectors but those of the line spanned by the image of v_{k+1}; for v_{k+3} the choice of $x - q^2$ vectors, all but the members of the plane spanned by the images of v_{k+1} and v_{k+2}; and so on. We see that the number of linear transformations whose null space is $N(x-1)(x-q)\cdots(x-q^{n-k+1})$. Let

$$\begin{bmatrix} n \\ k \end{bmatrix}$$

be the number of subspaces of V_n of dimension k; this number is called a *Gaussian coefficient*, and we shall study it later. Combining the two counts, we obtain the classical identity

$$x^n = \sum_{k \geq 0} \begin{bmatrix} n \\ k \end{bmatrix} (x-1)(x-q)\cdots(x-q^{n-k+1}). \tag{1}$$

As x may vary over infinitely many values, (1) is a polynomial identity in the rational variable x. This identity is the only combinatorial information we shall use. The number of subspaces of V_n is

$$G_n = \sum_{k \geq 0} \begin{bmatrix} n \\ k \end{bmatrix}, \tag{2}$$

and our purpose is to derive the properties of the sequences of these number using only identity (1) and elementary linear algebra. We call the G_n the Galois numbers.

3. The Symbolic Method

We now come to the basis of the symbolic method. Let P be the vector space over the rational numbers of all polynomials in the single variable x. The sequence of polynomials $\{p_n(x)\}$, $n = 0, 1, 2, \ldots$, defined by

$$p_0(x) = 1, \qquad p_n(x) = (x-1)(x-q)\cdots(x-q^{n-1}), \tag{3}$$

is a basis for P, since it is a sequence of polynomial of degrees $0, 1, 2, \ldots$. A linear functional L on P is uniquely defined by giving it values on the basis $\{p_n(x)\}$; we now define one as follows:

$$L(p_n(x)) = 1, \qquad n = 0, 1, \ldots.$$

We now come to the crucial step. Apply L to both sides of (1); this gives the identity

$$L(x^n) = \sum_{k=0}^{n} \begin{bmatrix} n \\ k \end{bmatrix}$$

or, in view of (2),

$$G_n = L(x^n). \tag{4}$$

This formula is the simplest and most explicit expression for the Galois numbers, as we shall demonstrate.

4. A Recursion Formula

We shall next use formula (4) to derive the following recursion formula for the Galois numbers:

$$G_{n+1} = 2G_n + (q^n - 1)G_{n-1}. \tag{5}$$

The first step is to rewrite the formula using the linear functional L, namely,

$$L(x^{n+1}) = 2L(x^n) + (q^n - 1)L(x^{n-1}), \tag{6}$$

and this is the formula we shall prove.

The appearance of the factor $(q^n - 1)$ on the right side of (6) suggests the old device of the Eulerian derivative. This is the linear operator D_q defined on all polynomials $p(x)$ as

$$D_q\, p(x) = \frac{p(qx) - p(x)}{x}. \tag{7}$$

For example, $D_q x^n = (q^n - 1)x^{n-1}$, which is what we want. Formula (6) can now be rewritten as

$$L(xx^n) = 2L(x^n) + L(D_q x^n). \tag{8}$$

But L is linear, hence (8) will hold for all n if and only if the identity

$$L(xp(x)) = 2L(p(x)) + L(D_q p(x)) \tag{9}$$

holds for all polynomials $p(x)$, since $\{x^n\}$ is also a basis for P.

We shall now work with the basis polynomials $p_n(x)$, introduced in (3). For these polynomials we have the evident recursion

$$p_{n+1}(x) = (x - q^n)p_n(x); \tag{10}$$

that is, $xp_n(x) = p_{n+1}(x) + q^n p_n(x)$. Hence, applying L to both sides,

$$L(xp_n(x)) = 1 + q^n. \tag{11}$$

One verifies trivially that $D_q p_n(x) = (q^n - 1)p_{n-1}(x)$, hence

$$1 + q^n = 2L(p_n(x)) + L(D_q p_n(x)). \tag{12}$$

Combining (11) and (12) we obtain

$$L(xp_n(x)) = 2L(p_n(x)) + L(D_q p_n(x)). \tag{13}$$

But the $p_n(x)$ are a basis for the vector space P. Therefore, (9) follows immediately from (13) and the linearity of L, and the verification of the recursion formula (5) is complete.

5. Generating Functions

Next we look at the *Eulerian generating function* of the Galois numbers, namely, the formal power series

$$G(t) = \sum_{n \geq 0} \frac{G_n t^n}{(1-q)(1-q^2)\cdots(1-q^n)}. \tag{14}$$

We shall prove that this series can be rewritten as an infinite product

$$G(t) = \prod_{i \geq 0} \frac{1}{(1-q^i t)^2}. \tag{15}$$

Convergence, as noted in the introduction, is in the q-adic norm. Replacing $p(x)$ by $D_q p(x)$ in identity (9) we obtain

$$L(xD_q p(x)) = 2L(D_q p(x)) + L(D_q{}^2 p(x)), \tag{16}$$

for all polynomials $p(x)$. Setting $p(x) = x^n$ in (16) gives

$$q^n L(x^n) = L(x^n) - 2(1 - q^n)L(x^{n-1}) + (1 - q^n)(1 - q^{n-1})L(x^{n-2}). \quad (17)$$

Multiply both sides by $t^n/(1 - q)(1 - q^2)\cdots(1 - q^n)$ and sum over $n \geq 0$, and obtain

$$G(qt) = G(t) - 2tG(t) + t^2 G(t) = (1 - t)^2 G(t), \quad (18)$$

or

$$G(t) = \frac{1}{(1 - t)^2} G(qt)$$

whence (15) follows immediately by iteration as

$$G(t) = \prod_{i \geq 0} \frac{G(q^i t)}{G(q^{i+1} t)}.$$

6. Gaussian Coefficients

Next, we shall briefly show how the symbolic method can be used to systematically derive the properties of Gaussian coefficients. Let a sequence L_k of linear functionals be given by

$$L_k(p_n(x)) = \delta_{kn},$$
$$\delta_{kn} = \begin{cases} 1, & k = n \\ 0, & k \neq n. \end{cases} \quad (19)$$

Applying L_k to both sides of (1) we obtain the identity

$$L_k(x^n) = \begin{bmatrix} n \\ k \end{bmatrix}. \quad (20)$$

The analog of the recursion formula (9) is

$$L_k((x - q^k)p(x)) = L_{k-1}(p(x)). \quad (21)$$

This is verified as follows. First, rewrite the formula as

$$L_k(xp(x)) = q^k L_k(p(x)) + L_{k-1}(p(x)).$$

As usual, it suffices to verify this formula as $p(x)$ ranges through the sequence $p_n(x)$. Use the identity $xp_n(x) = p_{n+1}(x) + q^n p_n(x)$ on the left side. If $n = k$, both sides equal q^k; if $n = k - 1$, both sides equal one; for all other n, both sides equal zero, as follows immediately from definition (19).

Also immediate from the definition is the identity

$$(q^k - 1)L_k(p(x)) = L_{k-1}(D_q(p(x))) \quad (22)$$

which is proved immediately by setting $p(x) = p_n(x)$.

Specializing $p(x) = x^n$ in (21) we get the classical recursion formula for the Gaussian coefficients

$$\begin{bmatrix} n \\ k \end{bmatrix} = \begin{bmatrix} n-1 \\ k-1 \end{bmatrix} + q^k \begin{bmatrix} n-1 \\ k \end{bmatrix}. \tag{23}$$

Similarly, from (22) we get

$$\begin{bmatrix} n \\ k \end{bmatrix} = \left(\frac{q^n - 1}{q^k - 1}\right) \begin{bmatrix} n-1 \\ k-1 \end{bmatrix}, \tag{24}$$

whence, by iteration, the classical expression

$$\begin{bmatrix} n \\ k \end{bmatrix} = \frac{(q^n - 1)(q^{n-1} - 1) \cdots (q^{n-k+1} - 1)}{(q^k - 1)(q^{k-1} - 1) \cdots (q - 1)}. \tag{25}$$

Identities (21) and (22), though formally similar to (23) and (24), are nevertheless much more pliable tools for guessing and proving identities for the Gaussian coefficients.

As an example, we prove a beautiful identity due to Gauss [2]:

$$\sum_{r=0}^{2m} (-1)^r \begin{bmatrix} 2m \\ r \end{bmatrix} = (1-q)(1-q^3) \cdots (1-q^{2m-1}). \tag{26}$$

To prove Gauss' identity it is necessary to prove simultaneously the identity

$$\sum_{r=0}^{2n+1} (-1)^r \begin{bmatrix} 2m+1 \\ r \end{bmatrix} = 0. \tag{27}$$

Note that in (26) and (27) we can replace the upper limit of summation by infinity. Now rewriting the formulas using the linear functionals $\{L_i\}$ we have

$$\begin{aligned} \sum_{r=0}^{\infty} (-1)^r L_r(x^{2m}) &= (1-q)(1-q^3) \cdots (1-q^{2m-1}), \\ \sum_{r=0}^{\infty} (-1)^r L_r(x^{2m+1}) &= 0. \end{aligned} \tag{28}$$

This suggests introducing the new linear functional

$$G = \sum_{r=0}^{\infty} (-1)^r L_r.$$

To prove (28) we must show that G satisfies the functional equations

$$\begin{aligned} G(x^{2m+2}) &= (1 - q^{2m+1}) G(x^{2m}) \\ G(x^{2m+3}) &= (1 - q^{2m+2}) G(x^{2m}) \end{aligned} \tag{29}$$

together with the boundary conditions

$$\mathrm{G}(x) = 0, \qquad G(x^2) = 1 - q. \tag{30}$$

Equations (29) can of course, be rewritten as the single equation

$$G(x^{n+2}) = (1 - q^{n+1})G(x^n) \qquad \text{or} \qquad G(x^2 x^n) = G(x^n) - qG((qx)^n) \tag{31}$$

But G is linear, thus (26) holds for all n if and only if

$$G(x^2 p(x)) = G(p(x)) - qG(p(qx)) \tag{32}$$

holds for all polynomials $p(x)$.

We now prove (32) by verifying it for the basis polynomials $p_n(x)$. From the definition of G it follows that

$$G(p_n(x)) = (-1)^n. \tag{33}$$

Applying this to the following identities

$$p_{n+1}(x) = (x - q^n)p_n(x) \qquad p_n(qx) = (qx - 1)q^{n-1}p_{n-1}(x),$$
$$p_{n+2}(x) = (x - q^{n+1})(x - q^n)p_n(x),$$

we get

$$\begin{aligned} G(xp_n(x)) &= (-1)^n(q^n - 1) \\ G(p_n(qx)) &= (-1)^{n-1}q^{n-1}(q^n - q - 1) \\ G(x^2 p_n(x)) &= (-1)^n(q^{2n} - q^{n+1} - q^n + 1). \end{aligned} \tag{34}$$

Using (33) and (34) we can now verify (32) for $p(x) = p_n(x)$ by direct substitution. Thus (32) is true.

To complete the proof of (26) and (27) we verify the boundary conditions (30). Now $Gp_0(x) = G(1) = 1$, and $Gp_1(x) = -1 = G(x - 1) = Gx - 1$. Therefore, $G(x) = 0$; $G(p_2(x)) = 1 = G((x - 1)(x - q)) = G(x^2 - qx - x + q) = G(x^2) + q$ and therefore $G(x^2) = 1 - q$.

7. Expansions

Next we derive an explicit representation, or, in classical language, a "summation formula" for the linear functional L, namely

$$L(p(x)) = \prod_{n \geq 0} (1 + q^n) \sum_{k \geq 0} \frac{[H_k(D_q)p(x)]_{x=1}}{(q^k - 1)(q^{k-1} - 1) \cdots (q - 1)} \tag{35}$$

where

$$H_k(x) = \sum_{i \geq 0} \begin{bmatrix} k \\ i \end{bmatrix} x^i \tag{36}$$

are the Eulerian analogs of the Hermite polynomials studied by Carlitz [3, 4].

We begin by remarking that for $k \geq n$

$$H_k(D_q)p_n(x) = \begin{bmatrix} k \\ n \end{bmatrix}(q^n - 1) \cdots (q - 1)$$
$$= (q^k - 1)(q^{k-1} - 1) \cdots (q^{k-n+1} - 1),$$

as follows immediately from (36) and the fact that $D_q p_n(x) = (q^n - 1)p_{n-1}(x)$. For $k < n$, the left side vanishes. It follows that

$$\sum_{i \geq 0} \frac{1}{(q^i - 1) \cdots (q - 1)} = \sum_{k \geq 0} \frac{[H_k(D_q)p_n(x)]_{x=1}}{(q^k - 1)(q^{k-1} - 1) \cdots (q - 1)}.$$

Now set

$$Q = \sum_{i \geq 0} \frac{1}{(q^i - 1) \cdots (q - 1)};$$

then

$$L(p_n(x)) = Q^{-1} \sum_{k \geq 0} \frac{[H_k(D_q)p_n(x)]_{x=1}}{(q^k - 1)(q^{k-1} - 1) \cdots (q - 1)},$$

since both sides are equal to one. By linearity it follows that

$$L(p(x)) = Q^{-1} \sum_{k \geq 0} \frac{[H_k(D_q)p(x)]_{x=1}}{(q^k - 1) \cdots (q - 1)}$$

for all polynomials $p(x)$. The proof of formula (35) will therefore be complete if we show that

$$Q^{-1} = \prod_{n \geq 0} (1 + q^n).$$

This follows from the well-known identity, due to Euler,

$$\sum_{k \geq 0} \frac{t^n}{(1 - q)(1 - q^2) \cdots (1 - q^n)} = \prod_{n \geq 0} \frac{1}{(1 - q^n t)}$$

upon setting $t = -1$.

We give a simple verification of this latter identity. It is immediate that the right side, call it $e(t)$, satisfies the functional equation

$$(1 - t)e(t) = e(qt).$$

By equating coefficients, one sees immediately that the left-hand side satisfies the same functional equation. As a special case of formula (35) we obtain the following expression for the Galois numbers:

$$G_n = \left[\prod_{m \geq 0} (1 + q^m)\right] \sum_{k \geq 0} \frac{[H_k(D_q)x^n]_{x=1}}{(q^k - 1)(q^{k-1} - 1) \cdots (q - 1)} \tag{37}$$

which can be used for asymptotic estimations.

References

1. G.-C. ROTA, The Number of Partitions of a Set, *Amer. Math. Monthly* **71**, No. 5 (1964), 498–504.
2. C. G. GAUSS, Summation Quarundam Serierum Singularium, *Werke* **2** (1876), 9–45.
3. L. CARLITZ, Some Polynomials Related to Theta Functions, *Ann. Math. Pura Appl.* **41** (1956), 359–373.
4. L. CARLITZ, Some Polynomials Related to Theta Functions, *Duke Math. J.* **24** (1957), 521–527.

GRAPHS, COMPLEXES, AND POLYTOPES

*Branko Grünbaum**

DEPARTMENT OF MATHEMATICS
UNIVERSITY OF WASHINGTON
SEATTLE, WASHINGTON

Dedicated to Hugo Hadwiger on his
60th birthday, December 23, 1968

In a long-past era a fierce battle raged between the "synthetic" and the "analytic" geometries; not only a battle, but a war, an epic confrontation, with frequent reverses of fortune, heroic stands at times when the cause seemed all but lost, marvelous feats of ingenuity, boldness, and daring. Though many exploits of those old campaigns are now the common heritage of all, most of the spirited sallies and stubborn defences—indeed almost all the moving forces of the struggle—would seem to us somewhat ridiculous if it were not for the mists of oblivion which mercifully shroud our view. However, one characteristic of this clash of spirits should not be permitted to escape our attention. It was not a difference in goals that was the cause of friction between the rival sects of geometers, but the willingness of each camp to let itself be led by its own tools, its readiness to tailor the strategic aims to suit the tactical weapons that each of them possessed.

History is said to tend to repeat itself. A simile to the dichotomy of geometry a century ago may be found today in that part of mathematics which is our common object of interest. What started as one discipline with Euler has in recent decades grown into two separate branches, with very little evidence of any common bond in all the thousands of papers published.

I am speaking, on the one hand, of topology, which managed to rid itself of the epithet "combinatorial" and of all the conceivable associations of this word, and which is following headlong after its successful algebraic tools. On the other hand, I have in mind graph theory, which, hiding behind a

* Research supported in part by Office of Naval Research contract N 00014-67-A-0103-0003.

number of key results and problems, grew too timid even to contemplate the fields beyond the one-dimensional pale erected by itself.

I would like to lure you into those fields, at least for a short inspection trip. My purpose in this florid introduction is to arouse your ire sufficiently to make you listen to my sales pitch for those fields. It is my hope that their inherent beauty will do the rest, persuading you to commit at least a part of your time and effort to their exploration.

Let me first state my aim more soberly. On the one side there is a large supply of combinatorial–geometric objects, complexes of different kinds—simplicial, cell, abstract, geometric, topological, piecewise linear—embedded or not, oriented or not, and so on. On the other side, there is a wealth of notions, properties, and operations that have evolved in graph theory, the theory of one-dimensional complexes. My whole objective may be stated as simply as this:

Why not apply those notions, properties, and operations to higher-dimensional complexes?

By some examples I will try to show that it is often possible to find meaningful generalizations of problems and results from graph theory and apply them to "combinatorial topology"—to use the contemptuously discarded term.

1. Everybody knows that a graph with n vertices has at most

$$\binom{n}{2}$$

edges, and that a graph with e edges has at least n vertices, where n is the least integer such that

$$\binom{n}{2} \geq e.$$

Only a few years ago Kruskal [8] found a generalization of these facts to arbitrary simplicial complexes. Kruskal's result, which would be a gem because of its elegance even it it were not important (though it is), may be stated as follows:

If the simplicial complex $\mathscr{C}$ has f_k k-faces, write f_k in the form

$$f_k = \binom{a_{k+1}}{k+1} + \binom{a_k}{k} + \binom{a_{k-1}}{k-1} + \cdots,$$

where a_{k+1} is first chosen as large as possible, then a_k is chosen as large as possible, etc. Then the number of m-faces of $\mathscr{C}$ is at least (if $m < k$) or most (if $m > k$)

$$\binom{a_{k+1}}{m+1} + \binom{a_k}{m} + \binom{a_{k-1}}{m-1} + \cdots.$$

These estimates are best possible for all k, f_k, and m. (For a simpler proof of Kruskal's theorem see Katona [7].)

To mention only one application of Kruskal's theorem, it may be used to settle certain (otherwise untractable) cases of the so-called "upper-bond conjecture" for convex polytopes (see Grünbaum [2]). That conjecture (and the whole area surrounding it) also forms a legitimate and appealing part of combinatorial topology.

If one is willing to add more information, other inequalities for the f_i's may be proved. For example:

For each cell complex and for $0 \leq m < k$,

$$\binom{k+1}{k-m} f_k \leq \binom{f_0 - m - 1}{k-m} f_m .$$

For each simplicial complex and for $1 \leq m < k$,

$$(k+1)\binom{k}{k-m} f_k \leq \binom{f_0 - m}{k-m} [(m+1)f_m - (k-m)f_{m-1}].$$

The inequalities are best possible for all m and k.

All these results lead to many unsolved problems. For example:

Is the last inequality valid for all cell complexes?

How should Kruskal's theorem be modified for cell complexes?

What are analogs of Kruskal's theorem involving three or more numbers f_i? (In other words, how should the last two results be modified to be best possible for each f_k?)

Characterize the $(k+1)$-tuples $(f_0, f_1, \ldots, f_k)$ which appear as numbers of faces of different dimensions for simplicial k-complexes (or cell k-complexes, or $(k+1)$-polytopes).

What better inequalities may be derived if the k-complex is embeddable in Euclidean d-space?

As a partial solution of the last problem, it may be proved easily that inequalities generalizing (and derivable from) the well-known inequalities for planar graphs hold for simplicial k-complexes embeddable in Euclidean $(k+1)$-space. They may be used to show that the 4-color problem (and 5-color theorem) for planar graphs have meaningful higher-dimensional analogs (Grünbaum [3]).

2. Two of the fundamental facts concerning planar graphs are the following:

Each graph not embeddable in the plane contains a subdivision of one of two "bad" graphs (Kuratowski [9]).

Each graph embeddable in the plane is geometrically (i.e., rectilinearly) embeddable in the plane (Wagner [14]).

I cannot tell you what the generalizations of those results to complexes are. But at least the case of simplicial n-complexes embeddable in $(2n)$-space does not seem completely hopeless. If we denote by $\mathscr{C}^n(v)$ the complete n-complex with v vertices, and if $\vee$ denotes the "join" operation, it is well known (since van Kampen and Flores in the 1930's) that $\mathscr{C}^n(2n+3)$ and $\mathscr{C}^0(3) \vee \cdots \vee \mathscr{C}^0(3)$ $(n+1$ times) are n-complexes not embeddable in E^{2n}. (For $n = 1$ they are the two "bad" graphs of Kuratowski [9].) But in general, they are only the two extremes of a whole series of analogous complexes: Whenever $n = n_1 + \cdots + n_p + p - 1$, the n-complex $\mathscr{C}^{n_1}(2n_1 + 3) \vee \cdots \vee \mathscr{C}^{n_p}(2n_p + 3)$ is not embeddable in E^{2n}. Moreover, each of those complexes is minimal in the same sense as the Kuratowski graphs. That is, for each of them, every proper subcomplex is not only embeddable in E^{2n} but is also geometrically embeddable in E^{2n} (Zaks [15], Grünbaum [5]).

I am willing to risk the following.

CONJECTURE. Every simplicial n-complex embeddable in E^{2n} is even geometrically embeddable in E^{2n}.

A companion to this conjecture is the following.

CONJECTURE. Every simplicial manifold topologically embeddable in some Euclidean space is even geometrically embeddable in the same Euclidean space.

3. Vizing's theorem (Vizing [13], Ore [10]) is well known. The simple case of it needed here claims that the edges of a k-valent graph may be colored by $k + 1$ colors.

Let a simplicial d-complex be called k-valent provided no $(d-1)$-face belongs to more than k d-faces; let us say that the complex is "properly colored" if its d-faces are assigned to color classes in such a way that no two d-faces of the same color have a common $(d-1)$-face. Some preliminary results suggest the following

CONJECTURE. Each k-valent simplicial d-complex may be properly colored by $d + k$ colors.

4. In analogy to the definition of interchange graphs of graphs, one may define interchange graphs of simplicial d-complexes and try to characterize the graphs obtainable in this fashion and to investigate their properties. (The

same idea was approached, in some special cases, by those interested in "association schemes.")[1]

A much more elegant and satisfying theory results, however, if one considers the *nerve* of a simplicial complex, where the complex is considered to be covered by its facets (maximal faces). Nerves of d-complexes, and of some other special families of complexes, may be characterized neatly. For a naturally restricted class of complexes (which includes manifolds, pseudomanifolds, and many other complexes) the nerve operation determines a dual pairing of complexes, leading to a number of interesting results and problems. Using a definite assignment of simplicial complexes to cell complexes, this duality is seen to generalize (among others) the notion of duality for convex polytopes and manifolds (Grünbaum [4]).

5. One of the more interesting open questions on graph coloring concerns the "total coloring" introduced by Behzad [1]. In a "total coloring" colors are assigned to vertices *and* edges, and no two incident or neighboring objects are permitted to have the same color. Behzad has conjectured that each k-valent graph may be totally colored by $k + 2$ colors.[2] Rosenfeld [12] and Gupta [6] have recently proved Behzad's conjecture for $k = 3$ and for some other special cases, but the general problem is still unsolved.

In a generalization of the total coloring to complexes it would probably be most interesting to assign colors to vertices and facets. A special case of this problem has recently been considered by Ringel [11] (for planar maps), but even here the solution is not complete.

The list of appealing, but practically unexplored domains of combinatorial topology, could be extended almost indefinitely. I will stop, however, in the hope that even the small sampler presented here will whet your appetite sufficiently to induce you to do something about it.

References

1. M. Behzad, Graphs and Their Chromatic Numbers, Ph. D. Thesis, Michigan State University, 1965.
2. B. Grünbaum, *Convex Polytopes*, Wiley, New York, 1967.

[1] Added in proof (March 6, 1969): A recent survey of results in this area is given in the author's paper "Incidence Patterns of Graphs and Complexes" (to appear in "The Many Facets of Graph Theory" (G. Chartrand and S. F. Kapoor, eds.), Springer Verlag Berlin, 1969).

[2] Added in proof (March 6, 1969): As mentioned by A. A. Zykov (Problem 12 in "Beiträge zur Graphentheorie" (H. Sachs, H. -J. Voss, and H. Walther, eds.), p. 228, Teubner, 1968). V. G. Vizing seems to have arrived independently at Behzad's conjecture concerning total colorings.

3. B. GRÜNBAUM, Higher-Dimensional Analogs of the Four-Color Problem and Some Inequalities for Simplical Complexes. *J. Combinatorial Theory* (to appear).
4. B. GRÜNBAUM, Nerves of Simplicial Complexes *Aequat. Math.* **2** (1968), 25–35.
5. B. GRÜNBAUM, Imbeddings of Simplicial Complexes (to appear).
6. R. P. GUPTA, The Cover Index of a Graph (to appear).
7. G. KATONA, A Theorem on Finite Sets, in *Theory of Graphs* (P. Erdös and G. Katona, eds.) (*Proc. Colloq. Tihany, September* 1966), pp. 187–207, Academic Press, New York, and Publ. House Hungar. Acad. Sci., Budapest, 1968.
8. J. B. KRUSKAL, The Number of Simplices in a Complex, in *Mathematical Optimization Techniques* (Symposium, Berkeley, 1960) (R. Bellman, ed.), pp. 251–278. Univ. of California Press, Berkeley, California, 1963.
9. C. KURATOWSKI, Sur le Problème des Courbes Gauches en Topologie, *Fund. Math.* **15** (1930), 271–283.
10. O. ORE, *The Four-Color Problem*, Academic Press, New York 1967.
11. G. RINGEL, Ein Sechsfarbenproblem auf der Kugel, *Abh. Math. Sem. Univ. Hamburg* **29** (1965), 107–117.
12. M. ROSENFELD, On the Total Chromatic Number of Certain Graphs (to appear).
13. V. G. VIZING, On an Estimate of the Chromatic Class of a p-Graph, *Diskret. Analiz* **3** (1964), 25–30.
14. K. WAGNER, Bemerkungen zum Vierfarbenproblem, *Jber. Deutsch. Math.-Ver.* **46** (1936), 26–32.
15. J. ZAKS, On a Minimality Property of Complexes, *Proc. Amer. Math. Soc.* (to appear).

ON THE STRUCTURE OF n-CONNECTED GRAPHS

R. Halin

UNIVERSITY OF KÖLN
KÖLN, GERMANY

1. PRELIMINARIES

The graphs considered in this paper are undirected and do not contain loops or multiple edges. By $V(G)$ and $E(G)$ we denote the set of the vertices and the set of the edges of the graph G, respectively, and we put $|V(G)| = \nu(G)$, $|E(G)| = \nu'(G)$; $\rho_G(v)$ denotes the valency of the vertex v in G, and by $\nu_n(G)$ we denote the number of vertices of valency n in G.

A section graph H of a graph G is a subgraph of G such that $(v, v') \in E(H)$ iff $(v, v') \in E(G)$ for any pair of distinct vertices v, v' of H. We write $H \subseteq G$ ($H \sqsubseteq G$) if H is a subgraph (section graph) of G. If $H \subseteq G$, we denote by $G - H$ the section graph of G generated by the vertices of G not in H.

If $a, b \in V(G)$ and $T \subseteq G$, we say that T *separates* a and b in G, and write

$$a.\ T.\ b(G),$$

if $a, b \notin V(T)$ and belong to different (connected) components of $G - T$. In general, we say that T separates G, or that T is a *cut* of G, if there exist $a, b \in V(G)$ such that $a.\ T.\ b(G)$ holds. An *n-cut* is a cut with n vertices.

If a, b are two distinct vertices of a graph G, we define as their *Menger number* $\mu_G(a, b)$ the maximal number of disjoint a, b-paths in G. A celebrated theorem by Menger states:

If a, b are distinct nonadjacent vertices of G, then

$$\mu_G(a, b) = \min\ \{\nu(T) \cdot/.\ a.T.b(G)\}.$$

A graph G will be called *n-vertex connected* if $\nu(G) \geq n + 1$ and G has no j-cut with $j < n$. The *vertex connectivity* of G, denoted by $c(G)$, is the maximal number n such that G is n-vertex connected. (It is easy to verify that this definition is also adequate in the case of infinite graphs and infinite connectivities.)

A theorem due to Whitney, closely related to Menger's theorem, states:

$$c(G) \geq n \qquad \textit{iff} \qquad \mu_G(a, b) \geq n$$

for every pair of distinct vertices a, b of G.

If $T \subseteq G$ and $a \in V(G - T)$, an *a, T-fan* of G is defined as the union of $v(T)$ paths in G, each beginning in a and ending in a vertex of T, such that any two of them have only a in common. Then we have the following theorem.

$c(G) \geq n$ *iff for every* $T \subset G$ *with* $v(T) = n$ *and every* $a \in V(G - T)$ *there is an a, T-fan in G.*

2. The Abundance of Structure of an n-Connected Graph

By definition of $c(G)$ we have

$$\rho_G(v) \geq c(G)$$

for all $v \in V(G)$. Hence

$$\frac{v'(G)}{v(G)} \geq \tfrac{1}{2} \cdot c(G).$$

W. Mader proved the following for finite G.
If

$$\frac{v'(G)}{v(G)} \geq 2^{\binom{n}{2}},$$

then G contains a subdivision of the complete n-graph K_n. Hence we have the following theorem.

Theorem 1 (Mader [8]). *For every integer k there exists an integer* $n(k)$ *such that* $c(G) \geq n(k)$ *implies the existence of a subdivision of* K_k *in G, provided that G is finite.*

This proves a conjecture of Grünbaum (written communication).

On the other hand, Jung [7] showed $n(k) \geq (k^2 - 1)/8$. Furthermore, every finite G with $c(G) \geq 6$ is nonplanar. It is a surprising observation by Mader [9] that there are infinite graphs of arbitrarily great finite vertex connectivity which are embeddable in the plane. [On the other hand, I [2] showed that every G with $v(G) > \aleph_0$ and $c(G) \geq 3$ contains a subdivision of one of the Kuratowski graphs; hence the planar graphs of vertex connectivity greater than 5 must be countable.]

3. Reductions of n-Vertex Connected Graphs

As a fundamental problem in the theory of finite n-vertex connected graphs one may note the following task. For a given natural number n, establish a complete list of all n-vertex connected graphs, starting with some initial elements of the class whose structure is known, and generating all members of the class step by step from those graphs which are already constructed through simple operations. This was done in the case $n = 3$ by Tutte [12, 13]. (For $n = 2$ see [6, Section 2], and Dirac [1].)[1] We should like to obtain an analog of Tutte's theory [12] in the case of general n. Thus the following question is of central importance: When is it possible to reduce an n-vertex connected graph by certain adequate operations, like the deletion or the contraction of an edge, in such a way that the resulting graph is again n-vertex connected?

We shall first study the structure of those n-vertex connected graphs which are not reducible with respect to the deletion of edges. If G is a graph and $e \in E(G)$, we denote by $G - e$ the graph $(V(G), E(G) - \{e\})$.

We call a graph G *minimally n-vertex connected* (or shortly, *n-minimal*) if $c(G) \geq n$ and $c(G - e) < n$ for every $e \in E(G)$.

Obviously, G is 1-minimal iff G is a tree. Note that every finite n-vertex connected graph can be reduced onto an n-minimal graph by the deletion of a sequence of edges. In the infinite case this is not always possible if $n > 1$.

We state two simple lemmas.

Lemma 1. *If $c(G) \geq n$, $e = (v, v') \in E(G)$ and $c(G - e) < n$, then for every j-cut T of $G - e$ with $j < n$ we have $j = n - 1$ and $v.T.v'(G - e)$.*

Lemma 2. *G is n-minimal iff $c(G) \geq n$ and for every edge $e = (v, v') \in E(G)$ we have $\mu_G(v, v') = n$.*

In [3] the following theorem is proved.

Theorem 2. *Every finite n-minimal graph G contains a vertex of valency n.*

In order to sharpen this result I shall do some studies on the n-cuts of a graph G with $c(G) = n$ and the parts into which G is decomposed by these subgraphs.

[1] For $n = 4$ see the thesis by Neil Robertson [11].

4. n-FRAGMENTS

If H is a proper section graph of G, according to Tutte I shall call a vertex v a *vertex of attachment* of H with respect to G if $v \in V(H)$ and there is a vertex $\bar{v} \in V(G) - V(H)$ such that v and $\bar{v}$ are adjacent in G. (I shall use the concept of vertex of attachment only for section graphs.) The set of vertices of attachment of H with respect to G will be denoted by $W(G, H)$. A vertex i of $V(H) - W(G, H)$ is called an *inner vertex* of H (with respect to G).

A section graph H of G with exactly k vertices of attachment and at least one inner vertex with respect to G will shortly be called *properly k-attached* (in G).

It is obvious that in G there exists a k-cut if G contains a properly k-attached section graph; and if T is a j-cut of G then there exists a properly k-attached section graph H of G with $k \leq j$ and $W(G, H) \subseteq V(T)$.

DEFINITION. A properly n-attached section graph of an n-vertex connected graph G will be called an *n-fragment of G*.

Now let us disregard the special graph G in which H is embedded as an n-fragment. Let H be a graph with $v(H) > n$ and $T \sqsubset H$, $v(T) = n$. Then H will be called an *n-fragment with respect to T* if there exists an n-vertex connected G with $H \sqsubset G$ and $W(G, H) = V(T)$.

THEOREM 3. *Let H be given with a proper section graph T consisting of n vertices. Then the following statements are equivalent.*

(i) *H is an n-fragment with respect to T.*
(ii) *For every vertex a of H not in T there exists an a, T-fan in H.*
(iii) *If $J \subset H$ with $v(J) < n$, then T meets every component of $H - J$.*
(iv) *There exists an integer r, $0 \leq r \leq n - 1$, such that $c(H \cup (T \times r)) \geq n$.* [Here we mean by $H \cup (T \times r)$ the graph which arises from H by adding r pairwise nonadjacent vertices $x_1, \ldots, x_r$ to H and joining each x_i to the vertices of T by edges.]
(v) *$c(G \cup K[T]) \geq n$, where $K[T]$ denotes the complete graph with vertex set $V(T)$.*

The proof of this theorem is based mainly on Menger's theorem.

The least integer r in (iv) may be used to classify the n-fragments (with respect to a given section graph of n vertices). It follows for n-fragments H:

$$\min r = \begin{cases} n - c(H) & \text{if} \quad c(H) < n \\ 0 & \text{iff} \quad c(H) \geq n. \end{cases}$$

It is seen from the above that particularly every $T \times 1$, where $v(T) = n$, is an n-fragment with respect to T.

5. n-Ends

In what follows we shall be interested mainly in those properly n-attached section graphs of an n-vertex connected graph G which do not contain a proper section graph which is also properly n-attached with respect to G. For these minimal n-fragments of an n-vertex connected G we introduce a new name.

Definition. If $c(G) \geq n$, a proper section graph H of G will be called an *n-end* of G provided that the following two conditions are fulfilled:

(i) H is an n-fragment of G;
(ii) no proper section graph of H is also an n-fragment of G.

Now let us again disregard the special G in which H is embedded as a section graph. Let H be a graph containing T with $v(T) = n$ as a proper section graph. We shall call H an *n-end with respect to T*, if there exists an n-vertex connected G containing H as an n-end with the vertices of T as vertices of attachment:

$$W(G, H) = V(T).$$

Theorem 4. *Let H be a graph, T a proper section graph of H with $v(T) = n$. If $v(H) = n + 1$, then obviously H is an n-end with respect to T iff H is of the form $T \times 1$. If $v(H) > n + 1$, then the following statements are equivalent.*

(i) *H is an n-end with respect to T.*
(ii) *For every $a \in V(H - T)$ the property holds that H is an $(n + 1)$-fragment with respect to the section graph of H generated by $V(T) \cup \{a\}$.*
(iii) *If J is a separating subgraph of H with $v(J) \leq n$, then T meets every component of $H - J$.*
(iv) $c(H \cup K[T]) \geq n + 1$.

The proof of this theorem also uses Menger's theorem as a principal tool.

6. n-Ends of n-Minimal Graphs

Now we shall study the n-ends of the finite n-minimal graphs. Our first aim is to give an estimate of the number of n-ends in these graphs.

LEMMA 3. *Let G be a finite n-minimal graph with $\nu(G) > n + 1$, $n \geq 1$. Let $v_1, \ldots, v_r$ be vertices of G and u a vertex of G different from $v_1, \ldots, v_r$ with valency $\rho_G(u) = n$ or $> r(r-1)$. Then there exists an n-end H of G which has none of the vertices $v_1, \ldots, v_r$ as an inner vertex.*

PROOF: Nothing is to be proved if $\rho_G(u) = n$; therefore we may exclude this case.

For every pair i, j with $1 \leq i < j \leq r$ in G there exist n disjoint v_i, v_j-paths $P_{ij}^{(\nu)}$, $\nu = 1, \ldots, n$, because $c(G) = n$. We now add to G every edge of the form (v_i, v_j) $(1 \leq i < j \leq r)$, provided that it is not yet contained in G. The resulting graph will be called G^*. Then $v_1, \ldots, v_r$ generate a complete graph in G^*.

CASE 1. If there is an n-cut T^* in G^*, T^* separates no two of the v_ρ in G^*; hence there is a component C^* of $G^* - T^*$ which contains none of the given vertices v_ρ. The section graph of G generated by $V(C^*) \cup V(T^*)$ obviously is an n-fragment of G with the elements of $V(T^*)$ as vertices of attachment and the elements of $V(C^*)$ as inner vertices. Hence it contains an n-end of G with the required property.

CASE 2. If there is no n-cut in G^*, then from $\nu(G^*) > n + 1$ we conclude $c(G^*) \geq n + 1$.

For each pair i, j, $1 \leq i < j \leq r$, there is at most one of the $P_{ij}^{(\nu)}$ which contains u, and therefore $\bigcup_\nu P_{ij}^{(\nu)}$ contains at most two edges incident with u. Hence $\bigcup_{i,j} \bigcup_\nu P_{ij}^{(\nu)}$ contains not more than

$$\binom{r}{2} \cdot 2 = (r-1) \cdot r$$

edges incident with u. Because $\rho_G(u) > r \cdot (r-1)$ there exists an edge e incident with u not contained in the union of the $P_{ij}^{(\nu)}$; $G^* - e$ is (at least) n-vertex connected and contains every $P_{ij}^{(\nu)}$. Now we delete successively the edges (e_i, e_j) not contained in $E(G)$. In every step we state that the resulting graph remains n-vertex connected, by Lemma 2, since none of the $P_{ij}^{(\nu)}$ is destroyed. Finally we conclude $c(G - e) \geq n$, which contradicts the fact that G was n-minimal. The proof is completed.

As an immediate consequence we have the following theorem.

THEOREM 5. *Let m be the maximal valency of the vertices of a finite n-minimal graph G with $\nu(G) > n + 1$. Then for the number $\varepsilon(G)$ of n-ends of G we have*

$$\varepsilon(G) > \sqrt{m}.$$

Particularly, every finite n-minimal graph G with $\nu(G) > n + 1$ *has* $\varepsilon(G) > (n + 1)^{1/2}$ *n-ends.*

PROOF: Let u be a vertex of valency m. Let $H_1, \ldots, H_r$ be $r \leq \sqrt{m}$ n-ends of G. In every H_i we choose an inner vertex v_i different from u. Then by Lemma 3 we get an n-end H_{r+1} which is different from each H_i because it does not contain v_i as an inner vertex.

We shall call an n-end H of an n-minimal graph G *normal* if it contains only one inner vertex which then necessarily has valency n in G; its n neighboring vertices are the vertices of attachment of H. The choice of this name (normal) is motivated by the following.

CONJECTURE 1. In an n-minimal graph every n-end is normal.

I was able to prove this conjecture only for $n \leq 4$. In what follows we shall show however that a great many of the n-ends of a finite n-minimal graph must be normal.

The next theorem shows that the assumption, that there exists a non-normal n-end H in an n-minimal graph G, has the curious consequence that H contains nearly the whole graph G, although, by definition, H is a minimal n-fragment.

THEOREM 6. *Let G be an n-minimal graph and H a nonnormal n-end of G with* $W(G, H) = V(T)$, $T \lhd H$. *Put* $G - H = B$, $H - T = A$. *Then*

(i) $\nu(B) \leq [n/2] - 1$;
(ii) $\nu(A) \geq \nu(B) + 3$;
(iii) $\nu(H) \geq \nu(G) - [n/2] + 1$.

OUTLINE OF THE PROOF: Let u, v be two vertices of A joined by an edge e. By Theorem 4, (iv) we have $c(H \cup K[T]) \geq n + 1$; hence there is a u, T-fan F_u and a v, T-fan F_v in H, both avoiding e. There exists a section graph T' with

$$u.\ T'.\ v(G - e), \qquad \nu(T') = n - 1.$$

We show that $B \subseteq T'$. For, if there were a vertex $b \in V(B), \notin V(T')$, then there exists a b, T-fan F_b in $G - A$. By $F_u \cup F_b$ and $F_v \cup F_b$ we get n disjoint u, b-paths and n disjoint v, b-paths in $G - e$. In each of these n-tuples of paths there exists one path which avoids T'; their union contains a u, v-path in $G - e$ avoiding T', which contradicts Lemma 2. Hence $B \subseteq T'$.

The rest of the theorem follows by considering the fact that, for every $t \in V(T)$, the union of the u, t-path of F_u and of the v, t-path of F_v must meet T'; I omit the details.

The section graph generated by $B \cup T$ is an n-fragment J of G. Every n-end $\subseteq J$ is normal, by Theorem 6, (ii). Hence we have the next theorem.

THEOREM 7. *If T is an n-cut of an n-minimal finite G, then there exists a vertex v of $G - T$ with $\rho_G(v) = n$.*

Furthermore one can prove (see [4 Section 6], for details):

LEMMA 4. *If H and H' are distinct nonnormal n-ends of the n-minimal graph G, then $(G - H) \cap (G - H') = \emptyset$.*

From Theorem 7 and Lemma 4 it follows that not more than half of the n-ends of a finite n-minimal graph can be nonnormal. Thus from Theorem 5 we conclude:

THEOREM 8. *If m is the maximal valency of the vertices of a finite n-minimal graph G, then G contains more than $\frac{1}{2}\sqrt{m}$ vertices of valency n. Particularly, if G is finite and n-minimal, then $v_n(G) > \frac{1}{2}(n+1)^{1/2}$.*

Theorem 8 does not seem to be the best possible at all. It states that $v_n(G)$ for finite n-minimal G tends to infinity with n, and, for constant n, with the maximal valency of G. But one should conjecture:

CONJECTURE 2. For every $n > 1$ there is a number c_n, $0 < c_n < 1$, such that for every finite n-minimal G

$$v_n(G) \geq c_n \cdot v(G).$$

In the case $n = 1$ this statement would be obviously wrong (as the path of length v shows). In [1] it is implicitly stated that $c_2 \geq \frac{1}{3}$, and one has $c_3 \geq \frac{2}{5}$, as is shown in [5].

An analogous conjecture may be stated for infinite graphs.

CONJECTURE 3. For every $n > 1$, in an infinite n-minimal graph G there are $v(G)$ vertices of valency n.

This conjecture is true for $n = 2$ (see [1]) and $n = 3$ (see [5]). This fact is somewhat surprising because there are infinite 1-minimal graphs (i.e., trees) whose minimal valency is arbitrarily great.

Slightly stronger results than Theorem 8 can be obtained under additional conditions, for instance:

THEOREM 9. *In an n-minimal graph G every n-end is normal if G contains a vertex of valency greater than or equal to $9n^2$.*

7. CRITICAL EDGES

In this section also contractions of edges will be considered as operations of reductions. If $e \in E(G)$ we denote by G/e the graph which arises from G by the contraction of e.

DEFINITION. We shall call an edge e of a graph G *critical* if both $c(G-e) < c(G)$ and $c(G/e) < c(G)$ hold.

DEFINITION. If $e = (v, v') \in E(G)$, we define $\min(\rho_G(v), \rho_G(v'))$ as the *order* of e in G and denote it by $\omega_G(e)$.

THEOREM 10. *Let $e = (v, v')$ be a critical edge of order greater than n in a graph G with $c(G) = n$. Then there is an n-cut T in G, containing at least one of the vertices v, v', such that one component of $G - T$ has at least $\nu(G) - [3n/2] + 1$ vertices (and hence the union of the other components of $G - T$ has less than or exactly $[n/2] - 1$ vertices).*

If now every edge of order greater than n is critical, then our n-vertex connected G is n-minimal, as is shown by a simple argument. Through Theorem 7 we then conclude that in a "small distance" from every vertex there must be a vertex of valency n in G. We find:

THEOREM 11. *For every $n \geq 1$ there exists $c_n' > 0$ such that for every G with $c(G) = n$, in which each edge of order greater than n is critical,*

$$\nu_n(G) \geq c_n' \nu(G).$$

(The analog of Theorem 11 holds also for infinite graphs.)

For instance we have $c_4' \geq \frac{1}{5}$, $c_5' \geq \frac{1}{7}$, $c_6' \geq \frac{1}{8}$, and it can be shown that every critical edge of order greater than 4 in a 4-vertex connected G is contained in a triangle of G whose third vertex has valency 4 in G. (Similar, but more involved results can be shown also for $n = 5$ and 6.)

8. 3-VERTEX CONNECTED GRAPHS

In this concluding section some remarks on the structure of 3-vertex connected graphs will be given, especially on Tutte's theory of reductions of these graphs. The proofs will be omitted (see [4, Section 7] and [5, Section 2]).

Because $[3/2] - 1 = 0$ we have as an immediate consequence of Theorem 10 that there cannot exist a critical edge of order greater than 3 in a 3-vertex connected graph. (From this statement we conclude the intermediate result of Tutte's theory that every essential edge of a 3-vertex connected graph is contained in a triangle or in a triad.) Furthermore one finds for every n: If a complete K_n is contained in an n-minimal graph G, then K_n contains at most one vertex of valency greater than n. Thus, in the case $n = 3$, we find that a triangle in a 3-minimal graph contains at least two vertices of valency 3.

There is a sharper result, implying the two results for $n = 3$ mentioned just now.

THEOREM 12. *If G is 3-minimal and $e \in E(G)$ with $\omega_G(e) > 3$, then e is not contained in a triangle, and G/e is again 3-minimal.*

From this theorem one finds:

THEOREM 13. *Every circuit of a 3-minimal graph contains at least two vertices of valency* 3.

The proof proceeds by induction on the length of the circuit under consideration. (Hence Theorem 13 is also valid for infinite graphs.)

Theorem 13 is used to prove the following two theorems.

THEOREM 14. *For every finite 3-minimal graph G one has* $\nu_3(G) \geq \frac{2}{5}(\nu(G) + 3)$.

THEOREM 14′. *For every infinite 3-minimal G one has* $\nu_3(G) = \nu(G)$.

We found by Theorem 12 that every finite 3-minimal graph can be reduced by a sequence of contractions of edges of order greater than 3 onto a 3-minimal graph in which every edge has order 3. But by these special reductions there is no possibility to come downstairs to the wheels, which are the final points of Tutte's reduction theory. [A wheel consists of a circuit C with one additional vertex which is adjacent to every $v \in V(C)$.] In order to refine Tutte's theory, I shall give another approach to the structure of 3-vertex connected graphs, avoiding completely contractions of edges of order greater than 3. For the rest of the paper, all the graphs under consideration shall be finite.

LEMMA 5. *If $\nu(G) > 4$ and $c(G) \geq 3$, to any four vertices of G there exists a circuit in G which contains exactly two of them.*

By the application of this lemma on a vertex of valency 3 and its neighbors in a 3-minimal graph, we easily find that in every triad (i.e., the set of the edges incident with a vertex of valency 3) of a 3-vertex connected graph G there is an edge e such that $c(G/e) \geq 3$.

Therefore we have:

To every 3-vertex connected G there exists a chain

$$G = G_0, G_1, \ldots, G_m = K_4$$

such that $c(G_i) \geq 3$ *and* $G_{i+1} = G_i - e_i$ *or* G_i/e_i *for some* $e_i \in E(G_i)$, $i = 0, \ldots, m-1$.

Tutte's stronger result says that the chain of the G_i's can be chosen in such a way that always exactly $v'(G_{i+1}) = v'(G_i) - 1$ holds, provided that G_i is not a wheel. In the case that e_i is contracted this equality is equivalent with the statement that e_i is not contained in a triangle of G_i.

An edge e of a 3-vertex connected G is called *essential* if no such reduction by means of e is possible, i.e., if $c(G - e) < 3$, and $c(G/e) < 3$ or e is contained in a triangle of G.

Tutte's result [12, 13] may be stated as follows:

A 3-vertex connected graph in which every edge is essential must be a wheel.

In [4] the following sharper result is proved.

THEOREM 15. *Let G be 3-minimal. If each edge of order 3 in G is essential, then G is a wheel.*

Using Theorem 15, one finds moreover:

THEOREM 16. *Let G be 3-minimal and not a wheel. By Theorem 15, there is a nonessential* $e \in E(G)$ *with* $\omega_G(e) = 3$. *Then it follows that* G/e *can be reduced onto a 3-minimal graph by the deletion of at most two edges (which can be chosen incident with the image of e in* G/e).

Thus we find that to every G with $c(G) \geq 3$ *there exists a sequence*

$$H_0, H_1, \ldots, H_r$$

such that the following statements hold:

(i) *each* H_i *is 3-minimal;*

(ii) $G \supseteq H_0$, $V(G) = V(H_0)$;

(iii) H_r *is a wheel;*

(iv) *for* $i = 0, \ldots, r-1$, H_{i+1} *arises from* H_i *by the contraction of a nonessential edge of order 3 in* H_i, *followed by the deletion of at most two edges of the contracted graph.*

Hence we have for $i = 0, \ldots, r-1$:

(v) $\nu(H_{i+1}) = \nu(H_i) - 1$,
(vi) $\nu'(H_i) - 1 \geq \nu'(H_{i+1}) \geq \nu'(H_i) - 3$.

This statement can be used to prove the following result of "Turán-type":

THEOREM 17. *If G is 3-minimal, then $\nu'(G) \leq 3 \cdot \nu(G) - 9$ for $\nu(G) \geq 7$; in the case $\nu(G) \geq 8$ equality holds if and only if G is isomorphic to the complete bipartite graph $K_{3, \nu(G)-3}$.*

Finally, we obtain through Theorem 16:

THEOREM 18. *Every 3-minimal graph G has chromatic number $\chi(G) \leq 4$.*

REFERENCES

1. G. A. DIRAC, Minimally 2-Connected Graphs, *J. Reine Angew, Math.* **228** (1967), 204–216.
2. R. HALIN, Ein Zerlegungssatz für unendliche Graphen mit Anwendung auf Homomorphiebasen, *Math. Nachr.* **33** (1967), 91–105.
3. R. HALIN, A Theorem on n-Connected Graphs (submitted to *J. Combinatorial Theory*).
4. R. HALIN, Zur Theorie der n-fach zusammenhängenden Graphen (submitted to *Abh. Math. Sem. Univ. Hamburg*).
5. R. HALIN, Untersuchungen über minimale n-fach zusammenhängende Graphen (submitted to *Math. Ann.*).
6. R. HALIN and H. A. JUNG, Über Minimalstrukturen von Graphen, insbesondere von n-fach zusammenhängenden Graphen, *Math. Ann.* **152** (1963), 75–94.
7. H. A. JUNG, Zusammenzüge und Unterteilungen von Graphen, *Math. Nachr.* **35** (1967), 241–267.
8. W. MADER, Homomorphieeigenschaften und mittlere Kantendichte von Graphen, *Math. Ann.* **174** (1967), 265–268.
9. W. MADER, Homomorphiesätze für Graphen (to appear in *Math. Ann.*).
10. O. ORE, *Theory of Graphs*, Am. Math. Soc., Providence, Rhode Island, 1962.
11. N. ROBERTSON, thesis (to appear).
12. W. T. TUTTE, A theory of 3-connected graphs, *Indag. Math.* **23**, 441–455 (1961).
13. W. T. TUTTE, *Connectivity in Graphs*, Chapt. 10, Univ. of Toronto Press, Toronto, 1966.

ON UNIONS AND INTERSECTIONS OF CONES*

A. J. Hoffman

IBM WATSON RESEARCH CENTER
YORKTOWN HEIGHTS, NEW YORK

1. Introduction

Let $m \leq n$ be given positive integers, and let $C_1, \ldots, C_n$ be closed, convex pointed cones in R^m (a cone is pointed if it contains no line). We tacitly assume that each C_i contains at least one nonzero vector. The problem considered is that of finding conditions on the pattern of intersections of the cones $\{C_i\}$ and $\{-C_i\}$ which will ensure that $\bigcup C_i \cup \bigcup -C_i = R^m$. We solve this problem only for certain values of m and n, as stated in Theorem 1.

A special case of a theorem of Ky Fan [1] on closed antipodal sets of the $(n-1)$ sphere is the inspiration for the present investigation.

Fan's Theorem. *If* $\bigcup C_i \cup \bigcup -C_i = R^m$, *then*

for every choice of $\varepsilon_n = \pm 1 (j = 1, \ldots, n)$ *and*

of a permutation σ *of* $\{1, \ldots, n\}$,

there exist indices $1 \leq i_1 < \cdots < i_m \leq n$ *such that* (1.1)

$$\bigcap_{h=1}^{m} (-1)^k \varepsilon_{\sigma ik} C_{\sigma ik} \neq 0.$$

The main result of the present paper is a partial converse.

Theorem 1. *If* $n = m$ *or* $n = m + 1$, *or if* $m = 2$ *or* $m = 3$, *then* (1.1) *implies* $\bigcup C_i \cup \bigcup - C_i = R^m$.

* This paper contains a section of the talk "Bounds on the Rank and Eigenvalues of Matrices, and the Theory of Linear Inequalities" given at the Third Waterloo Symposium on Combinatorial Mathematics, May, 1968. The research reported was sponsored in part by the Office of Naval Research under contract Nonr-3775(00).

We do not know if the theorem holds for other values of m. A corollary of the proof of Theorem 1 is the following curiosity.

COROLLARY. Let $P_1, \ldots, P_{t+3}$ be points in R^t in general position (i.e., no hyperplane contains $t + 1$ of the points). Let G be the graph whose vertices are the points, with two vertices P_i and P_j adjacent if and only if the $t + 1$ hyperplanes determined by the remaining points have the property that all separate the vertices P_i and P_j or none separate P_i and P_j. Then G is a simple polygon.

2. PRELIMINARIES

We first prove a theorem on the rank of real matrices, generalizing results given in [2] and [3], that may be of some independent interest.

THEOREM 2. *Let $\mathcal{M}$ be a set of matrices of order n, closed, convex, and not containing* 0, *and let* $1 \leq m \leq n$. *Then the following are equivalent:*

$$\begin{gathered}\text{For every real matrix } A, \operatorname{Tr} AM^T > 0 \text{ for all} \\ M \in \mathcal{M} \text{ implies rank } A > m - 1.\end{gathered} \tag{2.1}$$

$$\begin{gathered}\text{For every } n - m + 1 \text{ dimensional subspace} \\ L_{n-m+1} \subset R^n, \text{ there exists a matrix } M \in \mathcal{M} \\ \text{all of whose rows are in } L_{n-m+1}.\end{gathered} \tag{2.2}$$

PROOF: Suppose (2.2) holds and $\operatorname{Tr} AM^T > 0$. If rank $A \leq m - 1$, then there exists a subspace L_{n-m+1} such that $Ax = 0$ for all $x \in L_{n-m+1}$. Let M be the matrix whose existence is assured by (2.2), then $AM^T = 0$, contradicting $\operatorname{Tr} AM^T > 0$.

Conversely, assume (2.1). Let $\mathcal{L} = \mathcal{L}(L_{n-m+1})$ be the set of all matrices each of whose rows belong to a given L_{n-m+1}. Then $\mathcal{L}$ is obviously a linear subspace of the n^2-dimensional space of all matrices. If (2.2) is false, there exists an L_{n-m+1} such that $\mathcal{L} \cap \mathcal{M} = \varnothing$. By a hyperplane separation theorem there exists a linear function which is zero on $\mathcal{L}$ and positive on $\mathcal{M}$; i.e., there exists a matrix A such that

$$\begin{aligned} &\operatorname{Tr} AX^T = 0 \qquad \text{for all} \quad X \in \mathcal{L}, \\ &\operatorname{Tr} AM^T > 0 \qquad \text{for all} \quad M \in \mathcal{M}. \end{aligned}$$

The second statement is the hypothesis of (2.1); the first statement says that A annihilates L_{n-m+1}, which contradicts the conclusion of (2.1). Hence, if (2.2) is false, so is (2.1).

LEMMA 1. *Let D_i be the convex hull of $C_i \cap \{x \mid \|x\| = 1\}$ (note that D_i is nonvoid and does not contain zero). Let $\mathscr{M}_i$ be the set of all matrices of order n in which each entry is zero except for row i; in row i, the first m coordinates are given by an arbitrary vector in D_i, the remaining coordinates are zero. Let $\mathscr{M}$ be the convex hull of $\{\mathscr{M}_i\}$, $i = 1, \ldots, n$. Then $\mathscr{M}$ satisfies (2.2) if and only if $\bigcup C_i \cup \bigcup -C_i = R^m$.*

PROOF: Suppose $\mathscr{M}$ satisfies (2.2). Let x be any vector in R^m (where we identify R^m with the space spanned by the first m unit coordinate vectors of R^n), and $L^{(x)}_{n-m+1}$ be the space generated by x and the last $n - m$ coordinate vectors of R^n. From (2.2) it follows that some positive or negative multiple of x is in D_i; i.e., $\bigcup C_i \cup \bigcup -C_i$ contains x.

Conversely, assume $\bigcup C_i \cup \bigcup -C_i = R^n$. Let L_{n-m+1} be any $(n - m + 1)$-dimensional subspace of R^n; $L_{n-m+1} \cap R^m$ has dimension at least 1. Let $x \in L_{n-m+1} \cap R^m$, $x \neq 0$. Then some positive or negative multiple of x is contained in some D_i; hence, $\mathscr{L}(L_{n-m+1}) \cap \mathscr{M} \neq \varnothing$.

LEMMA 2. *Let $\mathscr{M}$ be a closed convex set of matrices not containing zero, and let $\mathscr{M}^T = \{M \mid M^T \in \mathscr{M}\}$. Then $\mathscr{M}^T$ satisfies (2.2) if and only if $\mathscr{M}$ satisfies (2.2).*

PROOF: Invoke from Theorem 2 the equivalence of (2.1) and (2.2), and recall that any matrix A and its transpose A^T have the same rank.

We define, for any real number x,

$$\operatorname{sgn} x \begin{cases} +1 & \text{if} \quad x > 0 \\ 0 & \text{if} \quad x = 0 \\ -1 & \text{if} \quad x < 0. \end{cases}$$

LEMMA 3. *If each L_{n-m+1} contains a vector $x = (x_1, \ldots, x_n)$, $x \neq 0$, such that*

$$\bigcap_{\operatorname{sgn} x_i \neq 0} \operatorname{sgn} x_i C_i \neq 0, \tag{2.3}$$

then $\bigcup C_i \cup \bigcup -C_i = R^m$.

PROOF: Let $\mathscr{M}$ be as in Lemma 1. Consider $\mathscr{M}^T$. Let

$$y \in \bigcap_{\operatorname{sgn} x_i \neq 0} \operatorname{sgn} x_i C_i, \qquad \|y\| = 1.$$

Then the matrix which consists entirely of zeros in the last $n - m$ rows, and in the first m rows has (i, j)th entry $y_i x_j / \sum_j |x_j|$ is clearly in $\mathscr{M}^T$. Hence, $\mathscr{M}^T$

satisfies (2.2). By Lemma 2, $\mathscr{M}$ satisfies (2.2), so the conclusion follows from Lemma 1.

3. Proof of Theorem 1

Case 1 ($n = m$). In this case, (1.1) says $\bigcap_i \varepsilon_i C_i \neq 0$ for all choices of $\varepsilon_i = \pm 1$. Further, in this case $n - m + 1 = 1$, so (2.3) is also the statement $\bigcap_i \varepsilon_i C_i \neq 0$ for arbitrary choice of $\varepsilon_i = \pm 1$. Hence, the theorem follows from Lemma 3.

Corollary. If $\bigcup_{i=1}^m C_i \cup \bigcup_{i=1}^m -C_i = R^m$, then there exist polyhedral subcones each with at most 2^{m-1} generators $E_i \subset C_i$ such that $\bigcup E_i \cup \bigcup -E_i = R^m$. [More generally $\bigcup_{i=1}^n C_i \cup \bigcup_{i=1}^n -C_i = R^m$ implies that there exist polyhedral subcones $E_i \subset C_i$ such that $\bigcup E_i \cup \bigcup -E_i = R^m$. But this we shall prove elsewhere. The case $n = m$ is of special interest, however, since it shows that the use of "round" cones to prove the nonsingularity of real square matrices (using Theorem 2 and Lemma 1) is superfluous.]

Proof: By Fan's theorem, for any subset $S \subset \{1, \ldots, n\}$ and its complement $\bar{S}$, there exists a vector $y_{(s;\,\bar{s})} \in \bigcap_{i \in s} C_i \bigcap_{i \in \bar{s}} -C_i$, with $y_{(\bar{s};\,s)} = -y_{(s\,;\bar{s})}$. The generators of E_i are all $y_{(s;\,\bar{s})}$, with $i \in S$.

Case 2 ($n = m + 1$). Then an arbitrary L_{n-m+1} becomes an arbitrary L_2. Assume that L_2 can be conceived as generated by two independent vectors $a = (a_1, \ldots, a_n)$ and $b = (b_1, \ldots, b_n)$, in which no $b_i = 0$. We choose a permutation σ so that

$$\frac{a_{\sigma 1}}{b_{\sigma 1}} \geq \cdots \geq \frac{a_{\sigma n}}{b_{\sigma n}}. \tag{3.1}$$

Define $\varepsilon_{\sigma t} = (-1)^t \operatorname{sgn} b_{\sigma t}$. We shall show that with this choice of σ and ε_i, (1.1) implies (2.3) for some x. Namely, we consider the n vectors x that are obtained from linear combinations of a and b of the form

$$x^{\sigma k} = b_{\sigma k} a - a_{\sigma k} b, \qquad k = 1, \ldots, n. \tag{3.2}$$

Note that $x_{\sigma k}^{\sigma k} = 0$ for all k, but $x^{\sigma k} = 0$ for no k. To verify (2.3) it is sufficient to show that, for at least one k,

$$\bigcap_{l \neq k} \operatorname{sgn} x_{\sigma l}^{\sigma k} C_{\sigma l} \neq 0. \tag{3.3}$$

But with our choice of σ and ε_i, (1.1) becomes the statement that, for at least one k,

$$
\begin{aligned}
0 &\neq \bigcap_{r<k} (-1)^r \varepsilon_{\sigma r} C_{\sigma r} \cap \bigcap_{r>k} (-1)^{r+1} \varepsilon_{\sigma r} C_{\sigma r} \\
&= \bigcap_{r<k} (-1)^{2r} \operatorname{sgn} b_{\sigma r} C_{\sigma r} \cap \bigcap_{r>k} (-1)^{2r+1} \operatorname{sgn} b_{\sigma r} C_{\sigma r} \\
&= \bigcap_{r<k} \operatorname{sgn} b_{\sigma r} C_{\sigma r} \cap \bigcap_{r>k} (-\operatorname{sgn} b_{\sigma r}) C_r .
\end{aligned}
$$

To show this implies (3.3), it is sufficient to prove that

$$
(\operatorname{sgn} x_{\sigma r}^{\sigma k})(\operatorname{sgn} x_{\sigma s}^{\sigma k}) = \operatorname{sgn}(r-k) \operatorname{sgn} b_{\sigma r} \operatorname{sgn}(s-k) \operatorname{sgn} b_{\sigma s} \quad \text{or} \quad 0 \qquad \text{for all} \quad r \neq s. \tag{3.4}
$$

But the left side of (3.4) is

$$
\begin{aligned}
&\operatorname{sgn} (a_{\sigma r} b_{\sigma k} - b_{\sigma r} a_{\sigma k})(a_{\sigma s} b_{\sigma k} - b_{\sigma s} a_{\sigma k}) \\
&= \operatorname{sgn} \frac{b_{\sigma r} b_{\sigma s}(a_{\sigma r} b_{\sigma k} - b_{\sigma r} a_{\sigma k})(a_{\sigma s} b_{\sigma k} - b_{\sigma s} a_{\sigma k})}{b_{\sigma k}^2 b_{\sigma r} b_{\sigma s}} \\
&= \operatorname{sgn} b_{\sigma r} b_{\sigma s} \left(\frac{a_{\sigma r}}{b_{\sigma r}} - \frac{a_{\sigma k}}{b_{\sigma k}} \right) \left(\frac{a_{\sigma s}}{b_{\sigma s}} - \frac{a_{\sigma k}}{b_{\sigma k}} \right) \\
&= \operatorname{sgn}(b_{\sigma r} b_{\sigma s}(r-k)(s-k)) \quad \text{or} \quad 0,
\end{aligned}
$$

from (3.1).

There remains to consider the case in which there is some index i such that $x_i = 0$ for all $x \in L_2$. But such an L_2 can be approximated by a sequence of two-dimensional subspaces for which this does not occur. Since each such subspace will satisfy (2.3), so will L_2.

CASE 3 ($m = 2$). For the study of this case, and also the case $m = 3$, it is convenient to prove first the following lemma.

LEMMA 4. *Let L_{n-m+1} be spanned by the row vectors of a matrix L with $n - m + 1$ rows and n columns, and assume that every $n - m + 1$ columns of L are linearly independent. Let P be a matrix with $m - 1$ rows and n columns, whose rows span the orthogonal complement of L_{n-m-1}, and whose columns are denoted by P_j.*

For each $S \subset \{1, \ldots, n\}$, with $|S| = m$, let $x \neq 0$, $x \in L_{n-m+1}$ satisfy $x_j = 0$ for all $j \notin S$. Then $\sum_{j \in S} x_j P_j = 0$.

PROOF: Let $x' = z'L$. Then, since $LP^T = 0$, it follows that $x'P^T = z'LP^T = 0$. But this is precisely $\sum_{j \in S} x_j P_j = 0$.

Note that our hypotheses on L ensure that an x satisfying the hypothesis exists for each S when $|S| = m$, is unique (up to multiplication by a constant), and $x_j \neq 0$ for any $j \in S$.

In applying Lemma 4, it may happen that L does not satisfy the hypothesis of Lemma 4 that every $n - m + 1$ columns are independent. Then replace L by a nearby matrix that does satisfy this hypothesis. It is clear that if we prove (2.3) for a sequence of linear spaces of dimension $n - m + 1$ approaching L_{n-m+1}, then (2.3) holds for L_{n-m+1}. So in both the cases $m = 2$ and $m = 3$, we make the assumption that the hypothesis of Lemma 4 holds.

In case $m = 2$, $n - m + 1$, and P consists of a single row $p = (p_1, \ldots, p_n)$ in which all p_i are different and no p_i is zero. Choose σ so that

$$p_{\sigma 1} > p_{\sigma 2} > \cdots > p_{\sigma n}. \tag{3.5}$$

Choose $\varepsilon_t = \operatorname{sgn} p_t$. With this choice of σ and $\{\varepsilon_i\}$, (1.1) becomes the statement that there exist indices $k \neq l$ such that

$$\varepsilon_{\sigma k} C_{\sigma k} \cap -\varepsilon_{\sigma l} C_{\sigma l} \neq 0 \tag{3.6}$$

$$x_{\sigma k} p_{\sigma k} + x_{\sigma l} p_{\sigma l} = 0. \tag{3.7}$$

Let $x = (x_1, \ldots, x_n)$ arise from Lemma 4 with $S = \{\sigma k, \sigma l\}$. To verify (2.3) we must show, in view of (3.6),

$$\operatorname{sgn} x_{\sigma k} = \varepsilon_{\sigma k}, \qquad \operatorname{sgn} x_{\sigma l} = -\varepsilon_{\sigma l} \tag{3.8}$$

or

$$\operatorname{sgn} x_{\sigma k} = -\varepsilon_{\sigma k}, \qquad \operatorname{sgn} x_{\sigma l} = \varepsilon_{\sigma l}$$

From (3.7), sgn $x_{\sigma k} x_{\sigma l} = -\operatorname{sgn} p_{\sigma k} p_{\sigma l} = -\varepsilon_{\sigma k} \varepsilon_{\sigma l}$ [from (3.5)], which proves (3.8).

CASE 4 ($m = 3$). In this case P is a matrix of two rows $a = (a_1, \ldots, a_n)$ and $b = (b_1, \ldots, b_n)$, and we may assume all b_i different from zero. Choose σ so that

$$\frac{a_{\sigma 1}}{b_{\sigma 1}} > \cdots > \frac{a_{\sigma n}}{b_{\sigma n}}. \tag{3.9}$$

Note that $b_i \neq 0$ and the strict inequality in (3.9) are consequences of the stipulations on L.

Let $\varepsilon_i = \operatorname{sgn} b_i$. From (1.1) we know there exist indices $j < k < l$ such that

$$\varepsilon_{\sigma j} C_{\sigma j} \cap -\varepsilon_{\sigma k} C_{\sigma k} \cap \varepsilon_{\sigma l} C_{\sigma l} \neq 0. \tag{3.10}$$

Let $x = (x_1, \ldots, x_n)$ arise from Lemma 4 with $S = \{\sigma j, \sigma k, \sigma l\}$. To verify (2.3) we must show, in view of (3.6), that if A_t denotes the tth column of P, then

$$x_{\sigma j} A_{\sigma j} + x_{\sigma k} A_{\sigma k} + x_{\sigma l} A_{\sigma l} = 0 \tag{3.11}$$

implies

$$\operatorname{sgn} x_{\sigma j} x_{\sigma k} = -\operatorname{sgn} b_{\sigma j} b_{\sigma k}, \qquad \operatorname{sgn} x_{\sigma k} x_{\sigma l} = -\operatorname{sgn} b_{\sigma k} b_{\sigma l}. \tag{3.12}$$

But (3.11) may be rewritten as

$$\frac{x_{\sigma j}}{x_{\sigma k}} A_{\sigma j} + \frac{x_{\sigma l}}{x_{\sigma k}} A_{\sigma l} = -A_{\sigma k}. \tag{3.13}$$

Using Cramer's rule on (3.13), and invoking (3.9), one demonstrates (3.12).

In case $m = 4$, $n = 6$, it is easy to make up examples which show that there exists L_{n-m+1} with the property that the intersection patterns guaranteed by (1.1) are by themselves insufficient to ensure (2.3). This remark, of course, is not yet a disproof of the converse of Fan's theorem.

PROOF OF COROLLARY: Let $n = t + 3$, $m = 3$, and L be the matrix with $n - m + 1 = t + 1$ rows and n columns, one of whose rows consists entirely of ones, the remainder of the matrix given by the column vectors P_j. Let σ be the permutation given in Case 4 of the proof of Theorem 1. Then it is easy to see from the proof of Case 4 that $P_{\sigma i}$ is adjacent to $P_{\sigma j}$ if and only if $|i - j| = 1$ or $n - 1$.

ACKNOWLEDGMENT

We are very grateful to Benjamin Weiss for stimulating conversations about this material. An alternative proof of the corollary has been kindly communicated to us by Professor H. D. Ursell.

REFERENCES

1. KY FAN, A Generalization of Tucker's Combinatorial Lemma with Topological Applications, *Ann. Math.* **56** (1952), 431–437.
2. A. J. HOFFMAN, On the Nonsingularity of Real Matrices, *Math. Comp.* **19** (1965), 56–61.
3. A. J. HOFFMAN, On the Nonsingularity of Real Partitioned Matrices, *ICC Bull.* **4** (1965), 7–17.

OPTIMAL DISTANCE CONFIGURATIONS

Leroy M. Kelly

DEPARTMENT OF MATHEMATICS
MICHIGAN STATE UNIVERSITY
EAST LANSING, MICHIGAN

1. Introduction

Problem 1. (Erdős *et al.* [3]). For $k = 2, 3, \ldots$, suppose

$$D_k \equiv \max_{\{w_1, \ldots, w_k\}} \prod_{\substack{u=1 \\ u \neq v}}^{k} \prod_{v=1}^{k} |w_u - w_v|$$

where $\{w_1, w_2, \ldots, w_k\}$ are all k element subsets of the complex plane and $|w_u - w_v| \leq 2$, $u, v = 1, \ldots, k$. If the w_u are roots of a polynomial then this is of course the discriminant of the polynomial. In [2] Erdős *et al.* conjectured that D_k would assume its maximum when the points $w_1, w_2, \ldots, w_k$ were the vertices of a regular k-gon.

Problem 2. (L. Moser [5]). If P_i, $i = 1, 2, \ldots, n$, are n points with mutual distances $\overline{P_i P_j} \geq 2$, $i \neq j$, then (according to Blichfeldt) for any point O, $\sum \overline{OP_i}^2 \geq 2(n-1)$ with equality possible for $n = 2, 3, 4$. What is the sharp inequality for somewhat larger values of n?

These two problems provided the impetus for the present investigation. They illustrate a class of very natural dual problems about which surprisingly little seems to be known. Problem 1 asks for those k-point configurations in E_2, with all distances *less* than or equal to 2, for which the product of the distances is a maximum.

I claim that Problem 2 asks for k-point configurations in, say, E_2 with all distances *greater* than or equal to 2, for which the sum of the squares is a minimum. Let me show, in fact, that Problem 2 can be phrased in this fashion.

A formula of Lagrange, which deserves to be better known, asserts that if P is any point in E_n, $\{P_i, m_i\}$, $i = 1, 2, \ldots, k$, a k-point mass system in E_n, and G its center of gravity, then

$$\overline{PG}^2 = \left(\sum_{i=1}^{k} \overline{PP_i}^2 m_i \Big/ \sum_{i=1}^{k} m_i\right) - \left[\sum_{i>j} \overline{P_i P_j}^2 m_i m_j \Big/ \left(\sum_{i=1}^{k} m_i\right)^2\right].$$

So in our case choosing $m_i = 1$, $i = 1, 2, \ldots, k$,

$$\left(\sum \overline{PP_i}^2/k\right) = \overline{PG}^2 + \left(\sum_{i>j} \overline{P_i P_j} \Big/ k^2\right)$$

or

$$\sum \overline{PP_i}^2 = k\overline{PG}^2 + (1/k)\sum_{i>j} \overline{P_i P_j}^2.$$

Thus for a fixed set of points $\sum_{i>1}^{k} \overline{PP_i}^2$ is minimum when $P = G$, at which point the minimum sum of squares is $(1/\mathrm{k}) \sum \overline{P_i P_j}^2$.

Thus to solve the Moser problem we must find the configuration of k points in the specified space, all of whose distances are at least 2, which minimizes the sum of the squares of the distances. Moser was not quite explicit about the space he had in mind though it seems clear he was thinking of E_3. The Blichfeldt minimum can clearly be obtained for any k if we are allowed to operate in E_{k-1}. We will concern ourselves here with the problem in E_2.

We find it convenient to normalize our maximum and minimum distances at 1 instead of 2. With this understanding a configuration in E_2 which *maximizes* D_k will be denoted by the symbol $C(k, \bar{\pi})$ while a configuration which solves the modified form of Moser's problem will be denoted $C(k, \underline{\Sigma}^2)$. In general $C(k, \bar{f})$ will stand for a k-point configuration in E_2 all of whose distances are less than or equal to 1 which maximizes some suitable function f of the distances. $C(k, \underline{f})$ is defined dually. We conclude this introduction with a summary of a few known results bearing on Problem 1.

The Erdős *et al.* problem was considered by Danzer and Pommerenke [1] who showed that the conjecture was false. More specifically they showed that D_{2k} never assumes its maximum on the vertices of a regular $2k$-gon. They were able to characterize the $C(4, \bar{\pi})$ and did obtain some interesting bounds in general. The present conjecture is that for odd k the vertices of the regular k-gon do provide the optimal set, and Danzer and Pommerenke have concocted a class that is rather hard to describe briefly as a candidate for the even case.

Some rather lengthy computations which I have recently carried out show that $C(5, \bar{\pi})$ is the vertex set of a regular pentagon; however, the possibility of extending these computations beyond this point seems unlikely. The organization of the computations closely parallels that for $C(5, \bar{\Sigma})$ which is described in Theorem 4.2.

We present here a simple argument to show that $C(6, \bar{\pi})$ is not the vertex set of a regular hexagon (see Fig. 1). The argument generalizes obviously to any even n.

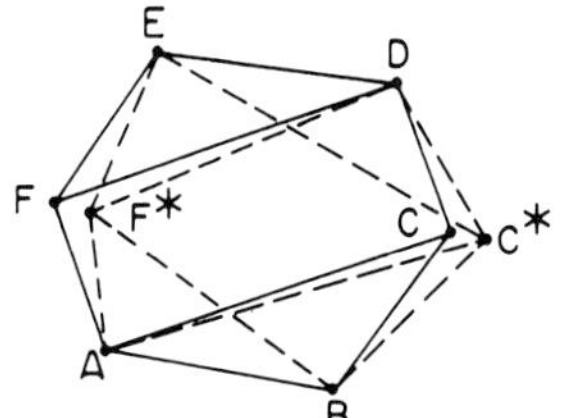

FIG. 1.

Suppose $ABCDEF$ is a regular hexagon and $\overline{FF^*} = \overline{CC^*} = \varepsilon$. The areas of $\Delta\ FDC$ and $\Delta\ F^*DC^*$ are equal. Therefore since $\angle\ FDC = 90°$, $\overline{FD} \cdot \overline{DC} = \overline{F^*D} \cdot \overline{DC^*} \sin \angle\ F^*DC$, and $F^*D \cdot DC^* > FD \cdot DC$.

Similarly $F^*E \cdot \overline{EC^*} > FE \cdot EC$, $F^*A \cdot AC^* > FA \cdot AC$, and $F^*B \cdot BC^* > FB \cdot BC$.

Thus $C(6, \bar{\pi})$ is not the vertex set of a regular hexagon.

A completely analogous argument shows that $C(2k, \bar{\pi})$ is not the vertex set of a regular $2k$-gon for any k.

2. Notation and General Theorems

Among the most useful tools in analyzing these optimal configurations are two graphs associated with a finite point set in a distance space known as the maximal distance graph (or the diameter graph) and the minimal distance graph. The maximal distance graph of a set S, denoted $M(S)$, has for vertices those points of S which are end points of at least one diameter of S and the edges of $M(S)$ are those pairs of vertices which are a maximal distance apart. The minimal distance graph, $m(S)$, is defined dually. The segments defined by the edges are called the edge segments.

REMARK 2.1. It is easy to show that for $S \subset E_2$ each two edge segments of $M(S)$ must intersect and that two edge segments of $m(S)$ have at most end points in common.

THEOREM 2.1 (Erdős). *$M(S)$, S a k-tuple in E_2, has at most k edges.*

THEOREM 2.2 (Erdős). *For S a k-tuple in E_2, $m(S)$ has at most $3k - 6$ edges.*

THEOREM 2.3 (Danzer and Pommerenke [1]). *$MC(k, \bar{\pi})$ is connected and spans $C(k, \bar{\pi})$.*

The proof of Theorem 2.3 is based on an interesting use of the maximum modulus principle and does not immediately extend to proving that $MC(n, \bar{\Sigma}^k)$, $k \geq 1$, is connected and spans $C(n, \bar{\Sigma}^k)$. However, an application of a generalized form of the maximum modulus principle (found in Pólya–Szego, for example) enables us to prove the following:

THEOREM 2.4. *$MC(n, \bar{\Sigma}^k)$, $k \geq 1$, is connected and spans $C(n, \bar{\Sigma}^k)$.*

PROOF: Suppose A is a component of $C(n, \bar{\Sigma}^k)$ and $C(n, \bar{\Sigma}^k) - A = B \neq \varnothing$. If we regard $C(n, \bar{\Sigma}^k)$ as a subset of the complex plane we may define the set $P(z) = A \cup (B + z)$. Since $|a_i - b_j| < 1$, $a_i \in A$, $b_i \in D$, there is a closed circular neighborhood, U, of the origin such that the diameter of $P(z) \leq 2$ for each z in U.

What we need to show is that $\sum_1^n \sum_1^n |b_i + z - a_j|^k$ is not maximal in U at $z = 0$. The theorem alluded to above assures us that the sum of the absolute values of n complex valued functions analytic in a closed domain assumes its maximum on the boundary. Hence

$$\sum_1^n \sum_1^n |b_i + z - a_j|^k = \sum_1^n \sum_1^n |(b_i + z - a_j)^k|$$

cannot be maximal at $z = 0$.

It is convenient to denote the class of k-tuples in E_2, by B_k, the class of k-tuples all of whose distances are at most 1 by $\bar{B}_k$ and the class all of whose distances are at least 1 by $\underline{B}_k$.

If the $\binom{n}{2}$ distances associated with a set S of n points in a distance space are arranged in a nonincreasing sequence, denoted $D(S)$, the result is called the ordered distance sequence of S. Sets of the same finite cardinality can then be partially ordered by the convention $P \geq Q$ iff each term of $D(P)$ is not less than the corresponding term of $D(Q)$. Under this ordering a maximal element in $\bar{B}_k$ is called an inner monotone k-tuple and a minimal element in $\underline{B}_k$ is an outer monotone k-tuple.

For $k = 2$, the end points of a unit segment are both inner and outer monotone and are the only such sets. For $k = 3$ the vertices of a unit equilateral triangle form the only monotone sets. For $k = 4$ the characterization of such sets is not so immediate.

Our interest in such sets arises, of course, from the fact that $C(n, \bar{\pi})$, $C(n, \bar{\Sigma}^k)$ are inner monotone and $C(n, \underline{\pi})$, $C(n, \underline{\Sigma}^k)$ are outer monotone.

THEOREM 2.5. *An outer monotone quadruple in E_2 is the vertex set of a rhombus of unit side length.*

PROOF: We first prove that the points of such a quadruple are the vertices of a strictly convex quadrilateral. For if $\{P_1, P_2, P_3, P_4\}$ is such a quadruple with P_4 in the convex hull of $\{P_1, P_2, P_3\}$, one of the angles $\{\angle P_i P_4 P_j\}$ must be at least 120° and hence the corresponding side of the $\Delta\, P_1 P_2 P_3$ must be at least $\sqrt{3}$. Now compare $\{P_1, P_2, P_3, P_4\}$ with $\{(Q_1, Q_2, Q_3, Q_4\}$ where $\overline{Q_1 Q_2} = \sqrt{3}$ and all other distances are 1. It should now be clear that $\{P_1, P_2, P_3, P_4\}$ is not outer monotone.

Now let $\{P_1, P_2, P_3, P_4\}$ be the vertices of a convex outer monotone set whose diagonals are $\overline{P_1 P_3}$ and $\overline{P_2 P_4}$. If $\overline{P_1 P_3} = 1$ then $\overline{P_2 P_4} > 1$ by Remark 2.1.

It is now an easy matter to see that $P_2 P_1 = P_2 P_3 = 1$ since otherwise a slight perturbation of P_2 will produce a "smaller" quadrilateral still in $\underline{B}_k$. A similar remark holds relative to P_4 and the quadrilateral is a rhombus with unit sides.

Suppose finally that neither diagonal is 1. An entirely similar argument shows that the sides of the quadrilateral are all equal to 1.

We include this rather pedestrian proof to show the elementary nature of most of the arguments.

By a similar but more involved analysis we can prove the following:

THEOREM 2.6. *The convex hull of an outer monotone 5-tuple is not a triangle.*

THEOREM 2.7. *If the convex hull of an outer monotone 5-tuple is a quadrilateral, then both diagonals and at least one side of the quadrilateral are greater than* 1.

THEOREM 2.8. *A $C(n, \bar{f})$ is an inner monotone set and a $C(n, \underline{f})$ is an outer monotone set where $f = \pi$, Σ^k or, in fact, any "monotone" function of the distances.*

THEOREM 2.9. *Both $C(n, \bar{\Sigma}^k)$ and $C(n, \bar{\pi})$ are the vertices of a convex n-gon in E_2.*

PROOF: Since $MC(n, \bar{\Sigma}^k)$ spans $C(n, \bar{\Sigma}^k)$, if any point of $C(n, \bar{\Sigma}^k)$ were interior to the convex hull of the others, one of the distances would have to exceed 1, which is impossible.

In fact suppose $\overline{P_e P_m} = 1$ with P_e interior to the convex hull of remaining points. The ray $\overrightarrow{P_m P_e}$ intersects the boundary of the convex hull in a point x. If x is a point of the set then $\overline{P_m x} > 1$, a contradiction. If x is on a side $P_j P_k$ not at a vertex then either $\overline{P_m P_j}$ or $\overline{P_m P_k}$ is greater than 1.

3. The Sum of the kth Powers, $k \geq 2$

The analysis of $C(n, \bar{\Sigma}^k)$ presents one feature which renders it conceptually more difficult than $C(n, \bar{\pi})$. In the latter case it is clear that the maximizing configuration consists of n distinct points while in the former it does not. Indeed we shall show that in many cases, $C(n, \bar{\Sigma}^k)$ is an n-tuple with precisely three *distinct points*. This prompts the following definition.

Definition 3.1. If S is an n-tuple of not necessarily distinct points let S^* be the set of pairwise distinct points of S. $M(S^*)$ is more conveniently denoted $M^*(S)$.

Definition 3.2. An n-tuple of points distributed as "evenly as possible" on the vertices of a unit equilateral triangle is denoted K_n.

More formally: $P_1, P_2, \ldots, P_n$ is denoted K_n iff $\overline{P_i P_j} = 0$, $i = j \bmod 3$; $\overline{P_i P_j} = 1$ otherwise.

We have already observed (Theorem 2.4) that each point of $C(n, \bar{\Sigma}^2)$ is a vertex of $MC(n, \bar{\Sigma}^2)$.

Theorem 3.1. *$M^*C(n, \bar{\Sigma}^2)$ is a cycle.*

We first show that no vertex is of order 1. Suppose, in fact, that P_1 is a vertex of order 1 in $M^*C(n, \bar{\Sigma}^2)$ and that $\overline{P_1 P_2} = 1$. From the Lagrange formula mentioned in the introduction, it follows that the locus of P such that

$$\sum_{i=3}^{n} \overline{PP_i}^2 = \sum_{i=3}^{n} \overline{P_1 P_2}^2$$

is a circle whose center is at the centroid G of $\{P_3, P_4, \ldots, P_k\}$. Now if the circle with center P_2 and radius $\overline{P_2 P_1} = 1$ cuts this circle then a slight perturbation of P_1 will produce an n point in $\bar{B}_n$ with a greater sum of squares than $C(n, \bar{\Sigma}^2)$. Thus the two circles must be tangent at P_1 and P_1, P_2, G are collinear. Now G is interior to the convex polygon $\{P_1, P_2, \ldots, P_n\}$ while P_1 and P_2 are vertices, so certainly G is between P_1 and P_2. But here again since the circle with center G and radius GP_1 is inside that with center P_2 and radius $\overline{P_2 P_1} = 1$, a slight perturbation of P_1 will produce an n point in $\bar{B}_n$ with a greater sum of squares than $C(n, \bar{\Sigma}^2)$.

Thus no vertex of $C(n, \bar{\Sigma}^2)$ is of order 1. Now each vertex is of order at least 2 and by Theorem 2.1 $M^*C(n, \bar{\Sigma}^2)$ has at most n edges. Hence each vertex must have precisely order 2 which together with the fact that $MC(n, \bar{\Sigma}^2)$ is connected implies that $M^*C(n, \bar{\Sigma}^2)$ is a cycle.

THEOREM. *A graph in E_2 with no three vertices collinear which is an even cycle has a pair of disjoint edge segments.*

PROOF: Suppose the theorem false and let $P_1, P_2, \ldots, P_{2n}$ be the vertices of such an even cycle with no pair of disjoint edge segments. Since each pair of edge segments intersect, the vertices $2i$, $i > 1$, must lie on one side of the line P_1P_2 and vertices $2i + 1$, $i \geq 1$, must lie on the other side. But this implies that $\overline{P_{2n} P_1} \cap \overline{P_2 P_3} = \varnothing$, contrary to the hypotheses.

THEOREM 3.2. $C(4, \bar{\Sigma}^2) = K_4$.

PROOF: Suppose $\{P_1, P_2, P_3, P_4\} = C(4, \bar{\Sigma}^2)$ and that all the points are distinct. Any 4-cycle on these points will include a pair of opposite sides of the convex quadrilateral P_1, P_2, P_3, P_4 which contradicts Remark 2.1. Hence at least two of the points coincide and it follows at once that $C(4, \bar{\Sigma}^2) = K_4$.

THEOREM 3.3. $C(5, \bar{\Sigma}^2) = K_5$.

PROOF: Suppose first that the five points P_1, P_2, P_3, P_4, P_5 are all distinct. They form the vertices of a convex pentagon; hence one of the angles, say that at P_1, is obtuse. Consider an associated $K_5 = \{Q_1, Q_2, Q_3, Q_4, Q_5\}$ where $Q_1 = Q_4$, $Q_2 = Q_5$.

Since $\{Q_2, Q_3, Q_4, Q_5\}$ is a K_4, $\sum_2^5 \overline{Q_i Q_j}^2 \geq \sum_2^5 \overline{P_i P_j}^2$. Also $\sum_2^5 \overline{Q_1 Q_i}^2 = 3$, while $\sum_2^5 \overline{P_1 P_i}^2 < 3$ since $\overline{P_1 P_2} + \overline{P_1 P_5}^2 < 1$. Thus the points of $C(5, \bar{\Sigma}^2)$ cannot all be distinct.

If $M^*C(5, \bar{\Sigma}^2)$ has four distinct vertices it must be a cycle, and by our previous observation must contain a pair of disjoint edges, which is impossible. Thus $M^*C(5, \bar{\Sigma}^2)$ has precisely three vertices and $MC(5, \bar{\Sigma}^2)$ is clearly K_5.

THEOREM 3.4. $C(6, \bar{\Sigma}^2) = K_6$.

PROOF: If all six points of $C(6, \bar{\Sigma}^2)$ are distinct, $MC(6, \bar{\Sigma}^2)$ is a 6-cycle and in E_2 such a cycle has at least one pair of disjoint edges. This is impossible and thus $C(6, \overline{\Sigma}^2)$ has at most five distinct points. If $C(6, \overline{\Sigma}^2)$ consists of five points, say, $P_1 = P_2, P_3, P_4, P_5, P_6$, then they form the vertices of a convex pentagon and at least two of the angles are obtuse.

If P_i is this vertex $\sum_{j=1}^{6} \overline{P_i P_j}^2 < 4$. Also by induction

$$\sum_{\substack{u \neq i \\ v \neq i}} \overline{P_u P_v}^2 \leq 8.$$

Hence $\sum \overline{P_i P_j}^2 < 12$. But for K_6, $\Sigma^2 = 12$. Hence $C(6, \overline{\Sigma}^2)$ has at most four points. But $M^*C(6, \overline{\Sigma}^2)$ with four vertices is impossible by the argument of Theoerm 3.2, and hence $C(6, \overline{\Sigma}^2) = K_6$.

Similar arguments allow us to assert:

THEOREM 3.5. $C(n, \overline{\Sigma}^2) = K_n$, $n \leq 9$.

The secret of the success of these arguments is the presence and distribution of a sufficient number of obtuse angles. However for $n = 10$ the number ceases to be sufficient for our purposes, all of which suggests that some other attack may be in order.

CONJECTURE. $C(n, \overline{\Sigma}^2) = K_n$ for all n.

THEOREM 3.6. *$C(4, \underline{\Sigma}^2)$ is the vertex set of a rhombus of unit edge each of whose diagonals is greater than or equal to* 1.

PROOF: $C(r, \underline{\Sigma}^2)$ must be an outer monotone quadruple by Theorem 2.5, and, of course, its diagonal must be at least 1. Since the sum of the squares of edges of any parallelogram is equal to the sum of the squares of the diagonals, the sum of the squares of the six distances in all such quadrilaterals is the same.

Let T be the union of points forming a triangular tesselation of the plane into unit equilateral triangles and let $T_3 = \{A_1, A_2, A_3\}$ be the vertices of any one of the triangles. Define $A_n \in T$ inductively by requiring that the distance from A_n to the centroid of T_{n-1} be minimal over all points of $T - T_{n-1}$. However, A_n is not uniquely defined by this scheme nor is it immediately clear that the various possible T_n, for fixed n, are all isometric. For low values of n the isometry is clear. The diagram of Fig. 2 gives some idea of the T_n sequence.

THEOREM 3.7. $C(5, \underline{\Sigma}^2) = T_5$.

The proof of this theorem is rather long and without any great generality. We suppress the details.

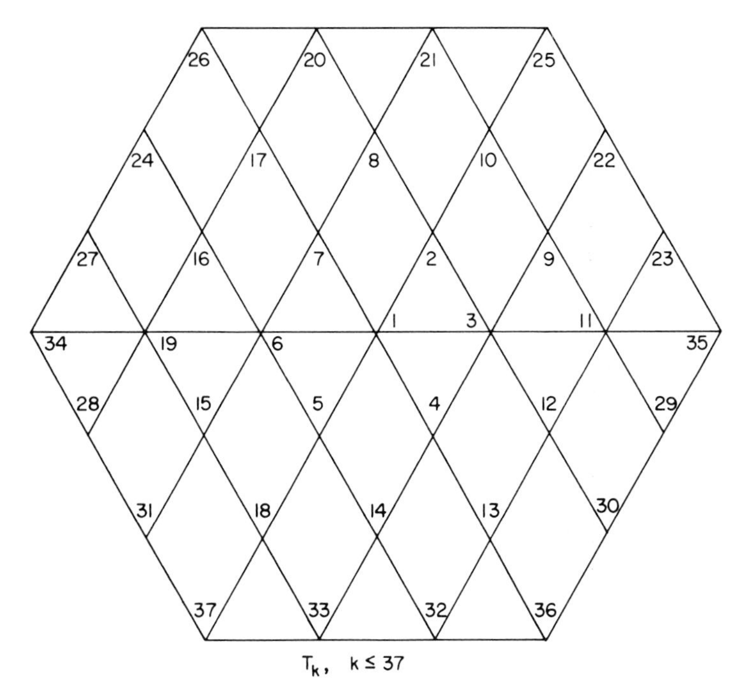

T_k, $k \leq 37$

FIG. 2.

CONJECTURE. $C(n, \underline{\Sigma}^2) = T_n$, $n \geq 3$.

THEOREM 3.8. $C(n, \overline{\Sigma}^k) = K_n$, $n \leq 9$, $k \geq 2$.

PROOF: Since the distances in $C(n, \overline{\Sigma}^k)$ are not greater than 1, the sum of the kth powers is less than or equal to the sum of the squares. Hence

$$C(n, \overline{\Sigma}^k) = C(n, \overline{\Sigma}^2) = K_n, \qquad n \leq 9.$$

CONJECTURE. $C(n, \overline{\Sigma}^k) = K_n$, $k \geq 2$.

This conjecture will be settled, of course, if it is answered affirmatively for $k = 2$.

THEOREM 3.9. *$C(4, \underline{\Sigma}^k)$ is the vertex set of a unit square $k > 2$.*

PROOF: $C(4, \underline{\Sigma}^k)$ is an outer monotone quadruple and hence is a rhombus with unit sides. It remains to minimize $x^k + y^k$ where $x^2 + y^2 = 4$, $x, y \geq 1$.

This is easily seen to occur when $x = y = \sqrt{2}$, that is to say when $C(4, \underline{\Sigma}^k)$ is the vertex set of a unit square.

However for $n > 4$ we make the following:

CONJECTURE. $C(n, \underline{\Sigma}^k) = T_n$, $k \geq 2$.

4. THE SUM OF THE DISTANCES

In general we are hoping that $C(n, \overline{\Sigma}^k) = K_n$ and that $C(n, \underline{\Sigma}^k) = T_n$. For $n = 4$ or 5, we have slight aberrations, but the other verified cases fit the pattern. However for $k = 1$ it again seems to take longer to settle down.

THEOREM 4.1. *$C(4, \overline{\Sigma}) = \{P_1, P_2, P_3, P_4\}$ with $P_1P_2 = P_1P_3 = P_1P_4 = P_2P_3 = 1$, $P_2P_4 = P_3P_4 = 2 \sin 15°$.*

PROOF: $C(4, \overline{\Sigma})$ constitutes the vertex set of a convex quadrilateral by Theorem 2.9. If either diagonal is not 1 then all the sides are seen to be 1, that is, the quadrilateral is a rhombus. But this is impossible since opposite unit sides do not intersect.

We may thus assume both diagonals are 1. Since $MC(4, \overline{\Sigma})$ is connected, at least one of the sides is 1. It is then clear that at least one more side must be 1 and that the two unit sides must be adjacent. We thus have a convex quadrilateral $ABCD$ with ABC a unit equilateral triangle, D on a unit circle centered at A (see Fig. 3). The point D on the circular arc BC which maximizes

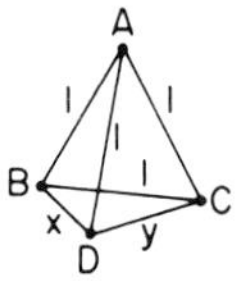

FIG. 3.

$x + y$ is easily seen to be at the middle point of the arc and the theorem is proved.

THEOREM 4.2. *$C(5, \overline{\Sigma})$ is the vertex set of a pentagon with unit diagonals.*

PROOF (outline): Since the points of $C(5, \overline{\Sigma})$ form a set of constant diameter and $MC(5, \overline{\Sigma})$ is connected there are three possibilities for $MC(5, \overline{\Sigma})$, as indicated by the diagrams of Fig. 4. Case (a) is easily eliminated as a possible $C(5, \overline{\Sigma})$ since the set $ABCD^*E$ of Fig. 5 is seen to have a larger sum of distances

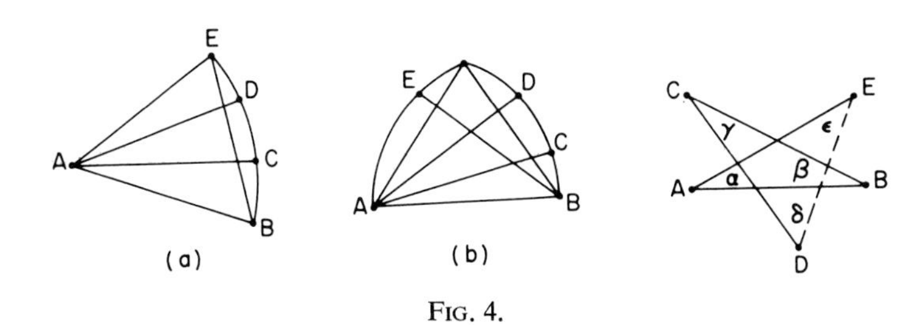

Fig. 4.

and to satisfy the boundary conditions. In Case (b), if E is "higher" than D the transformation indicated in Fig. 6 shows a set $ABCD^*E^*$ with a smaller sum of distances than $ABCDE$. We thus assume E is lower than D and $D \neq V$ (Fig. 7). Thus $\alpha \leq \beta < 60$, while $\gamma > 60$. Hence $\alpha < \gamma$. This implies

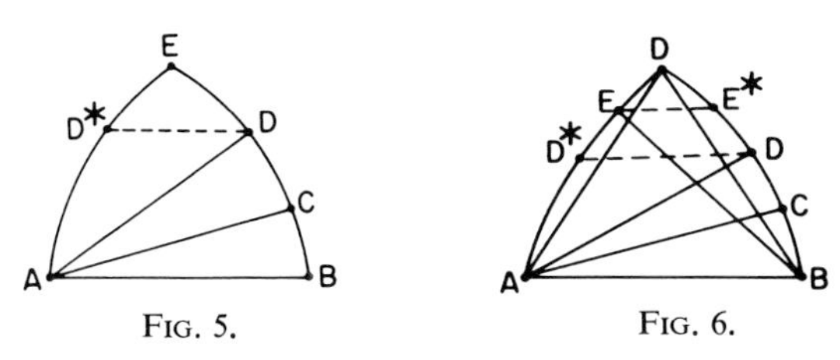

Fig. 5. Fig. 6.

that the normal to the ellipse with foci E and B at the point D intersects the segment AB and the point V must thus lie outside this ellipse. The point set $ABCVE$ has a larger distance sum than $ABCDE$ and we are led to consider the case shown in Fig. 8.

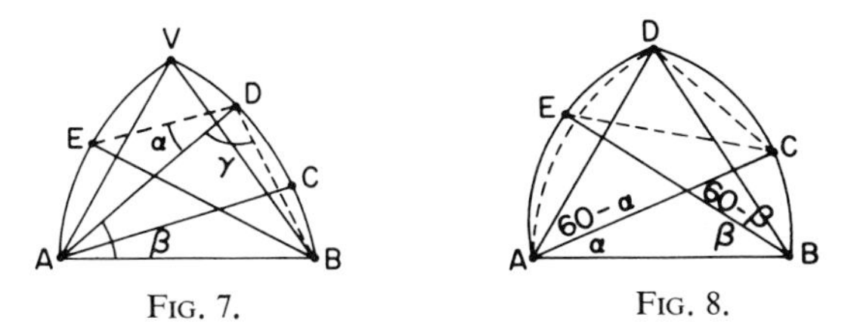

Fig. 7. Fig. 8.

In this case the problem is one in elementary calculus where we must maximize

$$2 \sin \tfrac{1}{2}\alpha + 2 \sin \tfrac{1}{2}\beta + 2 \sin \tfrac{1}{2}(60 - \alpha) + 2 \sin \tfrac{1}{2}(60 - \beta) + [(\sin \alpha - \sin \beta)^2 + (\cos \alpha + \cos \beta - 1)^2]^{1/2}.$$

A routine calculation shows that $\alpha = \beta$ and that the maximum occurs when $3^\circ < \alpha < 4^\circ$, leading to an upper bound for $\underline{\underline{\Sigma}}$ of 8.02. However, $\underline{\underline{\Sigma}}$ for the regular pentagon is 8.09, which eliminates Case (b).

In Case (c) a somewhat involved analysis reveals that DE must also be 1, after which an elementary calculus argument shows us that $\sin \frac{1}{2}\alpha + \sin \frac{1}{2}\beta + \sin \frac{1}{2}\gamma + \sin \frac{1}{2}\delta + \sin \frac{1}{2}\gamma$ where $\alpha + \beta + \gamma + \delta + \varepsilon = 180°$ is a maximum when all the angles are equal. Thus the optimal configuration is the vertex set of a regular pentagon.

Note that K_5 has a Σ of 8.0, so the pentagon is only slightly better. However $K_6 = 12$ while Σ for the regular hexagon is 11.2.

CONJECTURE. $C(n, \bar{\Sigma}) = K_n, n \geq 6$.

THEOREM 4.3. $C(4, \underline{\Sigma}) = T_4$.

PROOF: $C(4, \underline{\Sigma})$ is an outer monotone quadruple and hence must be the vertex set of a rhombus of unit side. We must then minimize the sum of the diagonals where their sum of squares is equal to 4. This minimum occurs when one diagonal is 1 and the other $\sqrt{3}$.

THEOREM 4.4. $C(5, \underline{\Sigma}) = T_5$.

We omit the proof.

CONJECTURE. $C(n, \underline{\Sigma}) = T_n, n \geq 3$.

REFERENCES

1. L. DANZER and CH. POMMERENKE, Uber die Deskriminante von Menge gegebenen Durchmessers, *Monatsh. Math.* **71/2** (1967), 100–113.
2. P. ERDŐS, On Sets of Distances of n Points, *Amer. Math. Monthly* **53** (1946), 240–250.
3. P. ERDŐS, F. HERZOG, and G. PIRANIAN, Metric Properties of Polynomials, *J. Analyse Math.* **6** (1958), 125–148.
4. P. ERDŐS, Some Unsolved Problems, *Michigan Math. J.* **4** (1957), 296.
5. L. MOSER, On Different Distances Determined by n Points, *Amer. Math. Monthly* **59** (1952), 85–91.
6. CH. POMMERENKE, On Metric Properties of Complex Polynomials, *Michigan Math. J.* **8** (1961), 97–115.

COMBINATORIAL DESIGNS AS MODELS OF UNIVERSAL ALGEBRAS

N. S. Mendelsohn

UNIVERSITY OF MANITOBA
WINNIPEG, MANITOBA

1. Introduction

This is the third in a series of papers by the author in which concepts of universal algebra are used in the construction of combinatorial designs. Algebraic systems such as groupoids, quasigroups, loops, and ternary algebras have been effectively used in the construction of designs. Some examples are Steiner triple systems [1], Room designs [1], generalized triple systems [2], and classes of directed graphs [3].

In this paper we apply these algebraic techniques to construct infinite classes of Balanced Incomplete Block Designs (BIBD) which have considerably more symmetry as compared with known designs; and to construct a class of latin squares which are orthogonal to their transposes. Furthermore, the algebraic formulation makes a study of the automorphism group of some of these designs a relatively simple procedure; it also enables us to study such concepts as simplicity of design and intersection properties of the blocks of a design. Because of the demonstrated effectiveness of these methods for a large number of problems, it is hoped that they will find a widespread acceptance for the solution of other combinatorial problems.

2. Nonassociative Fibonacci Numbers

In this section we are interested in the following type of design. Suppose we have a set S of v elements and b subsets of S each containing k elements. Suppose that each of the subsets (called blocks) are arranged cyclically and suppose that every ordered pair of elements of S appears exactly once in one of the blocks, e.g., the block $(a\ b\ c\ d\ e)$ contains the ordered pairs ab, bc, cd, de, and ea. For lack of a better term we designate such designs as "special block designs."

Consider any special block design. Let i be any element of S. Since there are $v-1$ ordered pairs of which i is the first member, each element appears in $r = v-1$ blocks. Also since each block contains k ordered pairs and there are altogether $v(v-1)$ ordered pairs we have $v(v-1) = kb$. If the unordered pair i, j appears in λ_{ij} blocks, we readily obtain

$$\sum \lambda_{ij} = b\,\frac{k(k-1)}{2}.$$

Again each $\lambda_{ij} \geq 2$, and if all the $\lambda_{ij} = \lambda$, we have a special BIBD with $\lambda = k-1$. Note also that if $k = 3$, every special block design is a BIBD with parameters v, $b = v(v-1)/3$, $\lambda = 2$, $k = 3$, $r = v-1$. These have been called by the author "generalized triple systems" and have been studied in [2]. Again, it is easy to display examples which show that not every BIBD with the appropriate parameters can have its blocks arranged cyclically so as to form a special block design.

To study the special block designs we extend the notion of a Fibonacci sequence to a nonassociative system. Using multiplicative notation and starting with two symbols, we generate an infinite sequence of words in which each word is the product of the two preceding words, viz., a, b, ab, $b(ab)$, $(ab)(b(ab))$, $(b(ab))((ab)(b(ab)))$, etc. Putting $W_1(a, b) = a$ and $W_2(a, b) = b$, the sequence is defined recursively by

$$W_{i+2}(a, b) = W_i(a, b)W_{i+1}(a, b). \tag{1}$$

Also it is clear that

$$W_{i+1}(a, b) = W_i(b, ab) = W_{i-1}(ab, b(ab)). \tag{2}$$

More generally, we can write

$$W_i(W_1(a, b), W_2(a, b)) = W_j(W_{i+1-j}(a, b), W_{i+2-j}(a, b))$$

where

$$1 \leq j \leq i. \tag{3}$$

We now construct a groupoid $\langle S, \cdot \rangle$ with a single identity

$$W_{k+1}(a, b) = a \qquad \text{for all} \qquad a, b \quad \text{in} \quad S. \tag{4}$$

It follows that

$$W_{k+2}(a, b) = W_{k+1}(b, ab) = b. \tag{5}$$

Hence the sequence $W_1(a, b), W_2(a, b), \ldots, W_k(a, b)$ repeats cyclically with cycles of length k, although there may be a smaller cycle length which is a divisor of k.

From (4) and (5)

$$W_k(a, b)a = W_k(a, b)W_{k+1}(a, b) = W_{k+2}(a, b) = b.$$

We note that the identity

$$W_k(a, b)a = b \tag{6}$$

could replace the identity $W_{k+1}(a, b) = a$ in the definition of the groupoid, though the proof is not trivial and will not be included here.

Now the right cancellation law will be proved. Suppose $ab = cb$. Then from (3), $W_{k+1}(a, b) = a$, and using (2), we obtain $W_k(b, ab) = a$. Also $W_k(b, cb) = c$. Hence $a = c$. The left cancellation law does not follow from (3), as it is easy to exhibit models in which the left cancellation law is not valid. However, there is much combinatorial significance for those systems in which the left cancellation law is valid.

3. Special Block Designs with $k = 4$

Consider the groupoid $\langle S, \cdot \rangle$ satisfying the identities

$$[b(ab)]a = b \tag{7}$$

$$a^2 = a \tag{8}$$

$$ca = cb \quad \text{implies} \quad a = b. \tag{9}$$

The first of these equations is simply (6) with $k = 4$. Since the right cancellation law also holds for such a groupoid, it follows that the system is a quasigroup. Now suppose a and b are two elements of S with $a \neq b$. It is readily verified that in the cyclic quadruple $\{a, b, ab, b(ab)\}$ all four elements are distinct. Furthermore, among the quadruplets so formed, if we have $\{a, b, c, d\}$ and $\{a, f, c, g\}$ then $b = f$ and $d = g$. This follows from $c = ab = af$ so that $b = f$. Hence we have a system of cyclic quadruplets in which given a pair a, b, these will appear exactly once in the forms $\{a, b, *, *\}$, $\{b, a, *, *\}$ and at most once in the form $\{a, *, b, *\}$. (Thus an unordered pair appears in at most three blocks.) Hence if S has v elements, we have a special block design in which $k = 4$, $b = v(v-1)/4$, and $r = v - 1$. Also $\sum \lambda_{ij} = 3v(v-1)/2$. As there are exactly $v(v-1)/2$ unordered pairs, the average value of $\lambda_{ij} = 3$. But the maximum value of $\lambda_{ij} = 3$. Hence $\lambda_{ij} = \lambda = 3$ for all i, j in S. Hence our special block design is a BIBD. The condition $b = v(v-1)/4$ implies $v \equiv 0 \bmod 4$ or $v \equiv 1 \bmod 4$. The following theorem yields a construction for a large infinite class of these designs.

THEOREM 1. *Let v be any odd integer which can be expressed as a sum of two squares. Then there exists a BIBD with parameters v, $b = v(v-1)/4$, $k = 4$, $\lambda = 3$, and $r = v - 1$. Also the blocks can be ordered cyclically in such a way that every unordered pair of elements a, b will appear in the blocks in all possible positions, viz.,*

$$\{a, b, *, *\}, \qquad \{b, a, *, *\}, \qquad \{a, *, b, *\}.$$

PROOF. The odd numbers which can be expressed as a sum of two squares are those in which a prime factor p, such that $p \equiv 3 \bmod 4$, appears with even multiplicity. Also, if a set of groupoids satisfies the same identities, then so does its products. Hence it will be sufficient to show that a groupoid satisfying (7), (8), and (9) can be constructed which has order p if p is a prime, $p \equiv 1 \bmod 4$, and has order p^2 if $p \equiv 3 \bmod 4$.

If $p \equiv 1 \bmod 4$, let the elements of the groupoid be the elements of $GF(p)$. Since -1 is a quadratic residue let $c \in GF(p)$ be such that $c^2 \equiv -1 \bmod p$. Define groupoid multiplication by $a \cdot b = ca + (1 - c)b$. One can verify immediately that (7), (8), and (9) are satisfied by this multiplication. If $p \equiv 3 \bmod 4$, then -1 is a quadratic nonresidue. Hence, $x^2 + 1$ is irreducible over $GF(p)$. Obtain the field $GF(p^2)$ by adjoining a root α of the equation $x^2 + 1 = 0$ to $GF(p)$. Again, letting the elements of the groupoid be those of $GF(p^2)$ and defining groupoid multiplication by

$$a \cdot b = \alpha a + (1 - \alpha)b,$$

it readily follows that (7), (8), and (9) are satisfied.

The following are the two simplest examples of designs so constructed.

$$\begin{aligned}
v = 5;\ &\{1, 2, 3, 4\}, \quad \{1, 3, 5, 2\}, \quad \{1, 4, 2, 5\},\\
&\{1, 5, 4, 3\}, \quad \{2, 4, 5, 3\}.\\
v = 9;\ &\{1, 2, 5, 8\}, \quad \{1, 3, 4, 6\}, \quad \{1, 4, 8, 9\},\\
&\{1, 5, 6, 7\}, \quad \{1, 6, 9, 2\}, \quad \{1, 7, 2, 4\},\\
&\{1, 8, 7, 3\}, \quad \{1, 9, 3, 5\}, \quad \{2, 3, 7, 5\},\\
&\{2, 6, 5, 4\}, \quad \{2, 7, 9, 8\}, \quad \{2, 8, 3, 6\},\\
&\{2, 9, 4, 3\}, \quad \{3, 8, 4, 5\}, \quad \{3, 9, 7, 6\},\\
&\{4, 7, 8, 6\}, \quad \{4, 9, 5, 7\}, \quad \{5, 9, 6, 8\}.
\end{aligned}$$

Note that to complete the proof of the existence of these designs for all $v \equiv 1 \bmod 4$, it is sufficient to prove the existence of the associated groupoid of order pq where p and q are primes congruent to 3 mod 4.

There are no known designs for $v \equiv 0 \bmod 4$. The following theorem gives a possible such construction with the use of a computer.

THEOREM 2. *If a* BIBD *of order V with parameters V, $b = V(V-1)/4$, $k = 4$, $\lambda = 3$,* and *$r = V - 1$ contains a subsystem of order v, then $V \geq 3v + 1$.*

PROOF: Let the subsystem be on the elements $\{1, 2, 3, \ldots, v\}$ and let the remaining elements be $\{v+1, v+2, \ldots, V\}$. Then there are $v(v-1)/4$ blocks of the type $\{i, j, k, l\}$ with $1 \leq i < j < k < l \leq v$. The remaining blocks each contain either four elements from the set $\{v+1, v+2, \ldots, V\}$ or three elements from this set and one from $\{1, 2, 3, \ldots, v\}$. Consider now blocks of the second type only. For each integer i, $1 \leq i \leq v$, there are the blocks $\{i, v+1, *, *\}$, $\{i, v+2, *, *\}$, $\ldots$, $\{i, V, *, *\}$. The totality of blocks of this type is $v(V-v)$. Hence the number of blocks is greater than or equal to $v(v-1)/4 + v(V-v)$. This implies that

$$\frac{V(V-1)}{4} \geq \frac{v(v-1)}{4} + v(V-v).$$

Hence $(V-(3v+1))(V-v) \geq 0$ or $V \geq 3v+1$.

In the case of equality, if $v \equiv 1 \bmod 4$ then $V \equiv 0 \bmod 4$. This suggests trying to construct a system of order $3v+1$ by extending a system of order v, possibly with the use of a computer.

THEOREM 3. *If p is a prime, $p \neq 5$, $p \equiv 1 \bmod 4$, the special* B.I.B.D. *of order p, constructed in Theorem* 1 *is such that no two blocks have three elements in common. If p is a prime $p \equiv 3 \bmod 4$, the corresponding design of order p^2 has the same property.*

PROOF. Both cases are treated in the same way. We have $a \cdot b = \lambda a + (1-\lambda)b$ where $\lambda^2 = -1$. The three blocks containing a and b are

(1) $\{a, b, \lambda a + (1-\lambda)b, (1+\lambda)a - \lambda b\}$;

(2) $\{b, a, \lambda b + (1-\lambda)a, (1+\lambda)b - \lambda a\}$; and

(3) $\left\{a, \frac{(1+\lambda)b}{2} + \frac{(1-\lambda)a}{2}, b, \frac{(1+\lambda)a}{2} + \frac{(1-\lambda)b}{2}\right\}$.

We must verify that no pair of these blocks have three elements in common. We check only the case for (1) and (2), the other cases following in the same way. If (1) and (2) have three elements in common then either

$$\lambda a + (1-\lambda)b = \lambda b + (1-\lambda)a$$

or

$$\lambda a + (1-\lambda)b = (1+\lambda)b - \lambda a$$

or

$$(1+\lambda)a - \lambda b = \lambda b + (1-\lambda)a$$

or

$$(1+\lambda)a - \lambda b = (1+\lambda)b - \lambda a.$$

The first of these equations implies that $(1-2\lambda)(a-b) = 0$ or $\lambda = \frac{1}{2}$. Hence $\frac{1}{4} \equiv 1 \bmod k$ or $p = 5$. The second or the third of these equations implies

$2\lambda(a - b) = 0$, which is impossible. The fourth equation implies that $(1 + 2\lambda)$ $(a - b) = 0$, or $\lambda = -\frac{1}{2}$, or $\frac{1}{4} \equiv -1 \bmod p$, or $p = 5$.

It is interesting to note that even when we have a BIBD with the appropriate parameters and such that no two blocks have three common elements, it is not necessarily possible to order the blocks cyclically so that each pair of elements occupies all possible positions. The following example is due to A. Gewirtz. Here $v = 16$, $b = 60$, $k = 4$, $r = 15$, and $\lambda = 3$. The blocks are:

(1, 2, 11, 12)	(2, 5, 8, 12)	(5, 6, 7, 8)
(1, 2, 13, 16)	(2, 5, 9, 13)	(5, 6, 13, 14)
(1, 2, 14, 15)	(2, 5, 10, 15)	(6, 6, 15, 16)
(1, 3, 7, 16)	(2, 6, 7, 12)	(7, 8, 13, 14)
(1, 3, 8, 15)	(2, 6, 9, 14)	(7, 8, 15, 16)
(1, 3, 10, 12)	(2, 6, 10, 16)	(7, 9, 11, 15)
(1, 4, 7, 14)	(3, 4, 9, 10)	(7, 9, 12, 14)
(1, 4, 8, 13)	(3, 4, 13, 15)	(7, 10, 11, 13)
(1, 4, 9, 12)	(3, 4, 14, 16)	(7, 10, 12, 16)
(1, 5, 8, 11)	(3, 5, 7, 9)	(8, 9, 11, 16)
(1, 5, 9, 16)	(3, 5, 11, 15)	(8, 9, 12, 13)
(1, 5, 10, 14)	(3, 5, 12, 14)	(8, 10, 11, 14)
(1, 6, 7, 11)	(3, 6, 8, 9)	(8, 10, 12, 15)
(1, 6, 9, 15)	(3, 6, 11, 16)	(9, 10, 13, 15)
(1, 6, 10, 13)	(3, 6, 12, 13)	(9, 10, 14, 16)
(2, 3, 7, 13)	(4, 5, 7, 10)	(11, 12, 13, 16)
(2, 3, 8, 14)	(4, 5, 11, 13)	(11, 12, 14, 15)
(2, 3, 10, 11)	(4, 5, 12, 16)	(13, 14, 15, 16)
(2, 4, 7, 15)	(4, 6, 8, 10)	
(2, 4, 8, 16)	(4, 6, 11, 14)	
(2, 4, 9, 11)	(4, 6, 2, 15)	

We will call a BIBD simple if it does not contain a design on a subset of its symbols. We now have the following theorem.

THEOREM 4. *The designs of order p and of order* p^2 *constructed in Theorem* 2 *are simple.*

PROOF: It is sufficient to show that any two elements generate the full groupoid. First, consider the case where $p \equiv 1 \bmod 4$. Let a and b be two distinct elements of $GF(p)$, and define $a \cdot b = \lambda a + (1 - \lambda)b$, $\lambda^2 = -1$. A direct computation shows that $b \cdot (b \cdot (a \cdot b)) = 2a - b$. Put $a = b + u$ where $u \neq 0$. Then $2a - b = b + 2u$. Hence from b and $b + u$ we can generate $b + 2u$ and iterating, we obtain b, $b + u$, $b + 2u$, $b + 2^2u, \ldots, b + 2^ru, \ldots$. Again, from

$b + 2^r u$ and $b + 2^{r+1}u$ we can generate $2(b + 2^{r+1}u) - (b + 2^r u) = b + 3(2^r u)$. From $b + 3(2^r u)$ and $b + 2^r u$ we can generate $2(b + 3(2^r u)) - (b + 2^r u) = b + 5(2^r u)$. Inductively, from $(b + (2k - 1)(2^r u))$ and $(b + (2k - 3)(2^r u))$ we can generate $2(b + (2k - 1)(2^r u)) - (b + (2k - 3)(2^r u)) = b + (2k + 1)(2^r u)$. Hence, taking $r = 0$ we have generated the elements b, $b + u$, $b + 3u$, $b + 5u$, $b + 7u$, ... Now $b + (2k + 1)u \equiv b + (2m + 1)u \bmod p$ if and only if $(2k + 1) \equiv (2m + 1) \bmod p$ (since $u \neq 0$). Hence $k = m$. Hence $b + u$, $b + 3u$, $b + 5u$, ..., give all elements of $GF(p)$. We also note that since $1, 2, 5, 7, 9, \ldots, 2p - 1$, and $1, 2, 3, 4, \ldots, p$ are both complete sets of residues mod p, then the elements generated by b and $b + u$ can be written as b, $b + u, b + 2u, b + 3u, \ldots, b + (p - 1)u$.

We now come to the case where $p \equiv 3$, mod 4. In this case the elements of our groupoid can be represented by $a + \lambda b$ where a and b range over $GF(p)$ and $\lambda^2 = -1$. Now from $a \cdot b = \lambda a + (1 - \lambda)b$, $\lambda^2 = -1$, $a = b + u$ we can generate as before b, $b + u$, $b + 2u$, ..., $b + (p - 1)u$. Again, since $(b + u) \cdot b = \lambda(b + u) + (1 - \lambda)b = b + \lambda u$, the same argument as in the first case with u replaced by λu allows us to generate b, $b + \lambda u$, $b + 2\lambda u$, ..., $b + (p - 1)\lambda u$. From $b + ru$ and $b + s\lambda u$ we can generate $2(b + ru) - (b + s\lambda u) = b + (2r - s\lambda)u$. As r ranges over all residues mod p, so does $2r$. Hence we can generate all the elements $b + (r + s\lambda)u$ where r and s range over all elements of $GF(p)$. Since $u \neq 0$, the set of all elements $b + (r + s\lambda)u$ are all the elements of $GF(p^2)$.

4. Extension for $k \geq 5$

In the general case, we consider the quasigroup with identities $a^2 = a$, $W_k(a, b) = b$.

The corresponding designs are still BIBD's for $k = 5$, but for $k > 5$ this is not necessarily the case. For $k > 5$, each ordered pair belongs to exactly one cyclic block but the λ_{ij}'s need not be distinct.

The solution for $k = 3$ and $k = 4$ generalizes. If we work modulo a prime p and define $a \cdot b = \lambda a + (1 - \lambda)b$, then the quasigroup identities are satisfied if and only if λ is a root of the polynomial $P_k(x) = 0$ where

$$P_k(x) = x^{k-1} - x^{k-2} + x^{k-3} - \cdots + (-1)^{k-1}.$$

The proof is a straightforward induction with some messy calculations.

If the equation has a solution in $GF(p)$ we have a design with $v = p$. Otherwise, we must go to $GF(p^r)$ where r is the degree of an irreducible divisor of $P_k(x)$. Note that these irreducible divisors are irreducible factors of cyclotomic polynomials. It is possible that the theory of cyclotomy might yield designs just as in the case of different sets.

5. Automorphism Groups

The large degree of symmetry of our designs indicates the possibility of a large automorphism group. The case where $v = p$, p prime, is an example where this is actually the case, as in the following theorem.

THEOREM 5. *Let p be a prime, k an integer, and suppose $P_k(x) = x^{k-1} - x^{k-2} + x^{k-3} - \cdots + (-1)^{k-1}$ has a solution in $GF(p)$. Then there exists the following design. The elements of $GF(p)$ can be arranged in cyclic blocks of length k such that every pair of elements lies in exactly one cyclic block. The design has an automorphism group which is doubly transitive on elements.*

PROOF: Take λ in $GF(p)$ so that $P_k(\lambda) = 0$. Define $a \cdot b = \lambda a + (1 - \lambda)b$. This yields a quasigroup satisfying $a^2 = a$, $W_k(a, b)a = b$. Given a and b in $GF(p)$ the unique cyclic block containing the ordered pair a, b is

$$\{a, b, W_3(a, b), W_4(a, b), \ldots, W_k(a, b)\}.$$

Consider the set of affine transformations of order $p(p - 1)$ given by

$$x \to x' = ux + v \qquad \text{where} \qquad u \not\equiv 0 \mod p$$

and v is any element of $GF(p)$. This group is doubly transitive on $GF(p)$. Now consider the effect of this mapping on the identity $x \cdot y = \lambda x + (1 - \lambda)y$. Suppose that $z = x \cdot y$. Now

$$\begin{aligned} x' \cdot y' &= \lambda x' + (1 - \lambda)y' \\ &= \lambda(ux + v) + (1 - \lambda)(uy + v) \\ &= u(\lambda x + (1 - \lambda)y) + v = uz + u = z'. \end{aligned}$$

Hence the mapping is an automorphism of the groupoid which implies that it is an automorphism of the design.

6. Orthogonal Latin Squares

In this section, a quasigroup will be used to construct a latin square which is orthogonal to its transpose. The construction yields an infinite class of such squares.

Let $\langle S, \cdot \rangle$ be a quasigroup with the single identity $a(ab) = ba$ for all a, b in S. Such a quasigroup is idempotent since $a(aa) = aa$ implies $aa = a$. Let M be the latin square which is the multiplication table of S, and let M^T be the transpose of M. When M^T is superposed on M, then the pair of elements in the cell with row index a and column index b is (ab, ba). Now suppose

$ab = cd$ and $ba = dc$. Then $a(cd) = a(ab) = ba = dc = c(cd)$. Hence $a = c$, and $b = d$. This means that M and M^T are orthogonal.

THEOREM 6. *Let $n = 4^k m$ where $k \geq 0$ and m is an odd integer such that every prime divisor p of m for which 5 is a quadratic nonresidue appears with even multiplicity in the decomposition of n into prime factors. Then there exists a matrix M of order n such that M and M^T are orthogonal.*

PROOF: Suppose $n = 4$. Take the elements of the quasigroup S to be those of $GF(2^2)$. If λ generates $GF(2^2)$ then $\lambda^2 = \lambda + 1$ and the multiplication defined by $a \cdot b = \lambda a + (1 + \lambda)b$ satisfies the identity $a(ab) = ba$ as well as the two cancellation laws. If $n = 5$ and we take the elements of S to be those of $GF(5)$ then $a \cdot b = 4a + 2b$ satisfies the identity $a(ab) = ba$ and the cancellation laws. If $n = p$, where 5 is a quadratic residue mod p, let the elements of S be $GF(p)$ and let $\lambda^2 = 5$. Now take

$$a \cdot b = \left(\frac{3 - \lambda}{2}\right)a + \left(\frac{\lambda - 1}{2}\right)b.$$

Again the identity $a(ab) = ba$ and the cancellation laws hold. Finally, if $n = p$ where 5 is a quadratic nonresidue mod p then $x^2 - 5$ is irreducible over $GF(p)$. Adjoin a root λ of $x^2 - 5 = 0$ to obtain the field $GF(p^2)$ and in this field define

$$a \cdot b = \left(\frac{3 - \lambda}{2}\right)a + \left(\frac{\lambda - 1}{2}\right)b.$$

Again, this multiplication yields an appropriate quasigroup. The theorem now follows by using direct products of quasigroups.

As an example, if $p = 3$ we obtain the following latin square of order 9 which is orthogonal to its transpose:

$$\begin{matrix} 1 & 7 & 8 & 6 & 9 & 3 & 4 & 5 & 2 \\ 4 & 2 & 9 & 7 & 3 & 8 & 5 & 1 & 6 \\ 5 & 6 & 3 & 2 & 8 & 9 & 1 & 4 & 7 \\ 3 & 5 & 6 & 4 & 7 & 2 & 8 & 9 & 1 \\ 2 & 9 & 4 & 8 & 5 & 1 & 6 & 7 & 3 \\ 8 & 1 & 7 & 5 & 2 & 6 & 9 & 3 & 4 \\ 6 & 3 & 5 & 9 & 1 & 4 & 7 & 2 & 8 \\ 9 & 4 & 2 & 1 & 6 & 7 & 3 & 8 & 5 \\ 7 & 8 & 1 & 3 & 4 & 5 & 2 & 6 & 9. \end{matrix}$$

References

1. R. H. Bruck, What is a Loop, in *M.A.A. Studies in Mathematics*, Vol. 2, Studies in Modern Algebra, (A. Albert, ed.), pp. 59–99. Prentice-Hall, Englewood Cliffs, New Jersey.
2. N. S. Mendelsohn, A Natural Generalization of Steiner Triple Systems (submitted for publication).
3. N. S. Mendelsohn, An Application of Matrix Theory to a Problem in Universal Algebra, in *Linear Algebra and its Applications* (to appear).

WELL-BALANCED ORIENTATIONS OF FINITE GRAPHS AND UNOBTRUSIVE ODD-VERTEX-PAIRINGS

C. St. J. A. Nash-Williams

DEPARTMENT OF COMBINATORICS AND OPTIMIZATION
UNIVERSITY OF WATERLOO, WATERLOO, ONTARIO

1. Introduction

This paper endeavors to explain, with such clarity and simplicity as can be achieved, the ideas underlying two closely related theorems in graph theory (Theorems 1 and 2 below) which were established by the author some years ago. Theorem 2 was discovered in the process of searching for a proof of Theorem 1 but seems also to be of some independent interest. Once Theorem 2 is proved, the deduction of Theorem 1 from it is a fairly easy exercise, as we shall see. On the other hand, the only proof of Theorem 2 which the author was able to discover involves a fairly intricate induction argument, which will only be sketched briefly here: full details are given in [4]. The two theorems indicate curious properties of graphs, and seem to be relatively deep in view of the comparative difficulty experienced by the author in proving them. They also do not seem particularly closely related to much other existing work in graph theory, and perhaps for this reason have not become particularly well known. On the other hand, these theorems seem to have a somewhat natural character which would suggest that there must ultimately be a place for them somewhere in the overall structure of graph theory.

2. Preliminary Definitions

Throughout this paper, the word "graph" means "undirected graph without loops," and, except in Section 8, it means "finite undirected graph without loops." Multiple edges in graphs are allowed. $V(G)$ denotes the set of vertices of a graph G and $E(G)$ denotes its set of edges. We shall also need to introduce and define notions of "directed graph" and "signed graph," abbreviated for convenience to "digraph" and "sigraph," respectively. As is well known, a digraph may be thought of intuitively as a graph in which each edge is assigned a direction (usually indicated diagrammatically by placing an arrow on the edge), so that one of its incident vertices becomes its "initial vertex" or "tail" and the other becomes its "terminal vertex" or "head." A "sigraph" is intuitively a graph with two kinds of edges, "positive" and "negative" edges, and one tends, for example, to take intuitively the view that, if two distinct vertices ξ and η of a sigraph are joined by 7 positive and 16 negative edges, then the total number of edges joining ξ to η is $7 - 16 = -9$.

Formally, we define a *digraph* to be an ordered pair (G, t), where G is a graph and t is a function from $E(G)$ into $V(G)$ such that $t(\lambda)$ is incident with λ for every $\lambda \in E(G)$. We shall tend to write $t(\lambda)$ as λt. This vertex is called the *tail* of the edge λ. The other vertex incident with λ is called the *head* of λ and denoted by λh. A *sigraph* is defined to be an ordered pair (G, Q), where G is a graph and Q is a subset of $E(G)$. If S denotes this sigraph, we call the elements of Q *positive* edges and the elements of $E(G)\backslash Q$ *negative* edges, and we write $Q = E_+(S)$, $E(G)\backslash Q = E_-(S)$. The graph G is the *underlying graph* of a digraph (G, t) or sigraph (G, Q). The letters G, D, S will always (without further introduction) denote, respectively, a graph, a digraph, and a sigraph. Since "graph" means "finite graph" in this discussion, it is of course understood that all digraphs and sigraphs considered here are finite in the sense of having finite underlying graphs.

All notation and terminology which are defined for graphs may be applied to digraphs and sigraphs with the understanding that they refer to the underlying graph; for instance, if $D = (G, t)$ then $V(D)$ denotes $V(G)$ and "vertices of D" are vertices of G, etc.

If $X \subseteq V(G)$, $\bar{X}$ denotes $V(G)\backslash X$. If $X, Y \subseteq V(G)$, $X \bigtriangledown Y$ denotes the set of all edges of G which join elements of X to elements of Y. If $X, Y \subseteq V(D)$, $X \rhd Y$ denotes $\{\lambda \in E(D) : \lambda t \in X, \lambda h \in Y\}$: thus $(X \rhd Y) \cup (Y \rhd X) = X \bigtriangledown Y$ and $(X \rhd Y) \cap (Y \rhd X) = (X \cap Y) \bigtriangledown (X \cap Y)$. If $L \subseteq E(S)$, L_+ denotes $L \cap E_+(S)$, L_- denotes $L \cap E_-(S)$ and $|L|_s$ denotes $|L_+| - |L_-|$. We might call $|L|_s$ the *signed cardinality* or *sicardinality* of L; i.e., it is, intuitively, the number of edges in L when a positive edge counts as one edge and a negative edge counts as minus one edge.

A *path* in G is a sequence

$$\xi_0, \lambda_1, \xi_1, \lambda_2, \ldots, \lambda_n, \xi_n \tag{1}$$

such that

$$n \geq 0; \qquad \xi_0, \xi_1, \ldots, \xi_n \in V(G); \qquad \lambda_1, \lambda_2, \ldots, \lambda_n \in E(G);$$

λ_i joins ξ_{i-1} to ξ_i $(i = 1, 2, \ldots, n)$; and $\lambda_1, \lambda_2, \ldots, \lambda_n$ are distinct. (When $n = 0$, the sequence (1) has just one term ξ_0.) A path $\xi_0, \lambda_1, \xi_1, \lambda_2, \ldots, \lambda_n, \xi_n$

in D (i.e., in the underlying graph of D) is a *directed path* or *dipath* if $\lambda_i t = \xi_{i-1}$ (and therefore $\lambda_i h = \xi_i$) for $i = 1, 2, \ldots, n$ (this condition being considered as being vacuously satisfied if $n = 0$). A *$\xi\eta$-path* [*$\xi\eta$-dipath*] is a path [dipath] with first term ξ and last term η; if $\xi = \eta$ the path is *closed.* An *Euler path* of G is a path in G which includes all the vertices and edges of G. Several paths in a graph are *edge-disjoint* if no edge appears in more than one of them.

The *edge-connectivity* $c(\xi, \eta)$ of two distinct vertices ξ and η of G is the minimum of the values of $|X \nabla \bar{X}|$ over all subsets X of $V(G)$ which include ξ but not η. Clearly $c(\xi, \eta) = c(\eta, \xi)$. Intuitively, $c(\xi, \eta)$ is the smallest number of edges whose removal from G would separate ξ from η. By a standard Menger-type theorem, $c(\xi, \eta)$ is also the maximum number of edge-disjoint $\xi\eta$-paths which can be found in G. The *edge-diconnectivity* $dc(\theta, \phi)$ of two distinct vertices θ, ϕ of D is the minimum of the values of $|X \rhd \bar{X}|$ over all subsets X of $V(D)$ which include θ but not ϕ. Intuitively, $dc(\theta, \phi)$ is the smallest number of edges whose removal from D would make it impossible to travel from θ to ϕ if the edges of D are regarded as one-way streets so that one is permitted only to travel along an edge in the direction from its tail to its head. A standard Menger-type theorem asserts that $dc(\theta, \phi)$ is the maximum number of edge-disjoint $\theta\phi$-dipaths which can be found in D. Note that $dc(\theta, \phi)$ is not necessarily equal to $dc(\phi, \theta)$. The *edge-siconnectivity* $sc(\alpha, \beta)$ of two distinct vertices α, β of S is the minimum of the values of $|X \nabla \bar{X}|_s$ over all subsets X of $V(S)$ which include α but not β. Clearly $sc(\alpha, \beta) = sc(\beta, \alpha)$.

G is *connected* if, for every ordered pair $(\xi\ \eta) \in V(G) \times V(G)$, there exists a $\xi\eta$-path in G, and G is *disconnected* if it is not connected. The *edge-connectivity* $c(G)$ of G is $\min\{|X \nabla \bar{X}| : \varnothing \subset X \subset V(G)\}$ (where $A \subset B$ means $A \subseteq B \neq A$); intuitively, $c(G)$ is the minimum number of edges whose removal from G disconnects G. D is *strongly connected* or *diconnected* if, for every $(\xi, \eta) \in V(D) \times V(D)$, there exists a $\xi\eta$-dipath in D. Of course, if (ξ, η) belongs to $V(D) \times V(D)$ then so does (η, ξ), and thus a diconnected digraph is one in which any two vertices ξ, η can be connected both by a $\xi\eta$-dipath and by an $\eta\xi$-dipath. The *edge-diconnectivity* $dc(D)$ of D is $\min\{|X \rhd \bar{X}| : \varnothing \subset X \subset V(D)\}$. Intuitively $dc(D)$ is the minimum number of edges whose removal from D would render it not diconnected. It is easily seen that G is connected if and only if $c(G) \geq 1$, and that D is diconnected if and only if $dc(D) \geq 1$.

An *orientation* of G is a digraph whose underlying graph is G.

3. Well-Balanced Orientations

In a paper entitled "A Theorem on Graphs with an Application to a Problem of Traffic Control," Robbins [8] considered the question "Which graphs have diconnected orientations?" The "application to traffic control" is more or less self-evident: if a network of streets is represented by a graph G, then asking whether G has a diconnected orientation is equivalent to asking whether we can convert all these streets into one-way streets in such a way that it shall be possible to travel from any point to any other. Certain graphs clearly have no diconnected orientations. For a start, disconnected graphs, i.e., graphs with edge-connectivity 0, obviously have no diconnected orientations. Moreover, if a graph G has edge-connectivity 1, then some subset X of $V(G)$ has the property that $X \nabla \overline{X}$ includes only a single edge λ (say), as illustrated in Fig. 1. Such a G has no diconnected orientation because, if

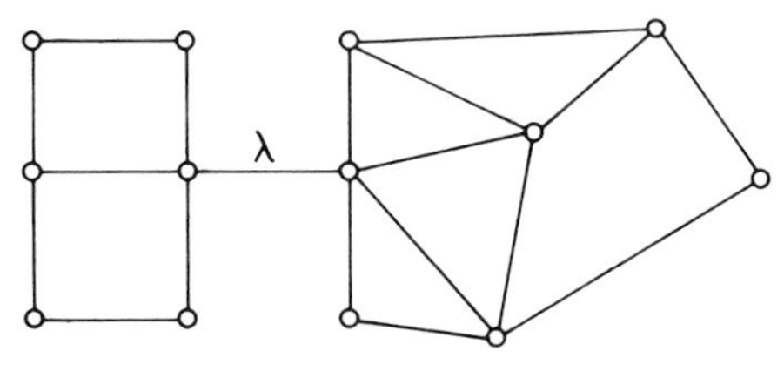

FIG. 1.

$D = (G, t)$ is any orientation of G and ξ is any element of X and η is any element of $\overline{X}$, then there is no $\xi\eta$-dipath in D if $\lambda t \in \overline{X}$ and there is no $\eta\xi$-dipath in D if $\lambda t \in X$, so that in either case D fails to be diconnected. Robbins' theorem asserts that *any graph with edge-connectivity* ≥ 2 *has a diconnected orientation*, thus assuring us that, apart from the obvious exceptions already noted, all graphs possess diconnected orientations. It is, in fact, not an unduly difficult exercise to prove Robbins' theorem by describing a definite procedure for orienting a graph of edge-connectivity greater than or equal to 2 so that the resulting digraph is diconnected.

Since D is diconnected if and only if $dc(D) \geq 1$, Robbins' theorem can also be expressed by saying that *if* $c(G) \geq 2$ *then* G *has an orientation* D *such that* $dc(D) \geq 1$. This immediately suggests the following generalization.

PROPOSITION 1. *If* k *is a positive integer and* $c(G) \geq 2k$ *then* G *has an orientation* D *such that* $dc(D) \geq k$.

More intuitively, Proposition 1 states that if, for every nonempty proper subset X of $V(G)$, X is joined to $\overline{X}$ by at least $2k$ edges, then G can be so oriented that, for every nonempty proper subset X of $V(G)$, X is joined to $\overline{X}$ by at least k edges in each direction. Thus we are concerned with orienting G to give a digraph in which there is, in a rough intuitive sense, a certain degree of "balance" between edges oriented in one direction and edges oriented in the opposite direction.

The author found it possible to develop a proof of Proposition 1, although this appeared to be considerably more difficult than proving the special case represented by Robbins' Theorem. It was then found that (after overcoming some further difficulties) a still more general result could be established. To see how Proposition 1 can be generalized further, we must once again express it in a different way, which is done by means of two trivial lemmas. For any set A, it is convenient to let $A * A$ denote $\{(x, y) \in A \times A : x \neq y\}$, i.e.,the set of all ordered pairs of *distinct* elements of A. Then we have

LEMMA 1. *$c(G) \geq r$ if and only if $c(\xi, \eta) \geq r$ for every $(\xi, \eta) \in V(G) * V(G)$.*

LEMMA 2. *$dc(D) \geq r$ if and only if $dc(\xi, \eta) \geq r$ for every $(\xi, \eta) \in V(D) * V(D)$.*

To prove Lemma 1, suppose that $c(G) \geq r$. Then $|X \nabla \overline{X}| \geq r$ for every X such that $\varnothing \subset X \subset V(G)$. Hence, if $(\xi, \eta) \in V(G) * V(G)$, then $|X \nabla \overline{X}| \geq r$ for every $X \subseteq V(G)$ such that $\xi \in X$, $\eta \in \overline{X}$, and so $c(\xi, \eta) \geq r$. Conversely, suppose that $c(\xi, \eta) \geq r$ for every $(\xi, \eta) \in V(G) * V(G)$. Then, if $\varnothing \subset X \subset V(G)$, we can select a $\xi \in X$ and an $\eta \in \overline{X}$ and we have $|X \nabla \overline{X}| \geq c(\xi, \eta)$ by the definition of $c(\xi, \eta)$, and $c(\xi, \eta) \geq r$ by hypothesis and consequently $|X \nabla \overline{X}| \geq r$. This shows that $c(G) \geq r$. Lemma 2 is proved similarly. These lemmas enable us to reformulate Proposition 1 in the following equivalent form.

PROPOSITION 1′. *If k is a positive integer and $c(\xi, \eta) \geq 2k$ for every $(\xi, \eta) \in V(G) * V(G)$, then G has an orientation in which $dc(\xi, \eta) \geq k$ for every $(\xi, \eta) \in V(G) * V(G)$.*

As usual, let $[x]$ denote the greatest integer less than or equal to the real number x. It is now immediately apparent that, if we prove that, for any graph G, there exists an orientation of G in which the inequality $dc(\xi, \eta) \geq [\frac{1}{2}c(\xi, \eta)]$ holds for all elements (ξ, η) of $V(G) * V(G)$, then this result will imply, and be stronger than, Proposition 1′, and hence it will also strengthen Proposition 1. We therefore call a digraph D *well-balanced* if $dc(\xi, \eta) \geq [\frac{1}{2}c(\xi, \eta)]$ for every $(\xi, \eta) \in V(D) * V(D)$, and our first theorem is

THEOREM 1. *Every graph has a well-balanced orientation.*

4. SOME LEMMAS

If $\xi \in V(G)$, the *valency* $v(\xi)$ of ξ is the number of edges incident with ξ. A vertex of odd [even] valency will be called an *odd vertex* [*even vertex*]. G is *Eulerian* if all its vertices are even. If $\alpha \in V(D)$, the *outvalency* $v_{\text{out}}(\alpha)$ of α is the number of edges of D with tail α and the *invalency* $v_{\text{in}}(\alpha)$ of α is the number of edges of D with head α. For some purposes, it seems natural to think of $v_{\text{out}}(\alpha) - v_{\text{in}}(\alpha)$ as a sort of algebraic measure of the total "flux out of α" in D and we may denote $v_{\text{out}}(\alpha) - v_{\text{in}}(\alpha)$ by $f(\alpha)$. If $v_{\text{out}}(\alpha) = v_{\text{in}}(\alpha)$ for every $\alpha \in V(D)$, D is *di-Eulerian* or *solenoidal*. The term "solenoidal" seems appropriate since in a solenoidal digraph the flux $f(\alpha)$ out of each vertex α is zero, which seems somewhat analogous to the fact that a solenoidal vector field $\mathbf{u}$ satisfies the condition div $\mathbf{u} = 0$, expressing the fact that the algebraic flux of $\mathbf{u}$ out of any infinitesimal element of space is zero. The term "di-Eulerian" seems appropriate since di-Eulerian digraphs often appear to play, in the theory of digraphs, a role analogous to that of Eulerian graphs in the theory of graphs, as is evidenced for instance by the well-known elementary theorems "every nonempty connected eulerian graph has a closed Euler path" and "every nonempty connected di-Eulerian digraph has a closed Euler dipath" (see [7, Chapter 3], and [2, Chapter II]). In [4] I was persuaded by a referee, on grounds of terminological standardization, to describe di-Eulerian digraphs as "quasi-symmetrical," but I do not favor this terminology so strongly.

LEMMA 3. *Every Eulerian graph has a di-Eulerian orientation.*

PROOF: Since every nonempty connected Eulerian graph has a closed Euler path ([7, Chapter 3], [2, Chapter II]), it follows that any nonempty connected Eulerian graph can be given a di-Eulerian orientation by orienting continuously along a closed Euler path of the graph. Hence, if G is any Eulerian graph, we can give a di-Eulerian orientation to each connected component of G, which clearly determines a di-Eulerian orientation of G itself.

LEMMA 4. *If G is Eulerian and $X \subseteq V(G)$ then $|X \nabla \bar{X}|$ is even.*

PROOF: The contribution of an edge of G to $\sum_{\xi \in X} v(\xi)$ is 2, 1, or 0 according as the edge belongs to $X \nabla X$, $X \nabla \bar{X}$, or $\bar{X} \nabla \bar{X}$, respectively. Therefore

$$\sum_{\xi \in X} v(\xi) = 2|X \nabla X| + |X \nabla \bar{X}|.$$

But $\sum_{\xi \in X} v(\xi)$ is even since G is Eulerian. Therefore $|X \nabla \bar{X}|$ must also be even.

COROLLARY 4a. *If S is Eulerian and $X \subseteq V(S)$, then $|X \nabla \bar{X}|_s$ is even.*

We first recall our convention that, when terminology defined only for graphs is applied to digraphs or sigraphs, it is to be interpreted as referring to the underlying graph; thus the hypothesis that S is "Eulerian" in Corollary 4a means simply that the underlying graph of S is Eulerian. To prove Corollary 4a, observe that the relations

$$|X \bigtriangledown \overline{X}| = |(X \bigtriangledown \overline{X})_{+}| + |(X \bigtriangledown \overline{X})_{-}|,$$
$$|X \bigtriangledown \overline{X}|_{s} = |(X \bigtriangledown \overline{X})_{+}| - |(X \bigtriangledown \overline{X})_{-}|$$

show that $|X \bigtriangledown \overline{X}|$ and $|X \bigtriangledown \overline{X}|_s$ have the same parity, and since the first of these quantities is even by Lemma 4, the second must also be even.

COROLLARY 4b. *If S is Eulerian and ξ, η are distinct vertices of S then $sc(\xi, \eta)$ is even.*

PROOF: $sc(\xi, \eta)$ is defined to be the minimum, over certain subsets X of $V(S)$, of $|X \bigtriangledown \overline{X}|_s$, and hence is by Corollary 4a the minimum of a collection of even numbers.

LEMMA 5. *If D is di-Eulerian and $X \subseteq V(D)$ then*

$$|X \rhd \overline{X}| = |\overline{X} \rhd X| = \tfrac{1}{2}|X \bigtriangledown \overline{X}|. \tag{2}$$

PROOF: The contribution of an edge λ of D to $\sum_{\xi \in X}(v_{\text{out}}(\xi) - v_{\text{in}}(\xi))$ is clearly 0 if $\lambda \in X \bigtriangledown X$, 1 if $\lambda \in X \rhd \overline{X}$, -1 if $\lambda \in \overline{X} \rhd X$, and 0 if $\lambda \in \overline{X} \bigtriangledown \overline{X}$. Hence

$$|X \rhd \overline{X}| - |\overline{X} \rhd X| = \sum_{\xi \in X} (v_{\text{out}}(\xi) - v_{\text{in}}(\xi)),$$

which is 0 since D is di-Eulerian. Therefore $|X \rhd \overline{X}| = |\overline{X} \rhd X|$; and since $|X \rhd \overline{X}| + |\overline{X} \rhd X| = |X \bigtriangledown \overline{X}|$ the result (2) follows.

REMARK. If $X \subseteq V(D)$, one could consider $|X \rhd \overline{X}| - |\overline{X} \rhd X|$ as algebraically measuring the "flux out of X in D," and Lemma 5 assures us that in a solenoidal digraph this flux is always zero. This could be thought of as analogous to the proposition that if $\mathbf{u}$ is a solenoidal vector field then the surface integral $\int_C \mathbf{u} \cdot d\mathbf{S}$ taken over any closed surface C vanishes, which expresses the fact that the algebraic total flux of a solenoidal vector field out of any bounded region vanishes. Lemma 4 can be considered as the analog of Lemma 5 for graphs, in the sense that the analog of a proposition concerning digraphs is often a corresponding "modulo 2" proposition concerning graphs.

COROLLARY 5a. *If D is di-Eulerian and ξ, η are distinct vertices of D, then $dc(\xi, \eta) = \frac{1}{2}c(\xi, \eta)$.*

PROOF: By definition

$$dc(\xi, \eta) = \min\{|X \rhd \overline{X}| : X \subseteq V(D), \xi \in X, \eta \in \overline{X}\},$$

$$c(\xi, \eta) = \min\{|X \bigtriangledown \overline{X}| : X \subseteq V(D), \xi \in X, \eta \in \overline{X}\},$$

and since $|X \rhd \overline{X}| = \frac{1}{2}|X \bigtriangledown \overline{X}|$ for every $X \subseteq V(D)$, by Lemma 5 it follows that $dc(\xi, \eta) = \frac{1}{2}c(\xi, \eta)$.

COROLLARY 5b. *Every di-Eulerian digraph is well-balanced.*

Corollary 5b follows at once from Corollary 5a. Combining Lemma 3 and Corollary 5b gives

LEMMA 6. *Every Eulerian graph has a well-balanced orientation.*

5. UNOBTRUSIVE ODD-VERTEX-PAIRINGS

A *partition* of a finite set A is a set $\{A_1, A_2, \ldots, A_n\}$ where $A_1, \ldots, A_n$ are nonempty subsets of A and $A_i \cap A_j = \varnothing$ whenever $i \neq j$ and $A_1 \cup A_2 \cup \cdots \cup A_n = A$. The set of odd vertices of G will be denoted by $V_o(G)$. It is well known that $|V_o(G)|$ is even for every graph G, and we define an *odd-vertex-pairing* of G to be a partition $\{A_1, A_2, \ldots, A_n\}$ of $V_o(G)$ such that $|A_1| = \cdots = |A_n| = 2$.

A *subgraph* of G is a graph H such that $V(H) \subseteq V(G)$, $E(H) \subseteq E(G)$, and each edge of H joins the same vertices in H as in G. If L is a subset of $E(G)$, $G - L$ denotes the subgraph H of G such that $V(H) = V(G)$, and $E(H) = E(G)\backslash L$. Analogously, a *subdigraph* of D is a digraph T such that $V(T) \subseteq V(D)$, $E(T) \subseteq E(D)$, and each edge of T has the same tail and head in T as in D; i.e., if $D = (G, t)$, then $T = (H, t')$ where H is a subgraph of G and t' is the restriction to $E(H)$ of the function t. If $M \subseteq E(D)$, $D - M$ denotes the subdigraph T of D such that $V(T) = V(D)$, and $E(T) = E(D)\backslash M$.

Let $|V_o(G)| = 2r$ and let $P = \{\{\xi_1, \eta_1\}, \{\xi_2, \eta_2\}, \ldots, \{\xi_r, \eta_r\}\}$ be an odd-vertex-pairing of G. Let us temporarily suppose that it is possible to find in G edges $\lambda_1, \lambda_2, \ldots, \lambda_r$ such that λ_i joins ξ_i to η_i for $i = 1, \ldots, r$. Then we might denote the graph $G - \{\lambda_1, \ldots, \lambda_r\}$ by $G - P$; i.e., it is obtained by removing from G edges one of which joins each pair of odd vertices which occurs in the odd-vertex-pairing P. Since P is an odd-vertex-pairing of G, we see at once that $G - P$ is Eulerian, and so has, by Lemma 6, a well-balanced orientation. One might reasonably ask whether the existence of a well-balanced orientation of G can in some way be deduced from the existence of a well-balanced orientation of $G - P$, and one might anticipate that this is most likely to be possible if, in some appropriate sense, the structure of $G - P$ is not too much unlike that of G, i.e., if the removal of the edges $\lambda_1, \lambda_2, \ldots, \lambda_r$ (joining the pairs of odd vertices which appear in P) does not alter too drastically the structure of G. If the removal of these edges does not alter too drastically the structure of G, one could say that the odd-vertex-pairing P has a certain quality of unobtrusiveness. There could, of course, be many criteria of whether the removal of $\lambda_1, \ldots, \lambda_r$ alters the structure of G too drastically; but the appropriate one for our purposes seems to be that the removal of these edges does not alter too drastically the structure of G if it does not reduce too greatly the edge-connectivity of any two vertices, and we shall therefore wish to consider P as being "unobtrusive" if there is no pair of distinct vertices whose edge-connectivity in $G - P$ falls very far below their edge-connectivity in G.

Of course, the notion of "unobtrusiveness" suggested in the preceding paragraph suffers from the disadvantage of requiring that, whenever a pair $\{\xi_i, \eta_i\}$ belongs to P, G must contain at least one edge joining ξ_i to η_i, which is not in general true for an arbitrary choice of G and P. To overcome this difficulty, we now replace the notion of *removing* an edge λ_i joining ξ_i to η_i by that of *adding* a *negative* edge λ_i joining ξ_i to η_i. Thus we redefine $G - P$ to be a *sigraph* obtained from G by adding edges $\lambda_1, \lambda_2, \ldots, \lambda_r$ such that λ_i joins ξ_i to η_i $(i = 1, 2, \ldots, r)$ and considering the added edges $\lambda_1, \lambda_2, \ldots, \lambda_r$ to be negative and the edges originally in G to be positive. The notion of "unobtrusiveness" proposed in the last paragraph has then to be amended by saying that we will call P "unobtrusive" if there is no pair of distinct vertices whose edge-*siconnectivity* in $G - P$ falls very far below their edge-connectivity in G. A precise definition along these lines will be given after we have proved an explanatory lemma.

Where the context makes it desirable to indicate explicitly the graph in which a symbol such as c is considered to be defined, this may be done by attaching the symbol denoting the graph to the other symbol as a suffix, e.g., $c_G(\xi, \eta)$ denotes the edge-connectivity of ξ and η in G. Of course the same convention can be used for digraphs or sigraphs. If $|V_o(G)| = 2r$ and $P = \{\{\xi_1, \eta_1\}, \ldots, \{\xi_r, \eta_r\}\}$ is an odd-vertex-pairing of G, then $G + P$ will denote a graph obtained from G by adjoining r new edges $\lambda_1, \ldots, \lambda_r$ such that λ_i joins ξ_i to η_i for $i = 1, \ldots, r$, and $G - P$ will denote the sigraph $(G + P, E(G))$, i.e., the sigraph with underlying graph $G+P$ in which the edges of G are considered as being positive and $\lambda_1, \ldots, \lambda_r$ are considered as being negative. Note that $V(G - P) = V(G + P) = V(G)$, since no vertices are added or removed in passing from G to $G + P$. If n is an integer, $n^\flat$ will denote the greatest even integer less than or equal to n, i.e., $n^\flat = n$ if n is even and $n^\flat = n - 1$ if n is odd.

LEMMA 7. *If P is an odd-vertex-pairing of G and ξ, η are distinct vertices of G, then $sc_{G-P}(\xi, \eta) \leq c_G(\xi, \eta)^\flat$.*

PROOF: Since $G - P$ has underlying graph $G + P$ and $E_+(G - P) = E(G)$, it follows that, for any $X \subseteq V(G) = V(G + P) = V(G - P)$,

$$(X \nabla_{G-P} \bar{X})_+ = (X \nabla_{G+P} \bar{X}) \cap E(G) = X \nabla_G \bar{X},$$

and consequently

$$\begin{aligned} |X \nabla_{G-P} \bar{X}|_s &= |(X \nabla_{G-P} \bar{X})_+| - |(X \nabla_{G-P} \bar{X})_-| \\ &\leq |(X \nabla_{G-P} \bar{X})_+| = |X \nabla_G \bar{X}|. \end{aligned}$$

But $sc_{G-P}(\xi, \eta)$ is the minimum, over subsets X of $V(G - P) = V(G)$ such that $\xi \in X$ and $\eta \in \bar{X}$, of $|X \nabla_{G-P} \bar{X}|_s$, and $c_G(\xi, \eta)$ is the minimum, over the same sets X, of $|X \nabla_G \bar{X}|$. Therefore $sc_{G-P}(\xi, \eta) \leq c_G(\xi, \eta)$. Moreover, since P is an odd-vertex-pairing of G, $G + P$ is Eulerian and so $G - P$ is an Eulerian sigraph, i.e., a sigraph with an Eulerian underlying graph, and consequently $sc_{G-P}(\xi, \eta)$ is even by Corollary 4b. Since $sc_{G-P}(\xi, \eta)$ is even and less than or equal to $c_G(\xi, \eta)$, it is less than or equal to $c_G(\xi, \eta)^\flat$.

We have indicated that we shall wish to consider an odd-vertex-pairing P of G to be "unobtrusive" if there are no two distinct vertices ξ, η for which $sc_{G-P}(\xi, \eta)$ is too far below $c_G(\xi, \eta)$. Since Lemma 7 shows that $sc_{G-P}(\xi, \eta)$ must always be at least as small as $c_G(\xi, \eta)^\flat$, the best we can hope for is that it shall never be any smaller. We therefore define an odd-vertex-pairing P of G to be *unobtrusive* if $sc_{G-P}(\xi, \eta) = c_G(\xi, \eta)^\flat$ for every pair ξ, η of distinct vertices of G; and our second main theorem is

THEOREM 2. *Every graph has an unobtrusive odd-vertex-pairing.*

6. Deduction of Theorem 1 from Theorem 2

In this section, we shall assume Theorem 2 and show how Theorem 1 can be deduced from this assumption. The difficult part of the theory is the proof of Theorem 2, which will be outlined briefly in Section 7.

We suppose, then, that we are given a graph G and wish to show that it has a well-balanced orientation. By Theorem 2, we can select an unobtrusive odd-vertex-pairing P of G. This enables us to form the graph $G + P$. Let E_P denote $E(G + P) \backslash E(G)$, i.e., the set of edges added to G to form $G + P$.

Since P is an odd-vertex-pairing of G, $G + P$ is clearly Eulerian and so by Lemma 3 we can select a di-Eulerian orientation Δ of $G + P$. Then $\Delta - E_P$, which we shall denote by D, is an orientation of G, and if we show that D is well-balanced our proof of Theorem 1 will be complete.

We note first that the sets $V(G)$, $V(G + P)$, $V(G - P)$, $V(\Delta)$, and $V(D)$ are identical, and so for simplicity we will denote each of these by V.

Suppose that $X \subseteq V$. Then, since $G - P$ and Δ both have underlying graph $G + P$, there is no difference in meaning between the symbols

$$X \nabla_{G+P} \bar{X}, \qquad X \nabla_{G-P} \bar{X}, \qquad X \nabla_{\Delta} \bar{X}.$$

Thus, using Lemma 5, we have

$$\begin{aligned} |X \rhd_{\Delta} \bar{X}| &= \tfrac{1}{2}|X \nabla_{\Delta} \bar{X}| = \tfrac{1}{2}|X \nabla_{G-P} \bar{X}| \\ &= \tfrac{1}{2}|(X \nabla_{G-P} \bar{X})_{+}| + \tfrac{1}{2}|(X \nabla_{G-P} \bar{X})_{-}|, \end{aligned} \tag{3}$$

and, since $E_P = E_{-}(G - P)$ by the definition of $G - P$, it follows that

$$\begin{aligned} |(X \nabla_{G-P} \bar{X})_{-}| &= |(X \nabla_{G-P} \bar{X}) \cap E_P| \\ &= |(X \nabla_{\Delta} \bar{X}) \cap E_P| \geq |(X \rhd_{\Delta} \bar{X}) \cap E_P|. \end{aligned} \tag{4}$$

Hence, since $D = \Delta - E_P$, we have

$$\begin{aligned} |X \rhd_D \bar{X}| &= |(X \rhd_{\Delta} \bar{X}) \backslash E_P| \\ &= |X \rhd_{\Delta} \bar{X}| - |(X \rhd_{\Delta} \bar{X}) \cap E_P| \\ &\geq \tfrac{1}{2}|(X \nabla_{G-P} \bar{X})_{+}| - \tfrac{1}{2}|(X \nabla_{G-P} \bar{X})_{-}| \end{aligned}$$

by (3) and (4). In other words, for every $X \subseteq V$, we have

$$|X \rhd_D \bar{X}| \geq \tfrac{1}{2}|X \nabla_{G-P} \bar{X}|_s. \tag{5}$$

If ξ and η are distinct elements of V, the definition of $sc_{G-P}(\xi, \eta)$ implies that $|X \nabla_{G-P} \bar{X}|_s \geq sc_{G-P}(\xi, \eta)$ for every $X \subseteq V$ such that $\xi \in X$, and $\eta \in \bar{X}$; therefore by (5), $|X \rhd_D \bar{X}| \geq \frac{1}{2}sc_{G-P}(\xi, \eta)$ for every such X, and so $dc_D(\xi, \eta) \geq \frac{1}{2}sc_{G-P}(\xi, \eta)$. But since P is unobtrusive, $sc_{G-P}(\xi, \eta) = c_G(\xi, \eta)^\flat$, which is of course the same thing as $c_D(\xi, \eta)^\flat$ since G is the underlying graph of D. Thus we have altogether $dc_D(\xi, \eta) \geq \frac{1}{2}sc_{G-P}(\xi, \eta) = \frac{1}{2}(c_G(\xi, \eta)^\flat) = \frac{1}{2}(c_D(\xi, \eta)^\flat) = [\frac{1}{2}c_D(\xi, \eta)]$ for every pair ξ, η of distinct elements of V, which shows that D is well-balanced, as required.

7. Sketch of the Proof of Theorem 2

If θ is an odd vertex of G and P is an odd-vertex-pairing of G, there will be a unique vertex ϕ such that $\{\theta, \phi\} \in P$; and it will be convenient to call ϕ the *P-partner* of θ.

We shall now outline briefly the main stages into which the proof of Theorem 2 can be divided, thus giving the reader an idea of the techniques involved. The full details of this proof are to be found in [4]. Let us say that a graph G' is *smaller* than a graph G if $|V(G')| + |E(G')| < |V(G)| + |E(G)|$. The proof of Theorem 2 is by induction and depends on showing that, if all graphs smaller than a graph G have unobtrusive odd-vertex-pairings, then G has one also. Therefore, in the remainder of this section, we assume the inductive hypothesis that all graphs smaller than G have unobtrusive odd-vertex-pairings.

To prove that this inductive hypothesis implies that G has an unobtrusive odd-vertex-pairing, we require two procedures for constructing odd-vertex-pairings of G from odd-vertex-pairings of smaller graphs. First, if $\lambda \in E(G)$, then $G - \lambda$ denotes the subgraph H of G such that $V(H) = V(G)$, and $E(H) = E(G)\backslash\{\lambda\}$. Suppose that λ joins the vertices ξ, η and that T is an odd-vertex-pairing of $G - \lambda$. Since $v_G(\xi) = v_{G-\lambda}(\xi) + 1$, it follows that ξ is odd or even in G according as it is even or odd, respectively in $G - \lambda$ and the same applies to η. Obviously all other vertices have the same valency and parity in G as in $G - \lambda$. From these remarks, it is easily seen that an odd-vertex-pairing $\tilde{T}$ of G is determined by the following rules.

(i) If ξ and η are odd in $G - \lambda$ and $\{\xi, \eta\} \in T$, let $\tilde{T} = T\backslash\{\{\xi, \eta\}\}$.

(ii) If ξ and η are odd in $G - \lambda$ and $\{\xi, \eta\} \notin T$ and θ, ϕ are the T-partners of ξ, η, respectively, let $\tilde{T}$ be obtained from T by replacing the pairs $\{\xi, \theta\}$, $\{\eta, \phi\}$ by the pair $\{\theta, \phi\}$.

(iii) If ξ is odd in $G - \lambda$ and η is even in $G - \lambda$ and θ is the T-partner of ξ, let $\tilde{T}$ be obtained from T by replacing the pair $\{\xi, \theta\}$ by the pair $\{\eta, \theta\}$.

(iv) If ξ is even in $G - \lambda$ and η is odd in $G - \lambda$ and ϕ is the T-partner of η, let $\tilde{T}$ be obtained from T by replacing the pair $\{\eta, \phi\}$ by the pair $\{\xi\ \phi\}$.

(v) If ξ and η are even in $G - \lambda$, let $\tilde{T} = T \cup \{\{\xi, \eta\}\}$.

In each of these cases, our rule consists essentially in making the smallest possible number of changes in T to turn it into an odd-vertex-pairing of G.

In the second place, if X is any nonempty subset of $V(G)$ and if X^* denotes the subgraph H of G such that $V(H) = X$ and $E(H) = X \nabla X$, then G_X will denote what may be thought of as the graph obtained from G by contracting its subgraph X^* to a single vertex $\hat{X}$. Thus, letting $\overline{X}$ denote $V(G)\backslash X$, we have $V(G_X) = \overline{X} \cup \{\hat{X}\}$, $E(G_X) = \overline{X} \nabla_G V(G)$, where $\hat{X}$ denotes some element not belonging to the set $V(G) \cup E(G)$. An element of $\overline{X} \nabla_G \overline{X}$ joins the same vertices in G_X as in G; and if λ is an edge joining an element α of X to an element β of $\overline{X}$ in G, then in G_X, λ joins the vertices $\hat{X}$ and β. This completes the formal definition of G_X, which will clearly be smaller than G if and only if $|X| \geq 2$.

Now suppose that $\varnothing \subset X \subset V(G)$ and that we are given an odd-vertex-pairing Q of G_X and an odd-vertex-pairing R of $G_{\overline{X}}$. Since the valency of $\hat{X}$ in G_X and the valency of $\hat{\overline{X}}$ in $G_{\overline{X}}$ are both equal to $|X \nabla_G \overline{X}|$, it follows that $\hat{X}$ will be an odd vertex of G_X and $\hat{\overline{X}}$ will be an odd vertex of $G_{\overline{X}}$ if $|X \nabla_G \overline{X}|$ is odd, and that $\hat{X}$ will be an even vertex of G_X and $\hat{\overline{X}}$ will be an even vertex of $G_{\overline{X}}$ if $|X \nabla_G \overline{X}|$ is even. Furthermore, all vertices in $\overline{X}$ clearly have the same valency and parity in G as in G_X, and all vertices in X clearly have the same valency and parity in G as in $G_{\bar{X}}$. From these remarks, it is easily seen that an odd-vertex-pairing $Q \vee R$ of G is determined by the following rules.

(i) If $|X \nabla_G \bar{X}|$ is even, let $Q \vee R = Q \cup R$.

(ii) If $|X \nabla_G \bar{X}|$ is odd and θ is the Q-partner of $\hat{X}$ and ϕ is the R-partner of $\hat{\bar{X}}$, let $Q \vee R$ be obtained from $Q \cup R$ by replacing the pairs $\{\hat{X}, \theta\}$, $\{\hat{\bar{X}}, \phi\}$ by the pair $\{\theta, \phi\}$.

Thus $Q \vee R$ is essentially what is obtained from $Q \cup R$ by making the smallest number of changes which will convert it into an odd-vertex-pairing of G.

Although the above definition of $Q \vee R$ makes sense if $|X| = 1$ or $|\bar{X}| = 1$, it should be noted that, if $|X| = 1$, then G_X is not smaller than G and so the existence of an *unobtrusive* odd-vertex-pairing of G_X does not follow from our inductive hypothesis; and likewise if $|\bar{X}| = 1$, we are not automatically entitled to assume that $G_{\bar{X}}$ has an unobtrusive odd-vertex-pairing. In view of the importance of this special situation, we call a subset X of $V(G)$ *vertical* if either $|X| = 1$ or $|\bar{X}| = 1$; i.e., "vertical" is to be read as meaning "pertaining to a vertex," in the sense that X is associated in a special way with a particular vertex if either X or its complement is the set whose sole element is that vertex.

We require the following further definitions. An odd-vertex-pairing P of G is *Y-unobtrusive*, where $Y \subseteq V(G)$, if $sc_{G-P}(\xi, \eta) = c_G(\xi, \eta)^\flat$ for every pair ξ, η of distinct elements of Y. (Thus P is unobtrusive if and only if it is $V(G)$-unobtrusive.) A subset X of $V(G)$ is *critical* if $|X \nabla \bar{X}|$ is even and there exist $\xi \in X, \eta \in \bar{X}$ such that $c(\xi, \eta) = |X \nabla \bar{X}|$. A subset X of $V(G)$ *divides* a subset Y of $V(G)$ if $Y \cap X \neq \varnothing$ and $Y \cap \bar{X} \neq \varnothing$. A subset X of $V(G)$ is *Y-minimal*, where $Y \subseteq V(G)$, if X divides Y and $|X \nabla \bar{X}| \leq |X' \nabla \bar{X}'|$ for every subset X' of $V(G)$ which divides Y. We note that, if $Y \subseteq V(G)$ and $|Y| \geq 2$, then there exist subsets of $V(G)$ which divide Y and so there exist Y-minimal subsets of $V(G)$ (at least two Y-minimal subsets, in fact, since the complement of a Y-minimal subset of $V(G)$ is Y-minimal). We note also that our definitions imply that, if a subset X of $V(G)$ is either critical or Y-minimal (for some $Y \subseteq V(G)$), then $|X| \geq 1$ and $|\bar{X}| \geq 1$, so that if in addition X is nonvertical then $|X|, |\bar{X}| \geq 2$ and G_X, $G_{\bar{X}}$ are graphs smaller than G which have unobtrusive odd-vertex-pairings by our inductive hypothesis.

After these preliminaries, our proof, assuming the specified inductive hypothesis, that G has an unobtrusive odd-vertex-pairing can be carried out in five stages.

STAGE I. If $V(G)$ has a nonvertical critical subset X, the inductive hypothesis implies that G_X, $G_{\bar{X}}$ have unobtrusive odd-vertex-pairings Q, R, respectively, and one can prove that $Q \vee R$ is, in this case, an unobtrusive odd-vertex-pairing of G.

STAGE II. Prove that, if $Y \subseteq V(G)$ and $\bar{Y} \bigtriangledown \bar{Y} \neq \varnothing$, then G has a Y-unobtrusive odd-vertex-pairing. The method of proof depends on whether or not $V(G)$ has a nonvertical critical subset. If $V(G)$ has a nonvertical critical subset, Stage I shows that G has an odd-vertex-pairing which is unobtrusive and therefore *a fortiori* Y-unobtrusive. If $V(G)$ has no nonvertical critical subset, select a $\lambda \in \bar{Y} \bigtriangledown \bar{Y}$, observe that by the inductive hypothesis $G - \lambda$ has an unobtrusive odd-vertex-pairing T, and (using the hypothesis that $V(G)$ has no nonvertical critical subset) show that $\tilde{T}$ is a Y-unobtrusive odd-vertex-pairing of G.

STAGE III. Prove that, if $Y \subseteq V(G)$ and $\bar{Y} \bigtriangledown \bar{Y} = \varnothing$ and there exists a nonvertical Y-minimal subset X of $V(G)$, then G has a Y-obtrusive odd-vertex-pairing. Here again, the inductive hypothesis implies that G_X, $G_{\bar{X}}$ have unobtrusive odd-vertex-pairings Q, R, respectively, and one can prove that $Q \vee R$ is a Y-unobtrusive odd-vertex-pairing of G.

STAGE IV. Prove that, if $Y \subseteq V(G)$ and $|Y| \geq 2$ and, for every $Z \subset Y$, G has a Z-unobtrusive odd-vertex-pairing, then G has a Y-unobtrusive odd-vertex-pairing. By virtue of the work already done in Stages II and III, we need only consider the case in which $\bar{Y} \bigtriangledown \bar{Y} = \varnothing$ and $V(G)$ has no nonvertical Y-minimal subset: we therefore make these two assumptions. The second assumption implies that $V(G)$ has a vertical Y-minimal subset X, and since X is vertical and divides Y, either X or $\bar{X}$ must be $\{\omega\}$ for some $\omega \in Y$. Since by hypothesis G has a Z-unobtrusive odd-vertex-pairing for every $Z \subset Y$, it has a $(Y \backslash \{\omega\})$-unobtrusive odd-vertex-pairing P, and using the assumption that $\bar{Y} \bigtriangledown \bar{Y} = \varnothing$ one can show that P is in fact Y-unobtrusive.

STAGE V. Since any odd-vertex-pairing of G is vacuously Y-unobtrusive if $Y \subseteq V(G)$ and $|Y| \leq 1$, repeated application of Stage IV yields a $V(G)$-unobtrusive odd-vertex-pairing of G, i.e., an unobtrusive odd-vertex-pairing of G.

The comparatively complicated nature of the foregoing proof, when written out in full, as contrasted with the comparatively simple and natural character of Theorems 1 and 2 might suggest that conceivably the most simple, natural, and insightful proof of these theorems has not yet been found and that a search for alternative proofs could be of interest. The only (very vague) suggestion I can offer in this direction is to observe that the theory of matroids has recently been found to shed considerable light on some areas of graph theory, and, since some parts of the proof of Theorem 2 seemed to me to have a somewhat matroid-like flavor, I have sometimes wondered whether there might be a way of using matroids, or something like matroids, to give a better and more illuminating proof of our two theorems.

8. Extension of Theorem 1 to Infinite Graphs

Robbins' theorem was extended to infinite graphs by Egyed [1]. It was stated in [4] that I had obtained an extension of Theorem 1 to infinite graphs, but I now feel this statement to have been an exaggeration and that a more accurate one would be that I had studied in some detail the problem of extending Theorem 1 to denumerably infinite graphs, and in somewhat less detail the problem of extending it to nondenumerably infinite graphs, and as a result I had arrived at the belief that I could very probably obtain such extensions of the theorem by doing the necessary amount of work. However, as the necessary arguments were never written out in full detail or published, I do not now think it fair to claim that any extension of Theorem 1 to infinite graphs is yet definitely established, and if anyone cares to work out and publish the necessary detailed arguments (before the present author finds time to do so), it would seem that he could quite properly claim this as a new, and interesting, contribution to graph theory. This modification of the claim made in [4] seems particularly necessary in view of the fact that in [5] I claimed on somewhat similar grounds to have obtained an extension to infinite graphs of a theorem proved in [5] for finite graphs, and recent rumors suggest that more careful investigations (I cannot at present recall by whom) have indicated that this theorem of [5] does not in fact extend to infinite graphs.

In considering how Theorem 1 might be able to be extended to *denumerably* infinite graphs, one is naturally tempted to think first of using König's "Unendlichkeitslemma." (For an account of this method see Chapter VI of [2] or Section 3 of [6].) Unfortunately, this method does not seem to work straightforwardly in the present problem, and the difficulties encountered are indeed very much of the general nature indicated by the discussion in Section 4 of [6] (which describes another problem in which an attempt to apply König's "Unendlichkeitslemma" does not seem to work very straightforwardly). However, as I mention in [6], difficulties in applying the "Unendlichkeitslemma" can sometimes be overcome by a little additional ingenuity, and my investigations have led me to believe that with a very high degree of probability Theorem 1 can in fact be extended to denumerably infinite graphs by a somewhat complicated argument based on the use of the following sharpened form of Theorem 1:

THEOREM 1#. *If G is a finite graph and H is a subgraph of G, then there exists a well-balanced orientation of G which induces a well-balanced orientation of H.*

Given Theorem 2, the proof of Theorem 1# is not unreasonably difficult. At a certain stage in the proof of Thoerem 1 (as described in Section 6) we had occasion to select an arbitrary di-Eulerian orientation Δ of the finite Eulerian graph $G + P$. Since most finite Eulerian graphs tend to have many di-Eulerian orientations, this in general leaves us a considerable degree of freedom in the choice of Δ, and the proof of Theorem 1# depends essentially on the idea of modifying this step in the proof of Theorem 1 by choosing Δ to be, not just *any* di-Eulerian orientation of $G + P$, but one which satisfies certain additional restrictions.

One is naturally tempted to conjecture that Theorem 1# is itself extensible to denumerably infinite (and perhaps also to nondenumerably infinite) graphs, but I have no idea how to prove or disprove this. The fact that the proof of Theorem 1 which was given in [4] does not seem to extend very easily and naturally to denumerably infinite graphs, and the fact that it seems to give us no clues as to how to extend the proof of Theorem 1# to such graphs, are perhaps additional reasons for conjecturing that the proof of Theorem 1 given in [4] might not be the best or most natural proof.

There seems to be a good chance that, after establishing an extension of Theorem 1 to denumerably infinite graphs, one could further extend it to nondenumerably infinite graphs by using techniques somewhat similar to those of [3, Section 3]; in fact I have already suggested in [6] that techniques of this general nature may be what are required in a number of situations in which one might wish to extend to nondenumerably infinite graphs results which have been established for finite and denumerably infinite ones.

REFERENCES

1. L. EGYED, Ueber die wohlgerichteten unendlichen Graphen, *Math. Fiz. Lapok* **48** (1941), 505–509 (Hungarian with German summary).
2. D. KÖNIG, *Theorie der endlichen und unendlichen Graphen*, Akademische Verlagsgesellschaft M.B.H., Leipzig, 1936; reprinted by Chelsea, New York, 1950.
3. C. ST. J. A. NASH-WILLIAMS, Decomposition of Graphs into Closed and Endless Chains, *Proc. London Math. Soc.* **10** (1960), 221–238.
4. C. ST. J. A. NASH-WILLIAMS, On Orientations, Connectivity, and Odd-Vertex-Pairings in Finite Graphs, *Canad. J. Math.* **12** (1960), 555–567.
5. C. ST. J. A. NASH-WILLIAMS, Edge-Disjoint Spanning Trees of Finite Graphs, *J. London Math. Soc.* **36** (1961), 445–450.

6. C. St. J. A. Nash-Williams, Infinite Graphs—a Survey, *J. Combinatorial Theory* **3** (1967), 286–301.
7. O. Ore, *Theory of graphs*, Colloquium Publ. Vol. 38, Am. Math. Soc. Providence, 1962.
8. H. E. Robbins, A Theorem on Graphs with an Application to a Problem of Traffic Control, *Amer. Math. Monthly* **46** (1939), 281–283.

THE PARTITION CALCULUS

R. Rado

DEPARTMENT OF MATHEMATICS
THE UNIVERSITY, READING, ENGLAND

1. Introduction

Dirichlet's pigeon-hole principle (chest-of-drawers principle, box argument, *Schubfachprinzip*) asserts roughly that if many objects are distributed over not too many classes then at least one class contains many of these objects. This mode of reasoning has interesting applications, for instance in number theory and algebra. The partition calculus aims at developing this principle for its own sake and studying its applications in combinatorial set theory. Some of its results have since been applied elsewhere, for instance in mathematical logic and in topology.

I use the customary notation for ordinal numbers (or ordinals)

$$0,\ 1,\ 2,\ \ldots,\ \omega(=\omega_0),\ \omega+1,\ \ldots,\ \omega^2,\ \ldots,\ \omega_1,\ \ldots,\ \omega_2,\ \ldots$$

and for cardinal numbers (or cardinals)

$$0,\ 1,\ 2,\ \ldots,\ \aleph_0,\ \aleph_1,\ \ldots.$$

The *initial ordinals* are

$$0,\ 1,\ 2,\ \ldots,\ \omega_0,\ \omega_1,\ \ldots.$$

If a is a cardinal then a^+ denotes the next-larger cardinal, and a^- the next-smaller cardinal if such a cardinal exists, and I put $a^- = a$ if there is no next-smaller cardinal to a. For $a \geq \aleph_0$ I denote by a' the least cardinal c such that a can be expressed as the sum of c cardinals each less than a. For convenience of notation I use the *obliteration operator* $\hat{}$ whose effect is to remove from a well-ordered sequence of objects the object above which it is placed. Other uses of this operator will be self-explanatory.

For every ordinal n and for cardinals a, $b_0, b_1, \ldots, \hat{b}_n$ the *partition relation*

$$a \to (b_0, \ldots, \hat{b}_n) \tag{1}$$

expresses, by definition, the truth of the following statement. Whenever a set A of cardinal $|A| = a$ is expressed as $A = K_0 \cup \cdots \cup \hat{K}_n$ then there are always a set $X \subseteq A$ and an ordinal $\nu < n$ such that $X \subseteq K_\nu$ and $|X| = b_\nu$. If $b_0 = \cdots = \hat{b}_n = b$ then (1) is also written either as $a \to (b)_n$ or $a \to (b)_{|n|}$, where $|n|$ denotes the cardinal of n. The logical negation of the relation (1), and similarly for all partition relations to be defined later, is denoted by $a \not\to (b_0, \ldots, \hat{b}_n)$.

A quantitative version of Dirichlet's principle states that if $1 \leq b_0, \ldots, \hat{b}_n < \omega$, then (1) *holds if and only if* $a > \sum (b_\nu - 1)$. It is easy to prove that for arbitrary cardinals a, $b_0, \ldots, \hat{b}_n$ *the relation* (1) *holds if and only if either*

(i) *there is* $\nu < n$ *such that* $b_\nu = 0$; *or*

(ii) $a < \aleph_0$; $1 \leq b_0, \ldots, \hat{b}_n \leq a$; $a > \sum (b_\nu - 1)$; *or*

(iii) $a \geq \aleph_0$; $1 \leq b_0, \ldots, \hat{b}_n \leq a$; $|\{\nu : b_\nu = a\}| < a'$;

$\sup\{b_\nu : b_\nu < a\} < a$; $|\{\nu : b_\nu \geq 2\}| < a$.

Here, for every set D of cardinals, sup D denotes the least cardinal s such that $s \geq d$ for every $d \in D$.

2. First Extension

The first extension of the scope of the partition calculus brings in structural features of sets other than their cardinals, such as the order type of an ordered set or the measure of a set of real numbers. It is concerned with relations

$$\mathbf{A} \to (\mathbf{B}_0, \ldots, \hat{\mathbf{B}}_n) \tag{2}$$

between classes $\mathbf{A}$, $\mathbf{B}_0, \ldots$ of structured sets. Thus $\mathbf{A}$ might denote the class of all ordered sets of order type $\omega_1{}^2$, or the class of all denumerable partially ordered sets which do not contain any infinite decreasing sequence of elements. The relation (2), by definition, holds if and only if, whenever $A \in \mathbf{A}$ and $A = K_0 \cup \cdots \cup \hat{K}_n$ then there are always a set $X \subseteq A$ and an ordinal $\nu < n$ such that $X \subseteq K_\nu$ and $X \in \mathbf{B}_\nu$. Here, the structure of X is that which is induced by considering X as a substructure of A. In practice one uses as notation for the classes $\mathbf{A}$, $\mathbf{B}_0, \ldots,$ those entities which characterize these classes, e.g., the symbol $\omega_1{}^2$ for the class of all sets ordered according to the type $\omega_1{}^2$, or $\aleph_2$ for the class of all sets of cardinal $\aleph_2$.

An instance of (2) is the relation (see [1], Theorem 1.33; also [2], corollary to Theorem 6)

$$\omega^\alpha \to (\omega^\alpha)_n \tag{3}$$

which holds for every ordinal α and every n such that $1 \leq n < \omega$. In the case of well-ordered sets the relation (2) has been completely analyzed [2] in the following sense.

(i) If $n < \omega$ and $\beta_0, \ldots, \hat{\beta}_n$ are any ordinals then there is an explicit expression, in terms of Cantor's standard representation of the β_ν as sums of powers of ω, for the least ordinal α for which $\alpha \to (\beta_0, \ldots, \hat{\beta}_n)$.

(ii) If $n, \beta_0, \ldots, \hat{\beta}_n$ are arbitrary ordinals then there is an algorithm, finite in a certain well-defined sense, which yields the least ordinal α for which $\alpha \to (\beta_0, \ldots, \hat{\beta}_n)$.

3. Finite Exponents

Ramsey [3] established a far-reaching generalization of Dirichlet's principle. For a set A and a cardinal r, put

$$[A]^r = \{X \colon X \subseteq A; |X| = r\}.$$

For cardinals $a, b_0, \ldots, \hat{b}_n, r$, let the relation

$$a \to (b_0, \ldots, \hat{b}_n)^r \tag{4}$$

denote the truth of the following statement. Whenever $|A| = a$ and $[A]^r = K_0 \cup \cdots \cup \hat{K}_n$ then there are a set $X \subseteq A$ and an ordinal $\nu < n$ such that $[X]^r \subseteq K_\nu$ and $|X| = b_\nu$. If $r = 1$ then (4) and (1) are logically equivalent. Ramsey proved that

$$\aleph_0 \to (\aleph_0)_n{}^r \qquad \text{for} \qquad n, r < \omega,$$
$$R^r(b_0, \ldots, \hat{b}_n) \to (b_0, \ldots, \hat{b}_n)^r$$

for arbitrary $n, b_0, \ldots, \hat{b}_n, r < \omega$, and some suitable number $R^r(b_0, \ldots, \hat{b}_n)$. The least value of $R^r(b_0, \ldots, \hat{b}_n)$, denoted here by $\bar{R}^r$, can be shown to lie between a $(r-2)$-fold exponentiation and a $(r-1)$-fold exponentiation. If we define the symbol $a_0 * a_1 * \cdots * a_k$ inductively by putting $a * b = a^b$ and $a_0 * a_1 * \cdots * a_l = a_0 * (a_1 * \cdots * a_l)$ for $3 \leq l < \omega$ then we have

$$\bar{R}^r(b_0, \ldots, \hat{b}_n) \leq n * (n^{r-1}) * (n^{r-2}) * \cdots * (n^2) * \left(\sum (b_\nu - r) + 1\right)$$

for $2 \leq n < \omega$; $2 \leq r \leq b_0, \ldots, \hat{b}_n < \omega$. If $r = 2$ then this estimate of $\bar{R}^r$ is to be interpreted as

$$\bar{R}^2(b_0, \ldots, \hat{b}_n) \leq n^{\sum (b_\nu - 2) + 1}.$$

For $r = 2$ and arbitrary cardinals $b_0, \ldots, \hat{b}_n$, the relation (4) has been almost completely analyzed, provided one assumes the "generalized continuum hypothesis"

$$2^a = a^+ \qquad \text{for} \qquad a \geq \aleph_0. \tag{GCH}$$

Considerable progress has been made in this direction without assuming GCH. For this and a detailed account of some parts of the calculus see [4].

In the case $r = 2$, relation (4) can be interpreted in terms of edge coloring of complete graphs. It states that if the edges of a complete graph of order a are colored with $|n|$ colors $c_0, \ldots, \hat{c}_n$, then there are always a color c_ν and a complete subgraph of order b_ν all of whose edges have color c_ν. Here is a result deduced on the assumption that GCH holds. We make the very natural assumptions that $a \geq \aleph_0$; $n \geq 2$; $2 \leq b_0, \ldots, \hat{b}_n \leq a$. Furthermore, we assume that if $a' > \aleph_0$, then $a' > (a')^-$ which means that a' is not one of the hypothetical inaccessible cardinals greater than $\aleph_0$. *Then* ([5], p. 130, Theorem I)

$$a \to (b_0, \ldots, \hat{b}_n)^2$$

holds if and only if $|n| < ((a')^-)'$ *and either*

(i) *there is* $\nu_0 < n$ *such that* $b_{\nu_0} = a$ *and* $b_\nu \leq ((a')^-)'$ *for* $\nu \neq \nu_0$; *or*

(ii) $b_0, \ldots, \hat{b}_n < a$, *and there is* $c < a^-$ *such that* $|\{\nu : b_\nu \geq c\}| < (a^-)'$.

4. Special Cases

As an example of a relation of the more general type (2), with "exponent" $r = 2$, I mention ([6], p. 432, Theorem 6)

$$\eta \to (\aleph_0, \eta)^2. \tag{5}$$

Here η denotes the order type of the rational numbers if ordered by magnitude. There are extensions of (5) involving cardinals $\aleph_\alpha > \aleph_0$ and Hausdorff's generalized rational order types η_α ([7]).

Let us now turn to relations for ordinals. It is shown in [8] that

$$\omega^{1+ph} \to (2^h, \omega^{1+p})^2 \qquad \text{for} \quad p < \omega_1;\ h < \omega, \tag{6}$$

and in [9] that

$$\omega_\alpha l(m, n) \to (m, \omega_\alpha n)^2 \qquad \text{for} \quad \alpha \geq 0 \quad \text{and} \quad m, n < \omega, \tag{7}$$

where $l(m, n)$ denotes a positive integer given explicitly as a simple function of m and n. If $l_\alpha(m, n)$ denotes the least number such that $\omega_\alpha l_\alpha(m, n) \to (m, \omega_\alpha n)^2$, then the number $l_0(m, n)$ can be computed for every m, n; it is characterized by a certain finite combinatorial property involving m and n. It is conjectured that $l_\alpha(m, n) = l_0(m, n)$ for all α but this has only been proved for some pairs m, n.

For any given values of $r, n < \omega$, and of ordinals $\alpha, \beta_0, \ldots, \hat{\beta}_n < \omega^\omega$ the truth of the relation $\alpha \to (\beta_0, \ldots, \hat{\beta}_n)^r$ can be decided in a finite number of steps. For ordinals of the form $\omega^t < \omega^\omega$ this has been shown by A. Hajnal and independently by F. Galvin.

5. Infinite Exponents

The question might be asked whether the exponent r can be taken to be infinite. The following result [10] shows that only trivial partition relations can hold for any such r.

Let $r \geq \aleph_0$ and let n be an ordinal such that $|n| = 2^r$. Let A be an arbitrary set. Then there is a partition $[A]^r = K_0 \cup \cdots \cup \hat{K}_n$, where $K_\mu \cap K_\nu = \varnothing$ for $\mu < \nu < n$, such that whatever set $X \subseteq A$ with $|X| = r$ we choose, the set $[X]^r$ will always contain elements from every one of the classes $K_0, \ldots, \hat{K}_n$. This implies that for every cardinal a and every $r \geq \aleph_0$ the relation $a \to (r)^r_{2^r}$ fails to hold and, indeed, does so in a very spectacular way. It is easy to deduce that, more generally, for every ordinal $n \geq 2$ and every choice of cardinals r, a, $b_0, \ldots, \hat{b}_n$, such that $r \geq \aleph_0$ and $b_0, \ldots, \hat{b}_n \geq r$, we have

$$a \nrightarrow (b_0, \ldots, \hat{b}_n)^r.$$

6. Dual Relations

The preceding section leads to the introduction of a partition relation which in a sense is dual to (4), namely the relation

$$a \to [b_0, \ldots, \hat{b}_n]^r. \tag{8}$$

By definition, (8) expresses the truth of the following statement. Whenever $|A| = a$ and $[A]^r = K_0 \cup \cdots \cup \hat{K}_n$; $K_\mu \cap K_\nu = \varnothing$ for $\mu < \nu < n$, then there are always a set $X \subseteq A$ and an ordinal $\nu < n$ such that $[X]^r \cap K_\nu = \varnothing$ and $|X| = b_\nu$. Hajnal's theorem quoted in Section 5 is expressed by the relation $a \nrightarrow [r]^r_{2^r}$ and holds for every cardinal a and every infinite cardinal r. The following result ([5], p. 150 Theorem 23) gives a complete analysis of the relation (8) for $r < \omega$ provided that $a > a' = \aleph_0$, or more generally, whenever $a > a' \to (a', a')^2$, and if GCH is assumed.

Let $1 \leq r < \omega$; $a > a' \to (a', a')^2$; $b_0, \ldots, \hat{b}_n \leq a$. Then (8) *holds if and only if either* (i) $n > 2^{r-1}$, *or* (ii) $n \leq 2^{r-1}$ *and the cardinals $a, a', b_0, \ldots, \hat{b}_n$*

satisfy one of a finite number of explicitly given sets of inequalities. The number of these inequalities depends on r only.

7. POLARIZED RELATIONS

Of various other types of partition relations which have been studied I mention one more kind, the *polarized* partition relation in the bipartite case. For cardinals $a, b, a_0, \ldots, \hat{a}_n, b_0, \ldots, \hat{b}_n$ the relation

$$\binom{a}{b} \to \binom{a_0, \ldots, \hat{a}_n}{b_0, \ldots, \hat{b}_n} \tag{9}$$

denotes the truth of the following statement. If $|A| = a$; $|B| = b$; $A \times B = K_0 \cup \cdots \cup \hat{K}_n$, then there are always sets $X \subseteq A$ and $Y \subseteq B$ and an ordinal $\nu < n$ such that $X \times Y \subseteq K_\nu$; $|X| = a_\nu$; $|Y| = b_\nu$. Clearly, (9) is a proposition concerning edge colorings of bipartite graphs. In the case $n = 2$ the relation (9) has been analyzed to a considerable extent. I mention the following results:

[5], p,166, 21.2)

$$\binom{a}{a} \nrightarrow \begin{pmatrix} a+1, & 1 \\ 1, & a \end{pmatrix} \quad \textit{for every} \quad a \geq 0. \tag{10}$$

([5], p.176 Corollary 17)

$$\binom{a^+}{a^+} \to \begin{pmatrix} a, & a \\ a, & a \end{pmatrix} \quad \textit{for every} \quad a \geq \aleph_0. \tag{11}$$

$$\textit{If} \quad a, b \geq \aleph_0 \quad \textit{then} \quad \binom{a}{b} \to \begin{pmatrix} a, & a \\ b, & b \end{pmatrix} \quad \textit{holds if and only if}$$

$$\{a, a^+, a', (a')^+\} \cap \{b, b^+, b', (b')^+\} = \phi. \tag{12}$$

([5], p. 187 Theorem 44)
In the proofs of (11) and (12) GCH was assumed.

8. APPLICATIONS

I will now mention some applications of partition relations to problems which are themselves fairly close to pure combinatorics.

(i) In [1] the relation $\omega_{\lambda+1}^3 \to (\omega_{\lambda+1}^3)_p{}^1$, which is deduced for every ordinal λ and every cardinal $p < \aleph_{\lambda+1}$, is used to construct, given any infinite cardinal a, *a graph* Γ_a *of order* a *which contains no triangle and has a chromatic number* $\chi(\Gamma_a) = a$.

(ii) Let m be an ordinal, and let us assume the following weak version of GCH. If $\mu < m$ then $2^{\aleph_\mu} \leq \aleph_m$. It is shown in [6] that under these assumptions

$$\beta \to (4, \omega_m + 1)^3$$

for every ordinal β such that $|\beta| > \sum (\nu < \omega_m) 2^{|\nu|}$. This relation is used in [9] to obtain the following result on transitivity domains of binary relations.

Let the binary relation R be defined on a set S and be such that for every $x, y \in S$ exactly one of the relations $x = y$; xRy; yRx holds. Let a be a cardinal. Then the relation R is transitive on some suitable subset X of S with $|X| = a$, provided that

$$\begin{aligned} |S| &\geq 2^{a-1} && \text{if} \quad a < \aleph_0, \\ |S| &\geq \aleph_0 && \text{if} \quad a = \aleph_0, \\ |S| &> \textstyle\sum (b < a) 2^b && \text{if} \quad a > \aleph_0, \end{aligned}$$

where the summation extends over all cardinals b less than a.

9. Methods of Proof

I will now describe in broad outline two of the general methods used in the discussion of partition relations.

(i) *The ramification method* ([5], p. 103, Lemma 1). Beginning with the proof of the first nontrivial partition relation, which is Ramsey's formula $\aleph_0 \to (\aleph_0)_2{}^2$, a mode of reasoning has frequently been employed which relies on a process of building up a tree whose vertices are associated with certain sets. By way of an illustration, suppose we want to prove a relation of the form $a \to (b, c)^2$. We take a set S such that $|S| = a$ and consider an arbitrary partition $[S]^2 = K_0 \cup K_1$. By Zorn's lemma we find a maximal set $R \subseteq S$ such that $[R]^2 \subseteq K_0$. Then, for $x \in S - R$, there is an element $f(x) \in R$ such that $\{x, f(x)\} \in K_1$. This means that we can write

$$S = R \cup \bigcup (x_0 \in R) S(x_0),$$

where

$$S(x_0) = \{x \in S - R : \{x_0, x\} \in K_1\}.$$

Now, for every $x_0 \in R$, we treat the set $S(x_0)$ in the same way S was treated. We select a maximal set $R(x_0) \subseteq S(x_0)$ such that $[R(x_0)]^2 \subseteq K_0$ and write

$$S(x_0) = R(x_0) \cup \bigcup (x_1 \in R(x_0)) S(x_0, x_1),$$

where

$$S(x_0, x_1) = \{x \in S(x_0) - R(x_0) \colon \{x_1, x\} \in K_1\}.$$

The process continues and leads to sets $R(x_0, x_1, \ldots)$, $S(x_0, x_1, \ldots)$, where the sequences of elements x_ν of S continue, under suitable definitions, to transfinite lengths. If at any stage one of the sets $R(\cdots)$ has a cardinal at least equal to b then we have found a set $X \subseteq S$ such that $[X]^2 \subseteq K_0$; $|X| \geq b$, as required for a proof a $a \rightarrow (b, c)^2$. If, on the other hand, every $|R(\cdots)| < b$, then the "ramification valency" of our tree is bounded by b and hence, if $|S|$ is sufficiently large, we would expect to find a "chain" of at least c nonempty sets S, $S(x_0)$, $S(x_0, x_1)$, ... For this particular sequence of elements x_ν the set $Y = \{x_0, x_1, \ldots\}$ satisfies $[Y]^2 \subseteq K_1$; $|Y| \geq c$, and the relation $a \rightarrow (b, c)^2$ is proved. To make the argument rigorous a considerable amount of care and attention to detail are needed. So far this method has only been successful when a is regular, i.e., when $a = a'$.

(ii) *The canonization method.* ([5], p. 111, Lemma 3). Suppose we want to discuss a relation of the form $a \rightarrow (b, c)^2$, where $a > a' = \aleph_k$. As in (i) we consider sets S, K_0, K_1. Now we can write $S = S_0 \cup \cdots \cup \hat{S}_m$, where $m = \omega_k$; $|S_\nu| < a$ for $\nu < m$; $S_\mu \cap S_\nu = \varnothing$ for $\mu < \nu < m$. Using GCH and the ramification method one can show that there is a sequence $T_0, \ldots, \hat{T}_m$ of sets and an increasing sequence of ordinals $\nu_0, \ldots, \hat{\nu}_m < m$ such that $T_\lambda \subseteq S_{\nu_\lambda}$ for $\lambda < m$ and $|T| = a$, where $T = T_0 \cup \cdots \cup \hat{T}_m$, so that the given partition of $[S]^2$, when restricted to $[T]^2$, is "canonical" in the following sense. If $x \neq y$, and $x \in T_\alpha$; $y \in T_\beta$, where $\alpha \leq \beta < m$, then the class K_κ to which $\{x, y\}$ belongs only depends on α and β but not on the particular choice of the elements x and y in T_α and T_β, respectively. This means that the partition of $[T]^2$ is completely described in terms of a partition of the smaller set $[\{0, 1, \ldots, \hat{m}\}]^2$. However, the case $\alpha = \beta$ has to receive extra attention here. Such reasoning leads to the result that $a \rightarrow (b, c)^2$ holds if and only if $a' \rightarrow (b, c)^2$ holds. The argument can be generalized. Thus, *if GCH is assumed then, for $a \geq \aleph_0$; $|n| < a$; $r < \omega$, and arbitrary cardinals $b_0, \ldots, \hat{b}_n$, the two relations $a \rightarrow (b_0, \ldots, \hat{b}_n)^r$, $a' \rightarrow (b_0, \ldots, \hat{b}_n)$ are equivalent.*

10. Problems

In conclusion I mention some problems whose solutions seem to be essential for progress to be made in certain directions.

Problem 1. To decide whether

$$\omega^\omega \rightarrow (3, \omega^\omega)^2.$$

Problem 2. Assuming GCH one can prove that

$$\omega_1\alpha \not\rightarrow (3, \omega_1\omega)^2 \qquad \text{for every} \qquad \alpha < \omega_1.$$

It is not known whether

$$\omega_1{}^2 \to (3, \omega_1\omega)^2.$$

PROBLEM 3. Obvious extensions of the notation (4) are relations such as $a \to ((b)_m, c)^r$ which denotes the relation (4) in the special case when $n = m + 1$; $b_0 = \cdots = \hat{b}_{mm} = b$; $b_m = c$. Using GCH one can prove that

$$\aleph_{\omega+1} = \aleph_\omega^{\aleph_0} \nrightarrow ((3)_{\aleph_0}, \aleph_{\omega+1})^2.$$

Without GCH only the weaker relation

$$\aleph_\omega^{\aleph_0} \nrightarrow ((\aleph_0)_{\aleph_0}, \aleph_{\omega+1})^2$$

has been proved. To what extent can the gap between these two results be bridged?

PROBLEM 4. Progress in a number of directions with the development of the partition calculus without GCH seems to be blocked by our present ignorance of the answer to the following question. Put $a^* = \sum (\nu < \omega) 2^{\aleph_\nu}$. Assume that $a^* > \aleph_\omega$ and that the series for a^* has no maximal term. Is it then true that

$$a^* \to (\aleph_\omega, \aleph_\omega)^2 ?$$

REFERENCES

1. G. H. TOULMIN, Shuffling Ordinals and Transfinite Dimensions, *Proc. London Math. Soc.* (3) **4** (1954), 177–95.
2. E. C. MILNER and R. RADO, The pigeon-Hole Principle for Ordinal Numbers, *Proc. London Math. Soc.* (3) **15** (1965), 750–68.
3. F. P. RAMSEY, On a Problem of Formal Logic, *Proc. London Math. Soc.* (2) **30** (1930), 71–83.
4. P. ERDŐS, A. HAJNAL, and R. RADO, *The Partition Calculus in Set Theory* (a monograph submitted to the Hungarian Academy of Sciences).
5. P. ERDŐS, A. HAJNAL, and R. RADO, Partition Relations for Cardinal Numbers, *Acta Math. Acad. Sci. Hungr.* **16** (1965), 93–196.
6. P. ERDŐS and R. RADO, A Partition Calculus in Set Theory, *Bull. Amer. Math. Soc.* **61** (1956), 427–89.
7. P. ERDŐS, E. C. MILNER, and R. RADO (to be published).
8. E. C. MILNER, Some Combinatorial Problems in Set Theory, Thesis, London University, London, 1962.
9. P. ERDŐS and R. RADO, Partition Relations and Transitivity Domains of Binary Relations, *J. London Math. Soc.* **42** (1967), 624–33.
10. A. HAJNAL (to be published).

TEACHING GRAPH THEORY TO A COMPUTER

Ronald C. Read

UNIVERSITY OF THE WEST INDIES
JAMAICA

1. Introduction

This is not going to be any kind of high-powered discussion of new and startling results in the application of computing techniques to graph theory. On the contrary, if anything, it will be directed more to those who have little or no acquaintance with computers rather than to the expert computer man. Although a few new results will be mentioned, the main purpose will be to show, by some simple examples, how the use of computers for the solution of graph-theoretical problems frequently necessitates an approach to these problems which is quite different from the usual theoretical approach—a completely new viewpoint, in fact.

Not only are the computer methods to be described different from those usually employed by graph theorists, they have, in general, little in common with the conventional techniques of numerical analysis. This is not always so, however, and we shall start by considering a problem of interest in graph theory which can be solved by standard numerical methods, although as we shall see, it turns out that the " best " method is not the most convenient.

2. Characteristic Polynomials of Graphs

The characteristic polynomial $\phi_G(\lambda)$ of a graph G, that is, of its adjacency matrix A, is defined as $|A + \lambda I|$. Now it is more usual, in numerical analysis, to be interested in the roots of the characteristic equation, i.e., in the eigenvalues, rather than in the polynomial itself, and every computing center has programs for finding eigenvalues. Nevertheless, there are standard methods for finding the coefficients in the polynomial, notably that due to Danielevsky (see [2], p. 382). At our computing center we wanted to compute the characteristic polynomials of all graphs on seven nodes. We had, for input, a deck of cards containing descriptions of all such graphs (one per card) and our first intention was to use Danielevsky's method. Eventually we found another method to be more convenient.

We can write

$$\phi_G(\lambda) = \lambda^7 + a_1\lambda^6 + a_2\lambda^5 + a_3\lambda^4 + a_4\lambda^3 + a_5\lambda^2 + a_6\lambda + a_7. \tag{1}$$

Since the diagonal elements of an adjacency matrix are zero, we have $a_1 = 0$ Further, the principal two-rowed minor

$$\begin{vmatrix} 0 & a_{ij} \\ a_{ji} & 0 \end{vmatrix}$$

is -1 if and only if nodes i and j are joined, and is 0 otherwise. Hence $-a_2$ is the number of edges of G. This number was already included on each input card, so with a_1 and a_2 known, only a_3 to a_7 remained to be found.

We therefore computed $\phi_G(\lambda)$ for $\lambda = 0, \pm 1, \pm 2$ by means of a standard program for evaluating determinants, and solved the resulting five equations for the a_i ($i = 3$ to 7) by an equation-solving program. In this way the coefficients were found for each graph in the input list.

This is clearly a method that could be used for the evaluation of the characteristic polynomial of *any* matrix, and is described in some textbooks of numerical analysis (see, for example, [2]); but it is not recommended on account of its instability and lack of accuracy. In our example, however, there is no risk; the matrices are small, and since all elements are 0 or 1 we know that the coefficients in the polynomial are integers. Thus we could not possibly accumulate enough errors to cast doubt on the correct values of these coefficients.

3. The Isomorphism Problem

Contrasting with those problems that are amenable to standard numerical methods are those which require a completely novel approach. Strangely enough, these often relate to the fundamental concepts of graph theory. For example, how can we tell whether two graphs are "the same," i.e., isomorphic? The definition of graph isomorphism is well known: two graphs G_1 and G_2 are isomorphic if there is a one-to-one correspondence between their nodes such that two nodes of G_1 are adjacent if and only if the corresponding nodes of G_2 are adjacent. But how are we to make use of this definition in practice?

In talking theoretically about graphs it is easy to overlook that any problem even exists here. When we think about graphical theorems, or illustrate them in papers or on the blackboard, we generally choose simple examples of graphs with comparatively few nodes. In these circumstances it is usually clear, either immediately or after a little manipulation, whether two graphs are isomorphic or not. If, on the other hand, we have to decide whether there is isomorphism between two graphs each on (say) 379 nodes (and this is something that may well arise in a computer application) then we have a very real problem on our hands. Clearly the brute force approach of testing each of the $p!$ possible one-to-one correspondences for preservation of adjacency is quite impracticable.

This "isomorphism problem" is one which has been extensively studied, but no really practical algorithm has yet been devised to solve it. Of course, it is often easy to show that two graphs are *not* isomorphic; it is sufficient to find a graph invariant which is different for the two graphs. Trivial invariants are p and q, the numbers of nodes and edges, respectively; less trivial are the partition (valency sequence) and the characteristic polynomial, and there are many others. Heuristic algorithms for the isomorphism problem have been proposed, based on the general idea that if you compute many invariants, and they all turn out to be the same for both graphs, then the graphs are *probably* isomorphic (see, for example, [4]); but if a definite *yes* or *no* answer is required, the heuristic approach will not be good enough.

It follows *a fortiori* that the more general problem of determining whether one graph is a *subgraph* of another is also unsolved in any practical sense. This goes to show what a host of unsuspected difficulties can lie concealed behind an apparently simple statement like "H is a subgraph of G."

4. What Is a Graph?

We have just considered the two extremes. Now let us look at graph-theoretical problems in general, and try to see them through the eyes of a computer. We first ask "What is a graph?" We are all familiar with the standard definition—a set V of nodes, and a set E of edges, i.e., pairs of nodes—but when we think of a graph we almost certainly have in mind some sort of diagram, drawing, or other geometrical realization of the graph. Now it is unlikely that a computer would be programmed to "see" a graph in this sort of way, so it is natural to ask just how one would store a graph in a computer so that it can be easily manipulated. There are three fairly obvious methods:

(a) We can number the p nodes of the graph in some arbitrary manner, and then list the pairs of nodes which make up the q edges. This gives us q

pairs of integers (say from 0 to $p-1$) which we can regard as $2q$ digits in the scale of p. This is equivalent to $2q \log_2 p$ bits of information.

(b) Having numbered the nodes in some way, we can store the adjacency matrix A. This consists of p^2 0's or 1's, i.e., p^2 bits of information.

(c) We can proceed as in (b) but store only the upper triangular part of *a*. No information is lost thereby, since A is symmetric and all its diagonal elements are zero. This requires $\frac{1}{2}p(p-1)$ bits of information.

Method (a) is more economical in storage space provided that the number of edges is not too large, and for many problems this method is by far the most convenient form for carrying out the required manipulations. Of the other two, (c) uses less space than (b) but introduces extra problems in the look-up of individual matrix elements.

5. Chromatic Polynomials

Having decided how to represent a graph in a computer, let us consider a slightly more sophisticated problem. The chromatic polynomial $M_G(\lambda)$ of a graph G is defined as the number of ways of coloring the nodes of G in λ or fewer colors in such a way that adjacent nodes receive different colors (for details, see [3]). If A and B are two nonadjacent nodes of G, denote by G' the graph obtained from G by adding the edge AB, and let G'' denote the graph obtained by identifying nodes A and B. Then, by dividing the colorings of G into those that allot different colors to A and B, and those that give A and B the same color, we easily show that

$$M_G(\lambda) = M_{G'}(\lambda) + M_{G''}(\lambda). \tag{2}$$

If we use the device, introduced by Zykov [5], of letting a drawing of a graph stand for its chromatic polynomial, we can illustrate Eq. (2) by the "equation" of Fig. 1.

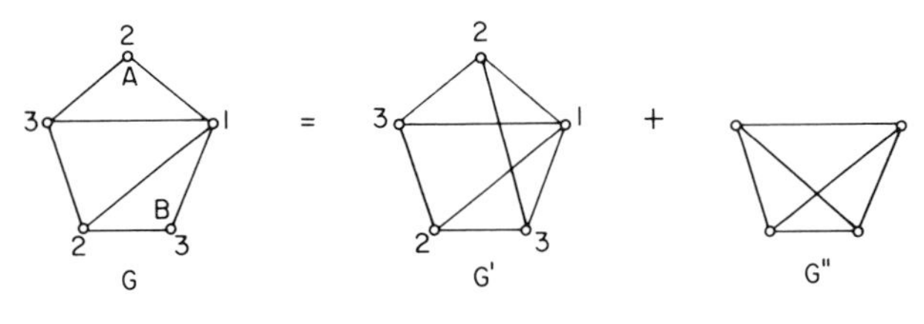

Fig. 1.

We can continue this process, as shown in Fig. 2, until all the graphs obtained are complete graphs. (In Figs. 1 and 2 ignore for the moment the numbers associated with some of the nodes.)

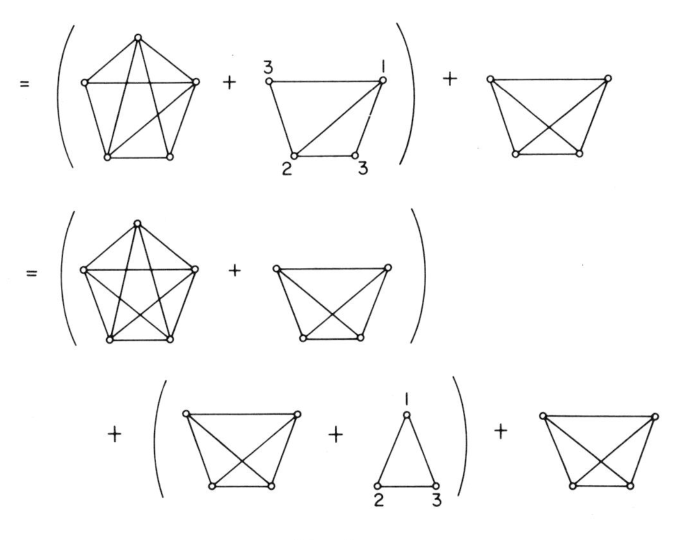

FIG. 2.

Since (as is easily verified) the chromatic polynomial of the complete graph with p nodes is $\lambda(\lambda-1)(\lambda-2)\cdots(\lambda-p+1)$, we see that the chromatic polynomial of the above graph is

$$\lambda(\lambda-1)(\lambda-2)(\lambda-3)(\lambda-4)+3\lambda(\lambda-1)(\lambda-2)(\lambda-3)+\lambda(\lambda-1)(\lambda-2)$$
$$=\lambda(\lambda-1)(\lambda-2)^3.$$

This illustrates an algorithm for finding the chromatic polynomial of a given graph, using a computer. We set up in the storage of the computer a list of graphs, initially consisting only of the given graph, and we zero an appropriate number of registers wherein to contain the coefficients. We then proceed as follows:

1. If the list is empty, stop. Otherwise transfer the last graph of the list to a working area.
2. Add an edge AB, not already present, to the graph in the working area.
3. If the resulting graph is complete, add the coefficients in

$$\lambda(\lambda-1)(\lambda-2)\cdots(\lambda-p+1)$$

to the coefficient registers; if not, add this graph to the end of the list.

4. Identify the nodes A and B.
5. If the resulting graph is not complete, repeat from Step 2; if it is complete, add the coefficients (as in Step 3) and repeat from Step 1.

It is easily verified that the list will never contain as many as p graphs, so that the storage requirements for this algorithm are quite modest.

In practice it is more convenient to use Eq. (2) in the alternative form

$$M_{G'}(\lambda) = M_G(\lambda) - M_{G''}(\lambda) \tag{3}$$

so that we successively *remove* edges AB already present, instead of adding them. The required polynomial is then expressed in terms of those of empty graphs, and these are simply powers of λ. Thus only one coefficient has to be increased in steps 3 or 5 of the algorithm. A program of this kind was used at our computing center to obtain the chromatic polynomials of all graphs on seven nodes

The original algorithm, using Eq. (2), has an advantage if, in addition to (or instead of) the chromatic polynomial, we require an actual coloring in a given number of colors. Thus if we want a coloring of the graph in three colors, we can find one as soon as we obtain a triangle in step 5 of the algorithm, as in Fig. 2. We color (or number) the nodes of this triangle, and then, so to speak, trace the "history" of these nodes during the course of the algorithm. This somewhat vague explanation is easily understood by referring to the numbering of the nodes in Figs. 1 and 2, and the process is summarized in Fig. 3.

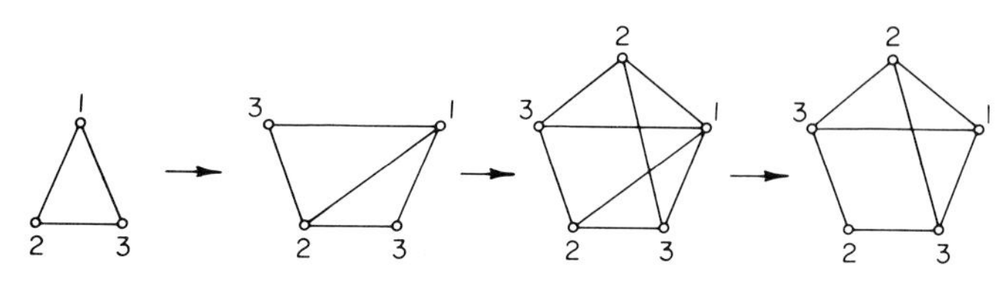

FIG. 3.

In the general case a coloring in λ colors can be constructed as soon as a complete graph on λ nodes appears during the algorithm.

6. Some Simple Properties of Graphs

Let us now consider three simple questions that one can ask concerning a graph G, namely:

(i) Is G connected?

(ii) Does G have a bridge, i.e., an edge whose removal disconnects G?

(iii) Does G have a cut node, i.e., a node whose removal (together with its incident edges) disconnects G?

One can always draw a diagram for a graph in a manner which makes clear at once the answers to these questions, and in illustrating a lecture or paper this is what one normally does. One can therefore easily fail to appreciate that there is any great difficulty associated with these problems; yet for a graph on a large number of nodes, given, shall we say, by its adjacency matrix, the answers to these questions are by no means easy to determine.

One possible approach uses more or less standard numerical matrix methods. If $(A^n)_{ij}$ denotes the (i, j)-element of the nth power of the adjacency matrix A of a graph G, then

$$(A^2)_{ij} = \sum_{k=1}^{p} a_{ik} a_{kj}$$

is the number of "walks" of length 2 between nodes i and j. For $a_{ik} a_{kj} = 1$ if and only if $a_{ik} = a_{kj} = 1$; there is then a walk from node i to node j via node k. (A walk may include a node or edge more than once.) Similarly

$$(A^r)_{ij} = \text{number of walks of length } r \text{ between nodes } i \text{ and } j. \tag{4}$$

A weaker consequence of (4) is that $(A^r)_{ij} > 0$ if and only if there is a walk of length r between nodes i and j. Consider now the matrix $A + I$. We easily see that $(A + I)_{ij} > 0$ if and only if there is a walk of length *r or less* between nodes i and j, because the 1's in the diagonal now enable a walk of length less than r to be "filled out" to a walk of length r by marking time at a suitable node. Now, if a graph is connected, there is a walk (in fact a path) between any two nodes which is of length not greater than $p - 1$. Hence we have the following result:

G is connected if and only if $(A + I)^{p-1}$ has no zero elements.

Since we are interested only in whether elements are zero or nonzero we can perform the required computation using (0, 1)-matrices throughout by modifying the arithmetic so that $1 + 1 = 1$.

Given a method such as this for testing for connectedness, we could use it to test for bridges. We remove each edge in turn and test the resulting graph for connectedness. In a similar way we can remove each node in turn to see whether it is a cut node. But these would be purely brute-force methods, liable to require much computer time. We shall now see how by a completely different approach, we can arrive at three algorithms, each an elaboration of the previous one, for tackling these three problems.

We first consider an operation which we can call "tree-felling." If a graph has nodes of valency 1 then it is either a tree or it has treelike appendages, as shown in Fig. 4.

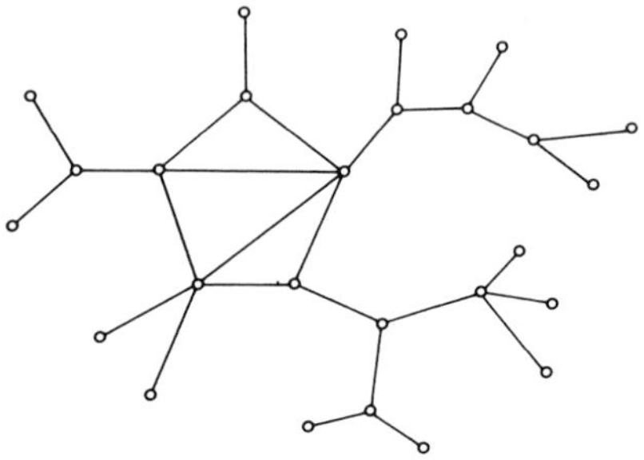

FIG. 4.

If we continually remove nodes of valency 1 (together with the incident edges) from the graph until no more remain, we shall have removed all the treelike appendages. Thus from the graph of Fig. 4 we would obtain that of Fig. 5. Although this method of felling trees, working from the top downwards,

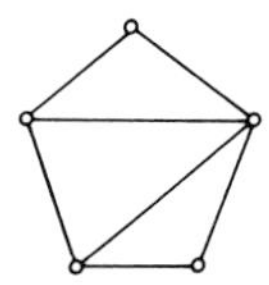

FIG. 5.

is, to say the least, rather unorthodox, it is easily and quickly carried out on a computer; we shall use it shortly.

Our test for connectedness closely resembles the well-known method for constructing a spanning tree of a graph (see, for example, [1]). We choose edges one at a time, subject only to the proviso that each new edge does not complete a circuit with those already chosen. If $p - 1$ edges can be so chosen, these form a spanning tree of the graph G, which is therefore connected. The only tricky part of this method is that of telling whether the new edge completes a circuit. A suitable algorithm is as follows (At a general stage of the algorithm some edges will have been chosen, and their incident nodes will have received labels 1, 2, 3, ..., etc.).

1. Choose a new edge if there is one. Otherwise go to Step 4.
2. Look at the end nodes of this edge.
3. (a) If neither end node is labeled, give the same, new label (i.e., one not already in use) to both of them.

 (b) If one end node only is labeled, give the same label to the other end node.

 (c) If both end nodes are labeled, and the labels are different, make these two labels equivalent.

This (3c) needs some explanation. Suppose that, starting with Fig. 6, we have reached the stage shown in Fig. 7, in which the unbroken lines denote the edges chosen so far. We now consider the dotted edge in Fig. 7. Its end nodes are labelled 1 and 2, so we make these labels equivalent. If we were doing this manually, we could just alter all the 1's to 2's (or *vice versa*) as in Fig. 8; but in a computer it is simpler merely to record that "1" and "2" are

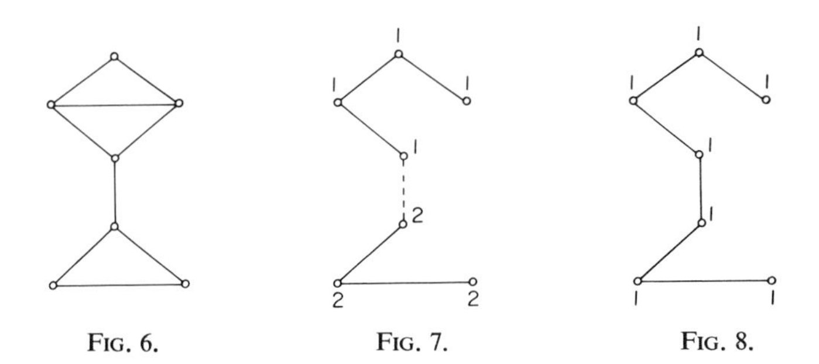

FIG. 6. FIG. 7. FIG. 8.

henceforth equivalent, i.e., are the "same" label. Thus, as the algorithm progresses, a table of equivalences is built up which can be referred to as necessary.

(d) If both end nodes are labeled, and the labels are equivalent, then the inclusion of this edge would complete a circuit. Hence we reject this edge and go to step 1.

4. (a) If $p - 1$ edges have now been chosen, G is connected.

(b) If not, and if there are no more edges to be considered, the graph is not connected.

(c) Otherwise, repeat from step 1.

If G is not connected, the equivalence classes of labels determine the components of G. In fact, the proof that the algorithm works is immediate once we note that this same remark holds, at any stage of the algorithm, for the subgraph H of G determined by the edges so far chosen.

We shall now elaborate this algorithm to make it detect bridges. It is well known that the circuits detected in Step 3d of the algorithm form a fundamental set of circuits of G. Any edge which is not a bridge will belong to at least one of these. We therefore modify Step 3d, as follows:

3. (d) If the new edge completes a circuit, we temporarily add it to the chosen set. The subgraph H determined by this set has just one circuit, which is most easily identified by carrying out a tree-felling operation on H, leaving the circuit. We note the edges forming this circuit, and *then* reject the new edge.

This time we must run through *all* edges of G, so we omit Step 4a. When the algorithm terminates, any edge which has not been noted is a bridge of a component of G. If G is connected (which the algorithm will discover in passing), then these edges are bridges of G.

A few preliminaries are needed before we devise a test for cut nodes. Consider Fig. 9, in which the edges of a spanning tree are drawn boldly, and the other edges lightly. The fundamental circuits determined [in Step 3d of the algorithm] by the nontree edges are BCD, ABD, $ADCPQR$, PSQ and QSR.

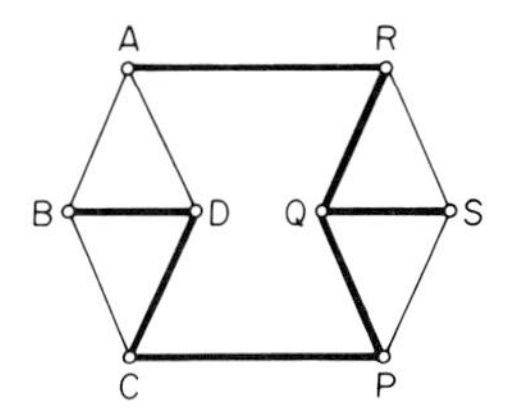

FIG. 9.

Let us call two edges which belong to a common fundamental circuit "cocyclic," and let us say that two edges are "equivalent" if they are the first and last edges of some sequence of edges in which every edge (except the first) is cocyclic with its predecessor. Thus in Fig. 9, edges AB and BD are cocyclic, and so are BD and BC. Hence, although AB and BC are not cocyclic, they *are* equivalent. The relation "equivalent" is an equivalence relation over the set of nonbridges, although "cocyclic" is not. Moreover, it is easy to show that two edges are in the same block of G if and only if they are equivalent. (A block is a maximal connected subgraph with no cut node.)

We therefore modify Step 3d again so that, instead of just noting that certain edges belong to a circuit, we keep track of the equivalences between these edges. The adjacency matrix A comes in handy for this purpose, and during the algorithm some of the 1's in it will be replaced by "labels" 2, 3, 4, ..., etc. (Note that we shall not count the original 1's as labels.) This replacing is done in such a way that, at all times, equivalent labels in A will represent equivalent edges.

When a circuit is detected in Step 3d we look at its edges and the elements of A which represent them. If these elements are all 1's we replace them all by a new label. Otherwise we make equivalent all the labels that occur, and replace by one of these equivalent labels all the 1's that occur among these elements. (For ease of exposition we shall assume that both the elements a_{ij} and a_{ji} are altered as the occasion arises. One could take always $i < j$ and manipulate only the upper triangular part of A, possibly with some advantage.)

When the algorithm terminates, we can make the following deductions:

(a) Any edge (i, j) for which $a_{ij} = 1$ is a bridge.

(b) The sets of equivalent labels in A define sets of equivalent edges, each set constituting a block of G.

(c) If all nonzero elements of row i of A are equivalent labels, then node i is not a cut node, for all edges incident to it are in the same block of G.

(d) If row i contains at least one 1, or if its nonzero elements are not all equivalent, then node i is a cut node, since it is incident with a bridge or is incident with edges of more than one block.

Thus bridges, blocks, and cut nodes can all be identified by this algorithm which, as far as I know, is original.

7. A Graph-Theoretical Programming Language

In implementing the algorithms given so far, we have been giving the computer string after string of "meaningless" instructions: odd manipulations of the adjacency matrix, labeling of nodes, and so on. The computer will carry out these instructions to the letter (as always) and may arrive at useful results. But (to speak anthropomorphically for the moment) the computer has no idea that it is solving graph-theoretical problems. If we really want to teach graph theory to a computer we must change this state of affairs. What we want is to be able to communicate with the computer in the same sort of way that we communicate with fellow graph theorists—by direct graph-theoretical statements. We want to be able to say to the computer "remove node i of G," without having to break this operation down into many separate instructions. In short, we need a programming language specifically designed for problems in graph theory.

Some progress has been made toward such a language at the University of the West Indies. Provisionally called "GTL," our language is a dialect of FORTRAN in which the usual statement set has been augmented by a large number of statements for calling graph-theoretical subroutines when required, and for handling graphs generally. Graphs and digraphs form two "types" additional to the familiar types (integer, real, etc.) of standard FORTRAN.

A complete description of the language is not yet available, and in any case would be too lengthy to detail here. The following few remarks and examples will give a general idea of how it goes.

One non-FORTRAN symbol is used, namely "@," which can usually be taken to mean "of." Thus we can write

```
N = NCOMPS@G
```

which causes the number of components of G to be computed, and stored as a fixed-point constant referred to as "N." Similarly

NC = NCYCLES@GX3

computes the number of circuits (cycles) in the graph GX3. The statement

DELETE(I)@ GA, *n*

(where n is a statement number) causes the deletion of node I from graph GA. If there is no node I, either because I is too large, or because there is a gap in the numbering (possibly as the result of a previous "DELETE" statement) control is transferred to statement number n. Similarly

DELETE(I, J)@G5, *n*

deletes the edge (i, j) of G5, or transfers control to statement n if there is no such edge.

There are also statements for transfer of control according to some graph-theoretical criterion. Thus

IF(G, PLANAR) 26, 152

transfers control to statement 26 if G is planar, and to statement 152 if it is not.

The following portion of a program, which tests whether a graph G has a node whose removal leaves a tree, will indicate how the language could be used.

```
  5  N = NNODES@G
     DO 61 I = 1, N
     GET G
 30  DELETE(I)@G, 20
     IF(G, TREE) 60, 61
 61  CONTINUE
106  ....
```

The fifth statement is the test for a tree after the deletion of node i. Statement 60 is reached if the answer to the problem is "yes," and Statement 106 is reached if the answer is "no." Statement 20 could be the start of an error routine, or perhaps a subprogram like

```
20  LABEL@G
    PUT G
    GO TO 30
```

which relabels G before continuing the DO-loop.

A word about the "GET" and "PUT" statements is in order here. Graphs are usually stored in the backing store of the computer, and brought into a working area as required. The GTL compiler keeps track of the *name* of the graph currently in this area. When a new name is mentioned the program automatically brings in the new graph from the backing store before performing the required manipulation, but if the graph name in a statement is the same as that of the graph already in the working area this is not done. This means, however, that if previous statements have modified this graph since it was originally brought in, then the operations will now be performed on the graph *as modified*. This may be what is wanted; but if, as in the above example, we want the next operation to be performed on the *original* graph, then we must call this graph from the backing store again. For this purpose a statement of the form "GET G" is used.

Statements of the form "PUT G" cause the graph currently in the working area to be stored in the backing store. If a graph with the same name is already there, this is overwritten.

A compiler for GTL is in process of being written for an IBM 1620 computer. The fact that this is a small, slow computer does not matter much in this context. The object of the exercise is to try out *some* realization of the language in order to show up deficiencies and suggest further improvements. This preliminary work should save much expensive computer time when, as we hope, GTL is later realized on a larger, modern machine. With such a language effectively developed it should become comparatively easy to persuade computers to take off our hands much of the hack work that crops up in the course of research in graph theory.

References

1. C. Berge, *Theorie des Graphes et ses Applications*, Dunod, Paris, 1958.
2. R. A. Buckingham, *Numerical Methods*, Pitman, New York, 1962.
3. R. C. Read, An Introduction to Chromatic Polynomials, *J. Combinatorial Theory* (4) **1** (1968), 52–71.
4. S. H. Unger, GIT—A Heuristic Program for Testing Pairs of Directed Line Graphs for Isomorphism, *Comm. ACM* (7) **1** (1964), 26–34.
5. A. A. Zykov, On Some Properties of Linear Complexes, Amer. Math. Soc. Transl. No. 79 (1952) (translated from *Matematicheskiĭ Sbornik* **24**, No. 66 (1949), 163–188.)

FINITE GRAPHS
(Investigations and Generalizations concerning the Construction of Finite Graphs Having Given Chromatic Number and No Triangles)

Horst Sachs

MATHEMATICAL INSTITUTE, TECHNICAL UNIVERSITY
ILMENAU, GERMAN DEMOCRATIC REPUBLIC

1. Introduction

A. *To every positive integer k there exists a graph containing no triangle and having chromatic number k.*[1]

This well-known theorem has been proved by several authors and has given rise to further interesting investigations and generalizations. To the author it seems worth while to compare systematically the different methods used; in addition, another generalization of A will be given.

2. Definitions and Notations

All graphs considered are finite, undirected graphs without loops and multiple edges. (For infinite graphs see Erdős and Rado [1a].)

A *p-clique* $\langle p \rangle$ is a complete graph with $p (p \geq 1)$ vertices.
$n(G)$ is the *number of vertices* of G.
$\chi(G)$ is the *chromatic number* of G.
$\omega(G)$ is the *density* of G, i.e., the maximum number of vertices of G generating a clique.

[1] Such graphs are characterized by $\omega(G) \leq 2$, $\chi(G) = k$ [with $\omega(G) = 2$ if $k > 1$]; they will be called "(2, k) graphs" for short.

$\sigma(G)$ is the *stability number* of G, i.e., the maximum number of vertices of G no two of which are joined by an edge.

$g(G)$ is the *girth* of G, i.e., the length of a shortest circuit of G.

G is called *critical* if the chromatic number decreases when an arbitrary edge of G is deleted.

3. k-Chromatic Graphs without Triangles

The first proofs of Theorem A were published by Descartes [1][2] in 1948 and, independently, by Zykov [2] in 1949. Zykov's very simple proof is by straightforward construction:[3]

B. *Let* $Z_1 = \langle 1 \rangle$ *and suppose* Z_k *to be already constructed. Take* k *copies* $Z_k^{\,i}$ $(i = 1, 2, \ldots, k)$ *of* Z_k *and* $[n(Z_k)]^k$ *additional vertices* v_j. *Consider all* $[n(Z_k)]^k$ k*-tuples* $T_j^{\,k}$, *each of them containing exactly one vertex out of each of the* $Z_k^{\,i}$. *For* $j = 1, 2, \ldots, [n(Z_k)]^k$, *join* v_j *with all vertices of* $T_j^{\,k}$ *by an edge. The resulting graph is* Z_{k+1}.

It is not difficult to show that for $k = 1, 2, \ldots, Z_k$ is a $(2, k)$-graph, which proves (A).

An easy computation yields

$$2^{(k-1)!} < n(Z_k) < 2^{2(k-1)!} \qquad \text{for} \quad k \geq 3. \tag{1}$$

The existence of $(2, k)$-graphs having been established one can ask for the structure of minimum $(2, k)$-graphs, i.e., of those $(2, k)$-graphs which have a minimum number n_k of vertices. The Zykov graphs are far from being minimum, but by slightly modifying Zykov's method, in 1967 Schäuble [3] constructed a sequence of $(2, k)$-graphs S_k which are minimal in a sense: They are critical (of course, the minimum graphs are critical *a fortiori*). Instead of using k copies of Z_k (as Zykov did) Schäuble takes one copy of each of the $S_1, S_2, \ldots, S_k$ for the construction of S_{k+1}. By an easy computation,

$$2^{2^{k-2}} < n(S_k) < 2^{2^{k-1}} \qquad \text{for} \qquad k \geq 3, \tag{2}$$

which is much better than (1), but still "bad enough." That for $k \geq 3$ the Z_k are not critical follows immediately from the fact that $S_k \subset Z_k$ and $S_k \neq Z_k$ for $k \geq 3$.

Another ingenious method of constructing $(2, k)$-graphs M_k was given Mycielski [4][4] in 1955:

[2] Unfortunately, the author had no access to [1].

[3] At the 3rd Waterloo Conference, 1968 the author learned that the same idea was given by J. B. Kelly and L. M. Kelly.

[4] As far as the author knows [4] was published by its author without having knowledge of [1, 2, 5, or 6].

C. *Let* $M_2 = \langle 2 \rangle$ *and suppose* M_k $(k \geq 2)$ *to be already constructed. Corresponding to every vertex* x_i *of* M_k, *take a new vertex* y_i *and join it with the neighboring vertices of* x_i *in* M_k $[i = 1, 2, \ldots, n(M_k)]$. *Join an additional vertex* z *with each of the* y_i.

The resulting graph is M_{k+1}.

It can easily be shown that for $k = 2, 3, \ldots, M_k$ is a $(2, k)$-graph with

$$n(M_k) = 3 \cdot 2^{k-2} - 1. \tag{3}$$

Equation (3) is considerably better than (2). Furthermore, M. Schäuble [3] proved:

(C′) M_k *is critical* $(k \geq 2)$.

The three methods of construction mentioned—together with another one of Descartes [5] and Kelly–Kelly [6], to be described later—rely on recursive algorithms and yield the graphs wanted in a straightforward manner. The number of vertices increases at least exponentially with k, as must be expected for recursively working procedures. The question may be raised as to whether there is any constructive method yielding much better upper bounds for n_k than the Mycielski construction does. That in fact there is such a method was shown by P. Erdős.

In 1957 and 1966, P. Erdős [7, 8] gave two nonrecursive procedures for constructing $(2, k)$-graphs.[5] Both methods, though very different in details, are based on very similar ideas. Illustrating in a beautiful way how propositions of metric geometry may serve for obtaining combinatorial statements, these methods deserve particular interest. Therefore, the first of them shall be developed here.

D. *To every pair of sufficiently large positive numbers* d *and* s (d *being an integer*) *a graph* $E(d, s)$ *is constructed in the following way: The vertices of* $E(d, s)$ *are those lattice points of the* d-*dimensional Euclidean space* $\mathbf{S}_d$ *whose distance from the origin does not exceed* $R = s\sqrt{d}$. *Any two distinct vertices of* $E(d, s)$ *are joined by an edge if and only if their Euclidean distance exceeds* $R\sqrt{3} = s\sqrt{3d}$. *Then, evidently,*

(a) $E(d, s)$ *contains no triangle, and*

(b) *the diameter of any stable set does not exceed* $s\sqrt{3d}$.

Establishing a one-to-one correspondence between lattice points and lattice cubes and making use of a well-known theorem of Kubota and

[5] The possibility of the second construction was already indicated in 1964 by Erdős and Hajnal [9].

Urysohn concerning convex bodies, one deduces from (b) that $\sigma(E(d, s))$ is not greater than the volume of a sphere of radius

$$\frac{1}{2}(R\sqrt{3} + 2\sqrt{d}) = \left(\frac{s}{2}\sqrt{3} + 1\right)\sqrt{d}.$$

On the other hand, $n(E(d, s))$ is not less than the volume of a sphere of radius

$$R - 2\sqrt{d} = (s - 2)\sqrt{d}.$$

If now s is chosen such that

$$\frac{s}{2}\sqrt{3} + 1 < s - 2$$

(which is certainly true for $s \geq 15$), then, obviously, $n(E(d, s))/\sigma(E(d, s))$ will exceed every bound when d increases without bound. Consequently, because $\chi(G) \geq n(G)/\sigma(G)$, we have:

D′. *To every positive integer k there exists a graph*

$$E(d_k, s) \qquad \text{with} \quad \chi(Ed_k, s)) \geq k.$$

By deleting (if necessary) suitably chosen vertices from $E(d_k, s)$ a $(2, k)$-graph can be derived. Minimizing $n(E(d, s))$ under the constraint

$$\chi(E(d, s)) \geq k$$

yields after due computation, the existence of a $(2, k)$-graph E_k with

$$n(E_k) < 82 \cdot k^{50} \quad (k = 1, 2, \ldots).^6 \tag{4}$$

The first method of Erdős however, does not allow us to replace the exponent 50 by a smaller integer. The second method of Erdös is based on the following discrete "cube space" S_d^*:

The 2^d vertices of S_d^* are in one-to-one correspondence with the 2^d vertices a d-dimensional unity cube D_d of S_d. The distance of two points is defined as the Euclidean length of the shortest way consisting of edges of C_d only and connecting the corresponding vertices of C_d.

The theorem of Kubota and Urysohn is replaced by a theorem of Kleitman [10]. After some calculation[6] one obtains:

E. *To every positive integer k the second* Erdös *construction yields a $(2, k)$-grap'ı E_k^* with*

$$n(E_k^*) < k^{13} \qquad \text{for sufficiently large} \quad k. \tag{5}$$

[6] Compare Schäuble [3].

These two constructive but nonrecursive methods give considerably better upper bounds for n_k than the recursive methods do. On the other hand, they are "less constructive" in as much as, in general, in order to reduce the chromatic number a subsequent amendment is necessary for which no effectively described algorithm is given. It is not very likely that much better bounds can be achieved by means of constructive methods at all. An essential improvement, however, is possible indeed if the search for effective constructions is abandoned altogether. Thus Erdős [11, 12], using certain ideas contained in [13] succeeded in developing very general and far-reaching counting methods (which he called "probabilistic arguments") by means of which he was able to show:

F. *To every $\varepsilon > 0$ an integer $k_0(\varepsilon)$ can be calculated such that for every positive integer $k > k_0(\varepsilon)$ there exists a $(2, k)$-graph G_k with less than $k^{2+\varepsilon}$ vertices*:

$$n(G_k) < k^{2+\varepsilon} \qquad \textit{for} \quad k \geq k_0(\varepsilon). \tag{6}$$

Even these pure counting methods may of course be called "constructive" in as much as they allow us to calculate an upper bound for the maximum number of steps to be performed in order to find a $(2, k)$-graph by some plain testing method.

4. k-Chromatic Graphs with Given Girth

For $k \geq 3$, the Z_k, S_k, M_k all contain circuits of length 4. Mycielski [4, Problem 131] raises the question as to whether to given positive integers $k \geq 2$ and $m \geq 3$ there exist k-chromatic graphs having girth greater than or equal to $m + 1$, i.e., containing no circuits of length less than or equal to m. In 1959, this question was answered affirmatively by Erdős [11] using probabilistic arguments.[7] To set up effective algorithms for constructing such graphs however, seems to be a very hard task—except for some special values of m. For $m = 5$, the problem was solved in 1947 by Descartes [1] and in 1955 independently by Kelly–Kelly [6], who invented an ingenious recursive algorithm which is, in a way, related to (B):

G. *Let T_2 be a circuit of length* 6 *and suppose T_k to be already constructed. Put*

$$A_k = \binom{(n(T_k) - 1) \cdot k + 1}{n(T_k)}.$$

[7] Analogous results concerning set systems (generalized graphs) were obtained in 1966 by P. Erdős and A. Hajnal [14].

Take A_k copies $T_k{}^i$ $(i = 1, 2, \ldots, A_k)$ of T_k and a set M of $(n(T_k) - 1)k + 1$ additional vertices. Then M contains A_k distinct subsets M_i $(i = 1, 2, \ldots, A_k)$ with $|M_i| = n(T_k)$. For $i = 1, 2, \ldots, A_k$, join $T_k{}^i$ with M_i by exactly $n(T_k)$ edges such that every vertex of $T_k{}^i$ is joined with exactly one vertex of M_1, and vice versa. The resulting graph is T_{k+1}.

Induction on k yields the result that each of the T_k has girth 6. Obviously, $\chi(T_2) = 2$; suppose $\chi(T_k) = k$ is already proved. Then $\chi(T_{k+1}) = k + 1$, for

(a) T_{k+1} can be admissibly colored with $k + 1$ colors: Color all $T_k{}^i$ with colors $1, 2, \ldots, k$ and color the vertices of M with color $k + 1$;

(b) T_{k+1} cannot be colored admissibly with k (or fewer) colors for otherwise (by virtue of Dirchlet's "Schubfach" principle) all vertices of at least one of the M_i would have the same color—which obviously leads to a contradiction.

The case $m = 7$ was treated in 1966 by Nešetřil [15] who found a recursive procedure for constructing k-chromatic graphs N_k having girth greater than or equal to 8. His method is related to (G) but is much more complicated, and in every step of the recursion the chromatic number of the graph constructed so far has to be determined.

The corresponding problem for set systems was solved in 1967 by Lovász [16], a young Hungarian student who developed a constructive procedure yielding, for any three positive integers $q \geq 2$, $m \geq 3$, and $k \geq 1$, a q-uniform k-chromatic set system having no nontrivial circuits of length less than or equal to m. Thus, in principle, the problem is also settled for graphs that correspond to 2-uniform set systems. Lovász, however, ends his paper with the following remark: "It is interesting that if I wanted to give the construction only for graphs I ought to use set systems, too."

Thus the question of an effective, purely graph-theoretic algorithm for constructing graphs of given chromatic number k and given girth $m + 1$ still remains open

5. A Generalization of (A)

In this section another generalization of the original problem will be considered.

Call a graph G *monochromatically colored* or, for short, *monocolored*, if each of its vertices has the same color. For $p \geq 2$, call a coloration of G *p-admissible* if G contains no monocolored p-clique. Denote by $\chi_p(G)$ the minimum number of colors with which G can be p-admissibly colored. Obviously,

$$\chi_2(G) = \chi(G).$$

Call G a (p, k)-graph if G contains no $(p+1)$-clique and $\chi_p(G) = k$. If G is a (p, k)-graph with $k > 1$, then evidently $\omega(G) = p$. The problem is to prove (or disprove) the existence of (p, k)-graphs with given $p \geq 2$, $k \geq 1$, and if possible, to do this by a straightforward construction.[8] In 1967, M. Schäuble and the author [17][9] gave a construction which is, in a sense, an extension of (B) yielding a little bit more, namely the following theorem:

H. *To every triple of integers* $p \geq 2$, $k \geq 1$, and $K \geq k$, *there can be constructed a graph* $G = G(p, k, K)$ *having the following properties:*

(1) G *contains no* $(p+1)$*-clique;*

(2) $\chi_p(G) = k$;

(3) *if* G *is p-admissibly colored with* K (*or fewer*) *colors then* G *contains* k (*or more*) *pairwise distinctly monocolored* $(p-1)$*-cliques.*

PROOF. If $K = 1$ then $k = 1$, and the problem becomes trivial. We may stipulate $G(p, 1, 1)$ to be a $(p-1)$-clique. Assume $K \geq 2$.

For $p = 2$, (H) reduces to (A); we may therefore choose $G(2, k, K) = Z_k$.

INDUCTION ON p. Suppose

(I) graphs $G(p, k, K)$ to be already constructed for $p = 2, 3, \ldots, P$ $(P \geq 2)$ and for all k, K with $1 \leq k \leq K$ $(K \geq 2)$.

By virtue of (I), there is a $G(P, K+1, K+1)$. Obviously, it is admissible to put

$$G(P+1, l, K) = G(P, K+1, K+1).$$

INDUCTION ON k. Suppose

(II) graphs $G(P+1, k, K)$ to be already constructed for

$$k = 1, 2, \ldots, l \; (1 \leq l \leq K-1).$$

We shall construct a graph $G(P+1, l, +1, K)$. Put for short

$$G_l \doteq G(P+1, l\, K).$$

Let q_l denote the number of distinct P-cliques contained in G_1. Take l copies $G_l{}^i$ $(i = 1, 2, \ldots, l)$ of G_l and a set M_l of $q_l{}^i$ additional vertices v_j. Consider all $q_l{}^i$ l-tuples $T_j^{l,P}$ of P-cliques, each of them containing exactly one P-clique

[8] The more existence may be established by using the "probabilistic arguments" of Erdős and Hajnal [14].

[9] A method of constructing (p, k)-graphs can also be deduced from [16], which is independent of [17].

out of each of the $G_l{}^i$. For $j = 1, 2, \ldots, q_l{}^i$, join v_j with all vertices of all P-cliques of $T_j^{l,P}$ by an edge. Call the resulting graph

$$\tilde{G}_l = \{G_l{}^1, \ldots, G_l{}^l; M_l\}.$$

Evidently, $\tilde{G}_l$ contains no $(P + 2)$-clique.

Put $H = G(P, K - l + 1, K - l + 1)$ [which, by virtue of (I), can be constructed] and denote the vertices of H by $x_1, x_2, \ldots, x_t$. Take t copies $\tilde{G}_l{}^\tau = \{G_l^{1\tau}, \ldots, G_l^{l\tau}; M_l{}^\tau\}$ $(\tau = 1, 2, \ldots, t)$ of $\tilde{G}_l$ and join two vertices of $M_l{}^\sigma$ and $M_l{}^\tau$ $(\sigma \neq \tau)$ by an edge if and only if x_σ and x_τ are joined in H.

Assertion. The resulting graph G^0 has properties (1), (2), and (3).

Proof of (I): This is obvious.

Proof of (II):

(a) Assume G^0 to be $(P + 1)$-admissibly colored with l colors $1, 2, \ldots, l$. Then also each of its subgraphs $\tilde{G}_l{}^\tau$ $(\tau = 1, 2, \ldots, t)$ and in these each of the, $\tilde{G}_l^{\lambda\tau}$ $(\lambda = 1, 2, \ldots, l)$, is $(P + 1)$-admissibly colored with colors $1, 2, \ldots, l$.

Being isomorphic with $G_l = G(P + 1, l, K)$, each of the $G_l^{\lambda\tau}$ has property (3) (with $p = P + 1$, $k = l$), i.e., it contains l pairwise distinctly monocolored P-cliques. Consequently, $M_l{}^\tau$ contains a vertex v which is joined with all vertices of a P-clique of $G_l^{\lambda\tau}$, all of them having color λ simultaneously for $\lambda = 1, 2, \ldots, l$. Since v has one of the colors $1, 2, \ldots, l$, we obtain a monocolored $(P + 1)$-clique, contradicting our assumption. Consequently,

$$\chi_{P+1}(G^0) \geq l + 1.$$

(b) G^0 can be $(P + 1)$-admissibly colored with $l + 1$ colors as follows. Color all of the $G_l^{\lambda\tau}$ $(P + 1)$-admissibly with colors $1, 2, \ldots, l$ and give color $l + 1$ to the remaining vertices. Each of the $(P + 1)$-cliques being contained in one of the $\tilde{G}_l{}^\tau$, no monocolored $(P + 1)$-clique will be created. Consequently,

$$\chi_{P+1}(G^0) \geq 1 + 1.$$

From (a) and (b) the proof of (2) follows.

Proof of (3): Assume G^0 to be $(P + 1)$-admissibly colored with K (or fewer) colors such that there are no $l + 1$ pairwise distinctly monocolored P-cliques. Being isomorphic with $G_l = G(P + 1, l, K)$, each of the $G_l^{\lambda\tau}$ contains l pairwise distinctly monocolored P-cliques. As a consequence of the assumption, each of these P-cliques has one out of exactly l different colors, $1, 2, \ldots, l$, say. The coloration being $(P + 1)$-admissible, in each of the $M_l{}^\tau$ there is a vertex x_τ^* having none of the colors $1, 2, \ldots, l$ and consequently having one of $K - l$ colors $l + 1, \ldots, K$. The vertices x_τ^* ($\tau = 1, 2, \ldots, t$) generate a subgraph H^* of G^0 isomorphic with $H = G(P, K - l + 1, K - l + 1)$ and being colored with $K - l$ colors $l + 1, \ldots, K$. Since $\chi_p(H) = K - l + 1$, H^* contains a P-clique monocolored with one of the colors $l + 1, \ldots, K$. Thus we still have found $l + 1$ pairwise distinctly monocolored P-cliques contradicting the assumption.

The assertion having been proved we may complete the construction by putting

$$G^0 = G(P + 1, l + 1, K).$$

By virtue of the twofold induction the theorem is now proved.

References

1a. P. Erdős and R. Rado, A Construction of Graphs without Triangles having Pre-assigned Order and Chromatic Number, *J. London Math. Soc.* **35** (1963), 445–448.
1. B. Descartes, A Three Colour Problem, *Eureka* (April 1947), Solution (March 1948).
2. A. A. Zykov, On Some Properties of Linear Complexes (Russian) *Math. Sb.* **24** (1949), 163–188.
3. M. Schäuble, Beiträge zum Problem der Existenz und Konstruktion endlicher Graphen gegebener Chromatischer Zahl, die gewissen Strukturbedingungen genügen, und zu verwandten Problemen, Dissertation, Ilmenau, 1968.
4. J. Mycielski, Sur le Coloriage des Graphes, *Colloq. Math.* **3** (1955), 161–162.
5. B. Descartes, Solution to Advanced Problem No. 4525, *Amer. Math. Monthly* **61** (1954), 532.
6. J. B. Kelly and L. M. Kelly, Paths and Circuits in Critical Graphs, *Amer. J. Math.* **76** (1954), 791–792.
7. P. Erdős, Remarks on a Theorem of Ramsey, *Bull. Res. Council Israel, Sect. F* **7**, No. 1 (1957).
8. P. Erdős, On the Construction of Certain Graphs, *J. Combinatorial Theory* **1**, No. 1 (1966).
9. P. Erdős and A. Hajnal, On Complete Topological Subgraphs of Certain Graphs, *Ann. Univ. Sci. Budapest. Eötvös Sect. Math.* **7** (1964).
10. D. J. Kleitman, Families of Nondisjoint Subsets, *J. Combinatorial Theory* **1** (1966) 153–155.
11. P. Erdős, Graph Theory and Probability, *Canad. J. Math.* **11** (1959), 34–38.
12. P. Erdős, Graph Theory and Probability II, *Canad. J. Math.* **13** (1961), 346–352.
13. P. Erdős and A. Renyi, On the Evolution of Random Graphs, *Publ. Inst. Hungar. Acad. Sci.* **5** (1960), 17–61.
14. P. Erdős and A. Hajnal, On Chromatic Number of Graphs and Set-Systems, *Acta Math Hungar.* **17** (1966) 1–2.

15. J. Nešetřil, On k-Chromatic Graphs without Circuits of a Length ≤ 7 (Russian), *Comm. Math. Univ. Carolinae* 7 (1966) 3.
16. L. Lovász, Graphs and Set-Systems, *Beiträge Graphentheorie, vorgetragen intern. Kolloq. Manebach*, 1967, Leipzig, 1968.
17. H. Sachs and M. Schäuble, Über die Konstruktion von Graphen mit gewissen Färbungseigenschaften, Beiträge Graphentheorie, vorgetragen *intern. Kolloq. Manebach*, 1967, Leipzig, 1968.

STRONGLY REGULAR GRAPHS

J. J. Seidel

DEPARTMENT OF MATHEMATICS
TECHNOLOGICAL UNIVERSITY EINDHOVEN
THE NETHERLANDS

INTRODUCTION

In Section 1, four introductory examples, described by their $(-1, 1)$ adjacency matrix, will be used to illustrate the concept of strong graphs (Section 2), a generalization of Bose's strongly regular graphs. The operation of complementation (Section 3) serves to relate certain distinct strongly regular graphs. Several results concerning the existence and the uniqueness of strongly regular graphs are collected and commented upon in Section 4. By geometric methods an additional graph will be constructed in Section 5.

1. EXAMPLES

The graphs considered are undirected, without loops and without multiple edges. We make use of adjacency matrices which have elements 0 on the diagonal, -1 and $+1$ elsewhere, according as the corresponding vertices are adjacent or nonadjacent, respectively (cf. [12, 13]). I denotes the unit matrix, J the all-one matrix, and $\mathbf{0}$ the all-zero matrix, of some order.

EXAMPLE 1.1. The *pentagon graph* (Fig. 1) consists of five vertices with cyclic adjacencies:

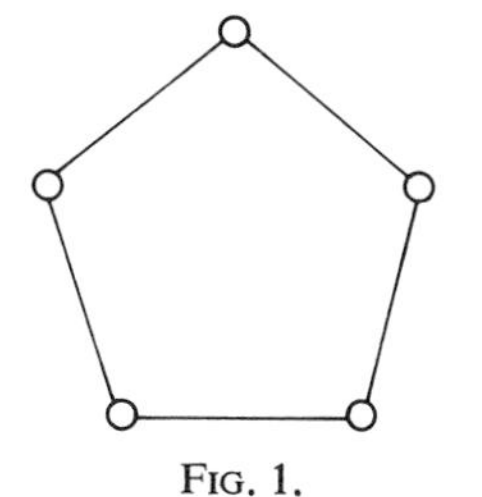

FIG. 1.

$$A_5 = \begin{bmatrix} 0 & - & + & + & - \\ - & 0 & - & + & + \\ + & - & 0 & - & + \\ + & + & - & 0 & - \\ - & + & + & - & 0 \end{bmatrix}.$$

Its adjacency matrix A_5 satisfies

$$A_5^2 = 5I - J, \qquad A_5 J = 0$$

and has the eigenvalues $0, \sqrt{5}, \sqrt{5}, -\sqrt{5}, -\sqrt{5}$.

EXAMPLE 1.2. The *Petersen graph* (Fig. 2) consists of the ten unordered pairs out of five symbols, any two pairs being adjacent if and only if they have no common symbol:

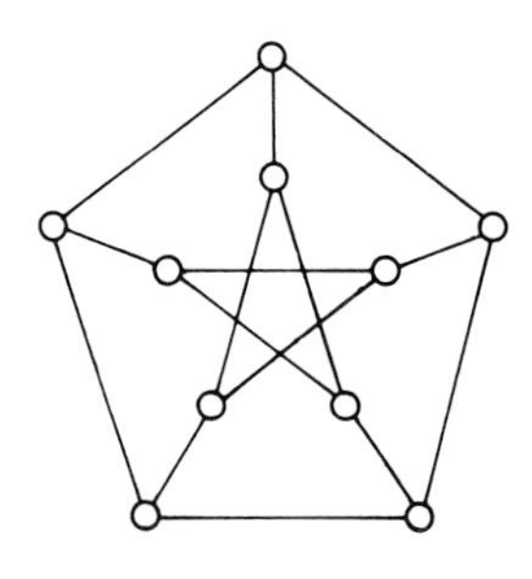

FIG. 2.

$$A_{10} = \begin{bmatrix} A_5 & J - 2I \\ J - 2I & -A_5 \end{bmatrix}.$$

Its adjacency matrix A_{10} satisfies

$$A_{10}^2 = 9I, \qquad A_{10} J = 3J$$

and has the eigenvalues -3 and 3, both of multiplicity 5.

EXAMPLE 1.3. The *complement of the Clebsch graph* (Fig. 3) consists of two subgraphs, a 5-claw and a Petersen graph on five symbols, whose adjacencies are described by inclusion if the end vertices of the 5-claw are taken as the symbols of the Petersen graph:

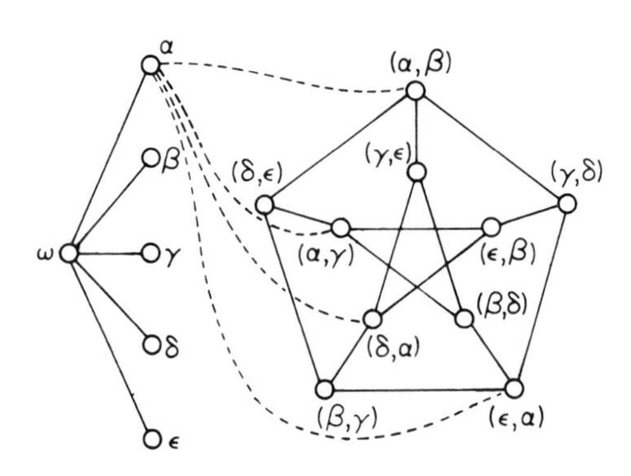

FIG. 3.

$$A_{16} = \begin{bmatrix} 0 & \text{——} & \text{—|—} & \text{—|—} \\ | & J-I & I+A_5 & I-A_5 \\ + & I+A_5 & A_5 & J-2I \\ + & I-A_5 & J-2I & -A_5 \end{bmatrix}.$$

Its adjacency matrix A_{16} satisfies

$$(A_{16}+3I)(A_{16}-5I)=\mathbf{0}, \qquad A_{16}J=5J$$

and has the eigenvalues -3 and 5, of multiplicities 10 and 6, respectively. The Clebsch graph itself has the adjacency matrix $-A_{16}$. Alternatively, as was remarked by Coxeter (cf. [14]), the Clebsch graph may be defined by the sixteen lines on the Clebsch quartic surface, any pair of lines being adjacent if and only if they are skew (cf. [5]).

EXAMPLE 1.4. The following *latin square graph* $L_3(5)$ [Fig. 4] consists of a latin square of order 5 superimposed onto the lattice graph $L_2(5)$:

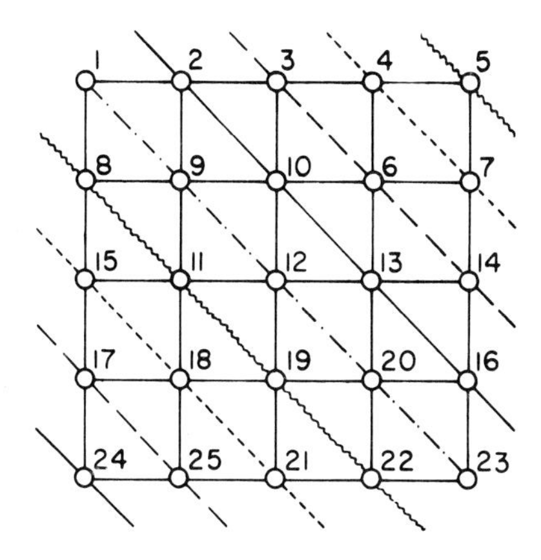

FIG. 4.

$$A_{25} = \begin{bmatrix} I-J & I-A_5 & I+A_5 & I+A_5 & I-A_5 \\ I-A_5 & I-J & I-A_5 & I+A_5 & I+A_5 \\ I+A_5 & I-A_5 & I-J & I-A_5 & I+A_5 \\ I+A_5 & I+A_5 & I-A_5 & I-J & I-A_5 \\ I-A_5 & I+A_5 & I+A_5 & I-A_5 & I-J \end{bmatrix}.$$

Its adjacency matrix A_{25} satisfies

$$A_{25}^2 = 25\,I - J, \qquad A_{25}\,J = \mathbf{0}$$

and has the eigenvalues 0, 5, -5 of multiplicities 1, 12, 12, respectively.

We recall the definition of the graphs $L_2(n)$ and $T(n)$.

The *lattice graph* $L_2(n)$ of order n, $n > 1$, consists of the n^2 ordered pairs out of n symbols, any two pairs being adjacent if and only if they have one symbol in common.

The *triangular graph* $T(n)$ of order n, $n > 3$, consists of the $\frac{1}{2}n(n-1)$ unordered pairs out of n symbols, any two pairs being adjacent if and only if they have one symbol in common.

2. Strong Graphs

A nonvoid and noncomplete graph on v vertices is called a *strong graph* (cf. [14]) if its adjacency matrix A satisfies

$$(A - \rho_1 I)(A - \rho_2 I) = (v - 1 + \rho_1\rho_2)J,$$
$$\rho_1 \text{ and } \rho_2 \text{ real numbers}, \qquad \rho_1 > \rho_2 .$$

If $v - 1 + \rho_1\rho_2 \neq 0$, then J is a linear combination of A^2, A, I. Hence these four matrices are simultaneously diagonalizable. The vector $(1, \ldots, 1)$ is an eigenvector of J belonging to the eigenvalue v, and hence an eigenvector of A belonging to the eigenvalue ρ_0, say. Therefore, in this case, the graph is regular. The only other eigenvalues of A are ρ_1 and ρ_2.

If $v - 1 + \rho_1\rho_2 = 0$, then there are exactly two distinct eigenvalues ρ_1 and ρ_2. Let their multiplicities be μ_1 and μ_2, respectively. From tr $A = 0$ it follows that

$$(\mu_1 - \mu_2)(\rho_1 - \rho_2) + (\mu_1 + \mu_2)(\rho_1 + \rho_2) = 0.$$

If $\mu_1 = \mu_2$, then $\rho_1 = -\rho_2 = (v-1)^{1/2}$. If $\mu_1 \neq \mu_2$ then $\rho_1 - \rho_2$ is rational. For strong graphs of this type, which are not any complete bipartite graph $K(k, v-k)$ or its complement, it readily follows (cf. [14]) that ρ_1 and ρ_2 are odd integers. For $v - 1 + \rho_1\rho_2 \neq 0$, analogous conclusions may be drawn from consideration of the multiplicities. Summarizing we have the following, partly overlapping, types of strong graphs:

1. $A^2 = (v-1)I$;
2. $(A - \rho_1 I)(A - \rho_2 I) = \mathbf{0}$, for odd integer ρ_1, ρ_2; and
3. $(A - \rho_1 I)(A - \rho_2 I) = (v - 1 + \rho_1\rho_2)J \neq \mathbf{0}$, $AJ = \rho_0 J$, for integer ρ_0 and odd integer ρ_1, ρ_2.

An equivalent set theoretic definition of strong graphs has been given in [14]. Let x and y be any pair of adjacent [nonadjacent] vertices of a graph, let p_1 [p_2] denote the number of vertices adjacent to x and nonadjacent to y, and let q_1 [q_2] denote the number of vertices nonadjacent to x and adjacent to y, (see Fig. 5).

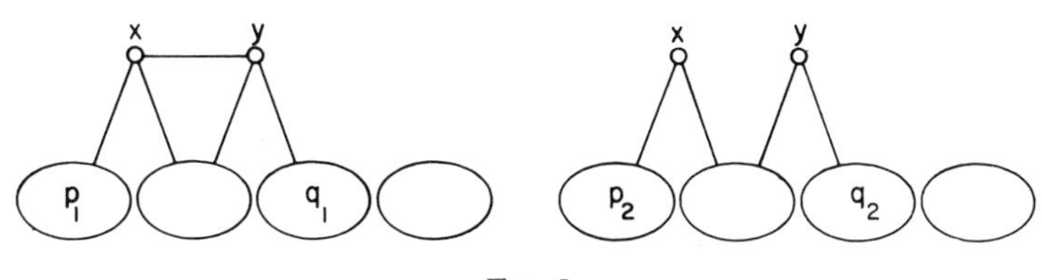

FIG. 5.

A graph is strong if and only if the integers

$$p_1 + q_1 \qquad \text{and} \qquad p_2 + q_2$$

are independent of the choice of x and y (cf. [14, Theorem 4]). These numbers are related to the eigenvalues ρ_1 and ρ_2 by

$$2(p_1 + q_1) = (\rho_1 - 1)(1 - \rho_2), \qquad 2(p_2 + q_2) = -(\rho_1 + 1)(1 + \rho_2).$$

Clearly strong graphs are regular if and only if each of the four integers p_1, q_1, p_2, q_2 is independent of the choice of x and y. The regular strong graphs are precisely the *strongly regular graphs* of Bose [2]. Therefore, the notion of a strong graph is a generalization of Bose's concept.

EXAMPLE 2.1. To the pentagon graph of Example 1.1 one isolated vertex (Fig. 6) is added:

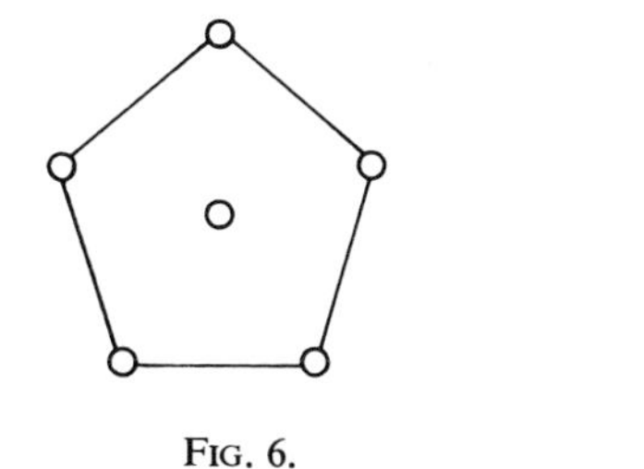

FIG. 6.

$$A_6 = \begin{bmatrix} 0 & -\!\!\mid\!\!- \\ \mid & A_5 \end{bmatrix}.$$

Then a strong graph of type 1 is obtained with

$$A_6^2 = 5I, \qquad p_1 + q_1 = 2, \qquad p_2 + q_2 = 2,$$

which is not regular.

EXAMPLE 2.2. The Petersen graph of Example 1.2 is a strong graph of types 1 and 2, which is regular.

EXAMPLE 2.3. The complement of the Clebsch graph of Example 1.3 is a strong graph of type 2, which is regular.

EXAMPLE 2.4. To the latin square graph of Example 1.4 one isolated vertex is added. Then a strong graph of types 1 and 2 is obtained with

$$A_{26}^2 = 25I, \qquad p_1 + q_1 = 12, \qquad p_2 = q_2 = 12.$$

This graph is not regular.

3. THE OPERATION OF COMPLEMENTATION

Let x be any vertex of any graph. *Complementation with respect to x* of the graph is defined to be the following operation: cancel all existing adjacencies to x and add all nonexisting adjacencies to x. The effect of complementation with respect to x on the adjacency matrix of the graph is that the row and the column corresponding to x are multiplied by -1. The operations of complementation with respect to any number of vertices generate an equivalence relation on the set of all graphs on v vertices (cf. [12], which contains tables of all classes of mutually equivalent graphs for $v = 2, 3, \ldots, 7$).

EXAMPLE 3.1 (Fig. 7).

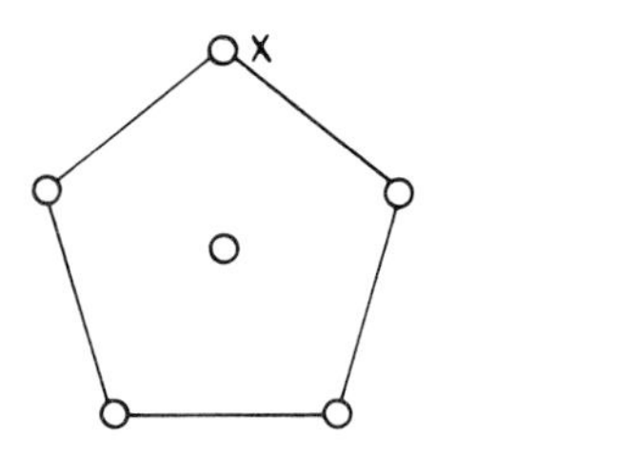

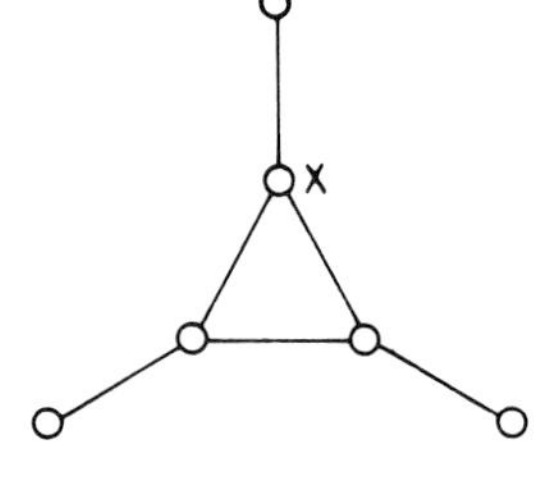

FIG. 7.

The right hand graph is the result of complementation with respect to x of the left hand graph. Both graphs satisfy the equation $A^2 = 5I$.

EXAMPLE 3.2 (Fig. 8). The complement of the Petersen graph is the triangular graph $T(5)$. The adjacency matrix of $T(5)$ is $-A_{10}$, where A_{10} is the adjacency matrix of the Petersen graph.

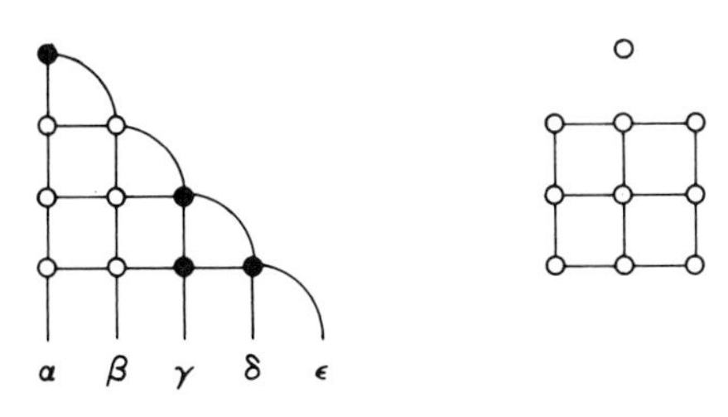

FIG. 8.

Complementation of $T(5)$ with respect to the dotted vertices yields the graph on the right hand side, which consists of the lattice graph $L_2(3)$ and one isolated vertex. The adjacency matrix A_9 of $L_2(3)$ satisfies

$$\begin{bmatrix} 0 & -|- \\ -|- & A_9 \end{bmatrix}^2 = 9I, \qquad A_9^2 = 9I - J, \qquad A_9 J = \mathbf{0}.$$

EXAMPLE 3.3. The adjacency matrix A of the Clebsch graph, explained in Example 1.3, satisfies

$$(A - 3I)(A + 5I) = \mathbf{0}, \qquad AJ = -5J.$$

This regular graph is equivalent to the graph consisting of the triangular graph $T(6)$ and one isolated vertex. This can be seen by complementation of the latter graph with respect to the dotted vertices (see Fig. 9):

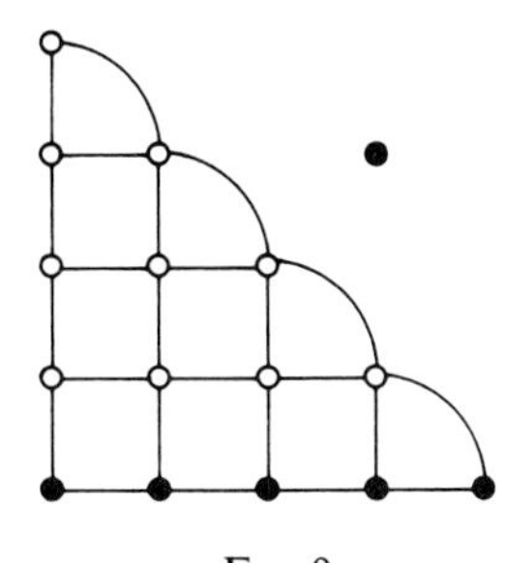

FIG. 9.

The latter graph is not regular. However, it is possible to perform complementations in two different ways so as to obtain regular graphs, which are not the Clebsch graph (Figs. 10 and 11).

The first way of doing so is by complementation with respect to the six dotted vertices (Fig. 10), which form two independent triangles. Then the lattice graph $L_2(4)$ is obtained, which is strongly regular with

$$(A - 3I)(A + 5I) = \mathbf{0}, \qquad AJ = 3J,$$

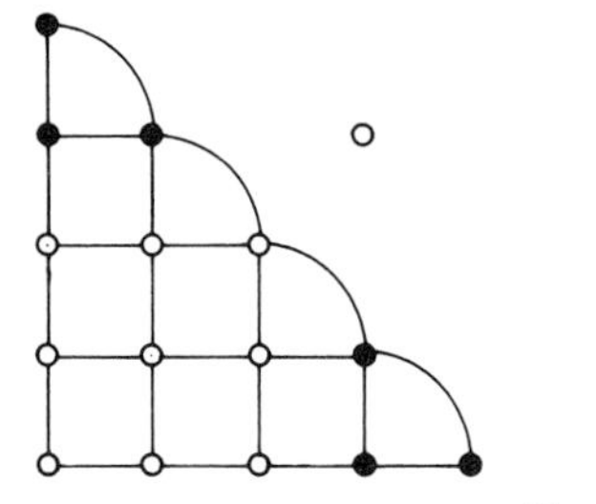

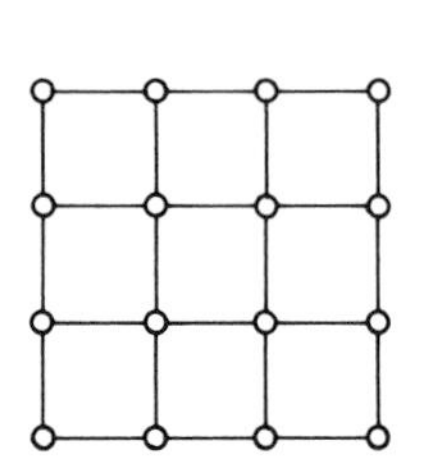

FIG. 10.

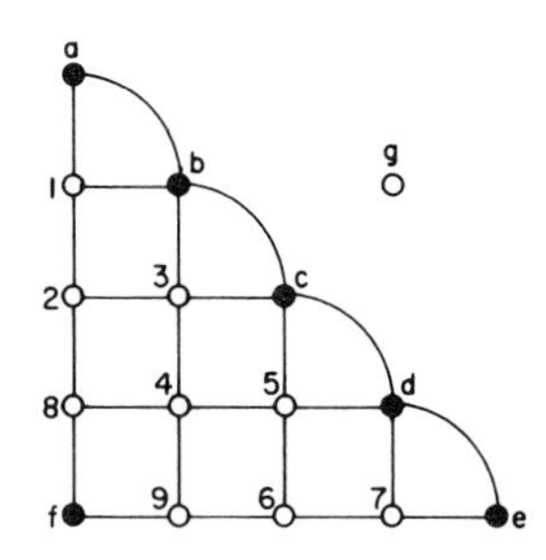

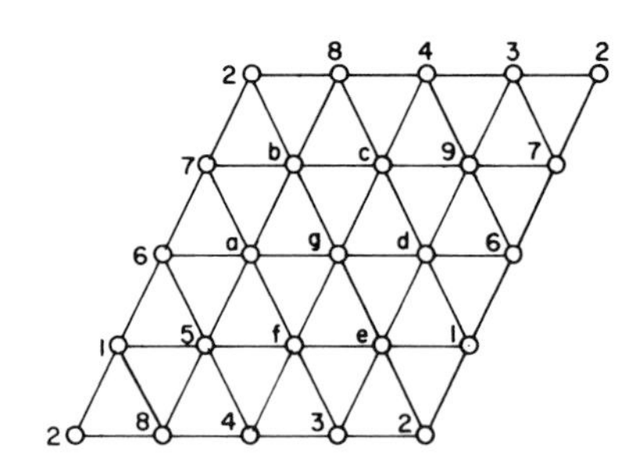

FIG. 11.

The second way of doing so is by complementation with respect to the dotted vertices a, b, c, d, e, f (Fig. 11), which form a hexagon subgraph. Then a graph of order 16 is obtained, which can be inbedded in the torus. This graph, which also is strongly regular with

$$(\mathrm{A} - 3I)(A + 5I) = \mathbf{0}, \qquad AJ = 3J,$$

is the complement of the graph obtained by superimposing onto $L_2(4)$ the nonextendable latin square of order 4.

EXAMPLE 3.4. The graph of Example 2.4, which is not regular, can be complemented in such a way that the resulting graph is regular. Indeed, the graph of order 26 with adjacency matrix

$$A_{26} = \begin{bmatrix} 0 & \text{—|—} & \text{—|—} & \text{———} & \text{———} & \text{———} \\ \text{—|—} & I-J & I-A & -I-A & -I-A & -I+A \\ \text{—|—} & I-A & I-J & -I+A & -I-A & -I-A \\ | & -I-A & -I+A & I-J & I-A & I+A \\ | & -I-A & -I-A & I-A & I-J & I-A \\ | & -I+A & -I-A & I+A & I-A & I-J \end{bmatrix}$$

is regular and satisfies

$$A_{26}^2 = 25I, \qquad A_{26}\,J = -5J.$$

4. Existence and Uniqueness

We summarize some results concerning the existence and the uniqueness of strongly regular graphs. The formulas stated in 4.2–4.9 may be derived by straightforward calculation, by use of the relations between the integers p_1, q_1, p_2, q_2 and the eigenvalues ρ_1, ρ_2, which were given in Section 2.

4.1 For the existence of strong graphs of order v of type 1 the following necessary conditions are known:

$$v \equiv 2 \pmod 4; \qquad v - 1 = a^2 + b^2, \qquad a \text{ and } b \text{ integers.}$$

These graphs have been constructed for the orders

$$v - 1 = p^\alpha \equiv 1 \pmod 4, \qquad p \text{ prime,}$$

and for some additional orders, e.g., for $v = 226$. The smallest order for which the existence is unknown is $v = 46$. For details the reader is referred to [7].

4.2 The graphs $H(n)$, $n > 3$, obtained from the complete graph on $2n$ vertices by deleting a 1-factor, constitute an infinite class of strongly regular graphs of type 3. These graphs satisfy

$$v = 2n, \qquad (A - 3I)(A + I) = 2(n - 2)J, \qquad AJ = (3 - 2n)J,$$

and are uniquely determined by the spectrum of their adjacency matrix (see [14, Theorem 11]).

4.3 The lattice graphs $L_2(n)$ are strongly regular with

$$v = n^2, \qquad (A - 3I)(A - 3I + 2nI) = (n - 2)(n - 4)J,$$
$$AJ = (n - 1)(n - 3)J.$$

These graphs, which are of type 3 unless $n = 2, 4$, are uniquely determined by the spectrum of their adjacency matrix unless $n = 4$. This follows from a result of Shrikhande [15], who also proved that the graph mentioned at the end of Example 3.3 provides the only exception for $n = 4$.

4.4 The triangular graphs $T(n)$ are strongly regular with

$$v = \tfrac{1}{2}n(n - 1), \qquad (A - 3I)(A - 7I + 2nI) = \tfrac{1}{2}(n - 5)(n - 8)J,$$
$$AJ = \tfrac{1}{2}(n - 2)(n - 7)J.$$

These graphs, which are of type 3 unless $n = 5, 8$, are uniquely determined by the spectrum of their adjacency matrix unless $n = 8$. This follows from results of Hoffman [10], Chang [4], and others. In addition, Chang proved that for $n = 8$ there exist exactly three exceptional graphs.

4.5 By the following theorems, for the proof of which we refer to Example 3.3 and to [13, 14], all strong and strongly regular graphs with $\rho_1 = 3$ are determined.

THEOREM. *Strong graphs with* $(A - 3I)(A - \rho_2 I) = \mathbf{0}$ *only exist for* $\rho_2 = -1, -3, -5, -9$. *Any such graph is equivalent to* $L_2(2)$, $T(5)$, $L_2(4)$, $T(8)$, *respectively.*

THEOREM. *The only strongly regular graphs with* $\rho_1 = 3$ *are the graphs* $H(n)$, $L_2(n)$, $T(n)$, *the* $1 + 3$ *exceptions to* $L_2(4)$ *and* $T(8)$, *and the graphs of Petersen, Clebsch, and Schläfli.*

The Schläfli graph mentioned in the theorem consists of the 27 lines on a general cubic surface, any pair of lines being adjacent if and only if they are skew.

4.6 A *Steiner triple system* of order n consists of $\frac{1}{6}n(n-1)$ unordered triples out of n symbols such that every unordered pair of symbols occurs in exactly one triple. Steiner triple systems exist for all $n \equiv 1, 3 \pmod 6$. A *Steiner graph* of order n consists of the $\frac{1}{6}n(n-1)$ unordered triples of a Steiner triple system, any two triples being adjacent if and only if they have one symbol in common. For $n \geq 9$ Steiner graphs are strongly regular with

$$v = \tfrac{1}{6}n(n-1), \qquad (A - 5I)(A - 8I + nI) = \tfrac{1}{6}(n-13)(n-18)J,$$
$$AJ = \tfrac{1}{6}(n-3)(n-16)J.$$

Recently, Aliev and Seiden [1] have characterized Steiner graphs provided that the number of vertices exceeds some constant. In fact, from [1], Theorem 2.1 it follows that, for sufficiently large n, any strong graph with

$$v = \tfrac{1}{6}n(n-1), \qquad \rho_1 = 5, \qquad \rho_2 = 8 - n$$

is isomorphic to some Steiner graph of order n. The graph mentioned in Example 3.4 of the present paper provides a counterexample to this statement for $n = 13$. Indeed, the two Steiner graphs of order 13 and the graph of Example 3.4 have the same spectrum but are nonequivalent in pairs (cf. [7, Theorem 3.4]).

4.7 A *latin square* of order n consists of n^2 ordered triples out of n symbols such that for each pair of coordinates every pair of symbols occurs exactly once. Latin squares exist for all n. A *latin square graph* of order n consists of the n^2 ordered triples of a latin square, any two triples being adjacent if and only if they have one symbol in common. For $n \geq 3$ latin square graphs are strongly regular with

$$v = n^2, \qquad (A - 5I)(A - 5I + 2nI) = (n-4)(n-6)J,$$
$$AJ = (n-1)(n-5)J.$$

It would be interesting to know whether methods analogous to those of Aliev and Seiden [1], who in turn use results of Bose and Laskar [3], would suffice to prove that, for sufficiently large n, any strong graph with

$$v = n^2, \qquad \rho_1 = 5, \qquad \rho_2 = 5 - 2n$$

is isomorphic to some latin square graph of order n. The complement of the Clebsch graph, mentioned in Examples 1.3 and 3.3, provides a counter-example to this statement for $n = 4$. The complements of the two other graphs mentioned in Example 3.3 furnish the only two distinct latin square graphs of order 4.

4.8 *Moore graphs* are regular graphs of diameter 2, which are maximal according to a certain definition. These have been investigated by Hoffman and Singleton [11]. From their results it follows that Moore graphs are strongly regular, that there exists a Moore graph of girth 5 with

$$v = 50, \qquad A^2 - 25I = 24J, \qquad AJ = 35J,$$

and that this graph is uniquely determined by the spectrum of its adjacency matrix.

4.9 Graphs of diameter 2, girth 4, with the property that any two non-adjacent vertices are both adjacent to a constant number of other vertices, have been investigated by Gewirtz [6]. These graphs appear to be strongly regular. The following graphs of this type have been found:

$$v = 56, \qquad (A + 5I)(A - 7I) = 20J, \qquad AJ = 35J,$$
$$v = 77, \qquad (A + 5I)(A - 11I) = 21J, \qquad AJ = 44J,$$
$$v = 100, \qquad (A + 5I)(A - 15I) = 24J, \qquad AJ = 55J,$$

the second graph being a subgraph of the third.[1]

Gewirtz [6] proved the existence of the first and the uniqueness of the first and the third of these graphs. Higman and Sims [9] gave a construction for the third graph.

4.10 A strongly regular graph with

$$v = 105, \qquad (A + 5I)(A - 19I) = 9J, \qquad AJ = 40J$$

will be constructed in the following section.

[1] For further investigations see J. M. Goethals and J. J. Seidel [16].

5. Equiangular Lines

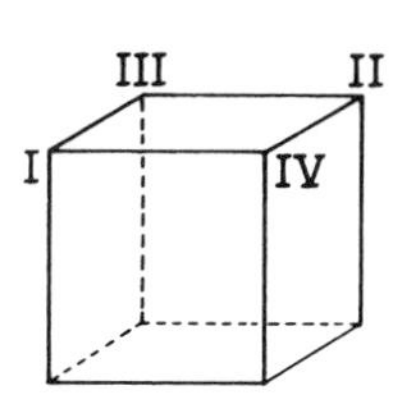

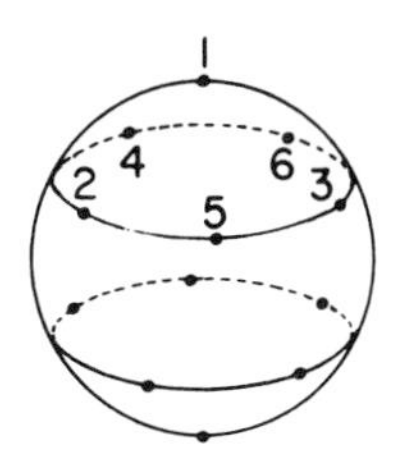

Fig. 12.

The four diagonals of the cube in Euclidean 3-space R_3 have the property that each pair has the same angle arccos (1/3). The six lines connecting the antipodal pairs of vertices of the icosahedron have the property that each pair has the same angle arccos $(1/\sqrt{5})$. More lines with this property do not exist in R_3. We consider the analogous problem in Euclidean r-space R_r. A set of lines in R_r is called *equiangular* if the angle between each pair of lines is the same. We investigate $n(r)$, that is, the *maximum number* of equiangular lines in R_r.

Theorem. *$n(2) = 3$, $n(3) = 6$, $n(4) = 6$, $n(5) = 10$, $n(6) = 16$, and $n(7) \geq 28$.*

For the proof of this theorem we refer to [8, 12]. The problem may be stated in terms of matrices as follows. A set of n lines, carried by the unit vectors $p_1, p_2, \ldots, p_n$, which span R_r, is equiangular if for each pair of vectors

$$\cos \varphi = \pm(p_i, p_j), \qquad 0 < \varphi < \tfrac{1}{2}\pi,$$

has the same value. Consider the matrices

$$P = [(p_i, p_j)] = \begin{bmatrix} 1 & & \pm\cos\varphi \\ & \ddots & \\ \pm\cos\varphi & & 1 \end{bmatrix}, \qquad A = \frac{1}{\cos\varphi}(P - I) = \begin{bmatrix} 0 & & \pm 1 \\ & \ddots & \\ \pm 1 & & 0 \end{bmatrix}.$$

The Gramian matrix P of the vectors $p_1, \ldots, p_n$ has order n, is symmetric and positive semidefinite of rank r, hence has the smallest eigenvalue 0 of multiplicity $n - r$. The matrix A has the smallest eigenvalue $-1/\cos\varphi$ of multiplicity $n - r$. Therefore, the number n of equiangular lines in R_r is large whenever a matrix A may be found, symmetric, with diagonal elements 0 and elements $+1$ and -1 elsewhere, whose smallest eigenvalue is of high multiplicity. In [8] it is proved that the optimal configuration in R_3 is provided by the six diagonals of the icosahedron, which corresponds to a matrix A which is the adjacency matrix occurring in Example 2.1. Furthermore, in [12] it is proved that the optimal configurations in R_5 and R_6 are provided by sets of lines whose matrix A is the adjacency matrix of the Petersen graph of Example 1.2 and of the complement of the Clebsch graph of Example 1.3, respectively.

We conclude this section by constructing 105 equiangular lines in R_{21} with angle arccos $\frac{1}{5}$. To that purpose we consider the incidence matrix of the 21 points and 21 lines of the projective geometry of order 4 in its cyclic form. Each row consists of 5 ones and 16 zeros and the inner product of each pair of distinct rows equals 1. We replace each row by the 5×21 matrix consisting of the columns of the matrix $J_5 - 2I_5$ at the 5 entries 1 of that row and of 16 zero columns elsewhere. Then the 105×21 matrix P is obtained, which satisfies

$$P^T P = 24I + J, \qquad PP^T = 5I + A,$$

where the symmetric matrix A has elements 0 on the diagonal and ± 1 elsewhere. The eigenvalues of $P^T P$ are 45 and 24, of multiplicities 1 and 20, respectively. Therefore, the eigenvalues of PP^T are 45, 24, and 0, and those of A are 40, 19, and -5, of multiplicities 1, 20, and 84, respectively. Hence A is the adjacency matrix of a strongly regular graph with

$$v = 105, \qquad (A + 5I)(A - 19I) = 9J, \qquad AJ = 40J,$$

and it follows that $n(21) \geq 105$.

References

1. J. S. O. Aliev and E. Seiden, Steiner Triple Systems and Strongly Regular Graphs, *J. Combinatorial Theory* **6** (1969), 33–39.
2. R. C. Bose, Strongly Regular Graphs, Partial Geometries, and Partially Balanced Designs, *Pacific. J. Math.* **13** (1963), 389–419.
3. R. C. Bose and R. Laskar, A Characterization of Tetrahedral Graphs, *J. Combinatorial Theory* **3** (1967), 366–385.
4. L. C. Chang, The Uniqueness and Nonuniqueness of the Triangular Association Scheme, *Sci. Record (Peking)* **3** (1959), 604–613.
5. A. Clebsch, Ueber die Flächen vierter Ordnung welche eine Doppelcurve zweiten Grades besitzen, *J. für Math.* **69** (1868), 142–184.
6. A. Gewirtz, Graphs with Maximal Even Girth, *Canad. J. Math.* (to appear).
7. J. M. Goethals and J. J. Seidel, Orthogonal Matrices with Zero Diagonal, *Canad. J. Math.* **19** (1967), 1001–1010.

8. J. Haantjes, Equilateral Point-Sets in Elliptic Two- and Three-Dimensional Spaces, *Nieuw Arch. Wisk.* **22** (1948), 355–362.
9. D. G. Higman and C. C. Sims, A Simple Group of Order 44,353,000, *Math. Z.* **105** (1968), 110–113.
10. A. J. Hoffman, On the Uniqueness of the Triangular Association Scheme, *Ann. Math. Statist.* **31** (1960), 492–497.
11. A. J. Hoffman and R. R. Singleton, On Moore Graphs with Diameters 2 and 3, *IBM J. Res. Develop.* **4** (1960), 497–504.
12. J. H. van Lint and J. J. Seidel, Equilateral Point Sets in Elliptic Geometry, *Koninkl. Ned. Akad. Wetenschap. Proc. Ser. A* **69** (or *Indag. Math.* **28**) (1966), 335–348.
13. J. J. Seidel, Strongly Regular Graphs of L-type and of Triangular Type, *Koninkl. Ned. Akad. Wetenschap. Proc. Ser. A* **70** (or *Indag. Math.* **29**) (1967), 188–196.
14. J. J. Seidel, Strongly Regular Graphs with $(-1, 1, 0)$ Adjacency Matrix Having Eigenvalue 3, *Linear Algebra and Applications* **1** (1968), 281–298.
15. S. S. Shrikhande, The Uniqueness of the L_2 Association Scheme, *Ann. Math. Statist.* **30** (1959), 781–798.
16. J. M. Goethals and J. J. Seidel, Regular Graphs Derived from Combinatorial Designs (to appear).

PROJECTIVE GEOMETRY AND THE 4-COLOR PROBLEM

W. T. Tutte

DEPARTMENT OF COMBINATORICS AND OPTIMIZATION
UNIVERSITY OF WATERLOO, WATERLOO, ONTARIO

At the Third Waterloo Conference on Combinatorics the author gave a course of three lectures entitled "On the Algebraic Theory of Graph-Colorings." This was based on a paper of the same title that appeared in the first issue of the *Journal of Combinatorial Theory* [1]. The present paper is based in turn on the course of lectures. However it differs from them and from the original paper in being purely expository in nature. It attempts to describe what the journal paper is about, and it states the results; but for proofs of the main theorems the reader is referred to the original paper.

1. Introduction

It is often of interest to represent a graph as an example of a more general mathematical system. For example a graph can be represented as a set of points in a finite projective geometry over $GF(2)$, as we shall explain below. When such a representation has been discovered we may hope to use the known properties of the general system to obtain theorems about graphs. Alternatively, we may try to generalize theorems about graphs so as to obtain new theorems about the more general structures.

In this paper we adopt the second approach, but we generalize not a theorem but a conjecture. We obtain a conjectural proposition about sets of points in finite projective geometries over $GF(2)$, and this new conjecture implies the 4-Color Conjecture for planar graphs.

2. Cycles and Coboundaries

Let G be a finite graph with no loops or multiple edges. Let its edge-set be $E(G)$ and its vertex-set $V(G)$. We define a 1-*chain* on G as a mapping of $E(G)$

into the field $GF(2)$. This field has only two elements, 0 and 1, and its arithmetical properties may be summarized as

$$0 + 0 = 1 + 1 = 0, \qquad 0 + 1 = 1;$$
$$0 \times 0 = 0 \times 1 = 0, \qquad 1 \times 1 = 1.$$

In a 1-chain f each edge A of the graph is associated with a single element $f(A)$ of $GF(2)$. This is called the *coefficient* of A in f.

There is a *zero* 1-chain on G in which the coefficient of each edge is zero. This is conveniently denoted by the symbol 0. Given two 1-chains f and g we may speak of their sum $f + g$. This is the 1-chain h on G in which the coefficient of each edge is the sum of its coefficients in f and g. This rule can be extended to the sum of any number of 1-chains. The addition thus defined is evidently commutative and associative.

A set of 1-chains on G is *linearly dependent* if one of its nonnull subsets sums to the zero 1-chain, and *linearly independent* otherwise. The *rank* $r(S)$ of a set S of 1-chains is the maximum number of members of S constituting a linearly independent subset. The usual algebraic rules of linear dependence apply. Thus if we define a *basis* of S as a set of $r(S)$ linearly independent members of S, then each linearly independent subset of S is contained in a basis of S. Moreover each member of S has a unique expression as a sum, or "linear combination" of members of any given basis.

We define the 0-*chains* on G as the mappings of $V(G)$ into $GF(2)$. We define addition and linear dependence for them just as for 1-chains.

We write $L_0(G)$ for the set of all the 0-chains of G, and $L_1(G)$ for the set of all the 1-chains. These sets are Abelian groups with respect to the operation of addition. They are indeed finite vector spaces.

We now describe some correspondences between $L_0(G)$ and $L_1(G)$ whose definitions involve the incidence relations of G.

The *boundary* ∂f of a 1-chain f is the 0-chain g with coefficients defined as follows: if $v \in V(G)$ then $g(v)$ is the sum of the coefficients in f of the edges incident with v. Analogously the *coboundary* δh of a 0-chain h is a 1-chain k with coefficients defined thus: if $A \in E(G)$ then $k(A)$ is the sum of the coefficients in h of the two ends of A.

We can regard a 1-chain on G as a coloring of the edges in two colors 0 and 1. A vertex has coefficient 1 in the boundary of this chain, if and only if the number of incident edges of color 1 is odd. Similarly a 0-chain h is a coloring of the vertices in two colors 0 and 1, and an edge has coefficient 1 in δh if and only if it has one end of each color.

As a simple exercise in graph theory we can establish the following identities:

$$\partial(f_1 + f_2) = \partial f_1 + \partial f_2,$$
$$\delta(g_1 + g_2) = \delta g_1 + \delta g_2.$$

The second of these shows that the set $\Delta(G)$ of coboundaries of 0-chains of G is an Abelian group with respect to addition. We call it the *coboundary-group* of G.

A *cycle* of G is a 1-chain whose boundary is the zero 0-chain. The first of the above identities shows that the set $\Gamma(G)$ of cycles of G is an Abelian group with respect to addition. We refer to it as the *cycle-group* of G.

A graph is said to be *Eulerian* if the valency of each of its vertices is even. Evidently a 1-chain k of G is a cycle if and only if the edges with coefficient 1 in k are the edges of an Eulerian subgraph of G.

Let us define the *unit* 1-chain of G as the 1-chain in which each edge has coefficient 1. We observe that this chain is a cycle if and only if G is Eulerian. It is a coboundary if and only if G is *bipartite*, that is, if $V(G)$ can be partitioned into two disjoint subclasses such that each edge of G has one end in each subclass.

Two 1-chains f and g are said to be *orthogonal* if

$$\sum_{A \in E(G)} f(A)g(A) = 0.$$

It can be shown, as another exercise in graph theory, that a 1-chain is a cycle if and only if it is orthogonal to every coboundary, and a coboundary if and only if it is orthogonal to every cycle.

3. Geometrical Representations

Let P be a d-dimensional projective geometry over $GF(2)$. By definition its points are in one-to-one correspondence with the vectors

$$(x_1, x_2, \ldots, x_{d+1}),$$

where the $d + 1$ components x_1 are elements of $GF(2)$ that are not all zero. Linear relations between these "coordinate vectors" are spoken of also as linear relations between the corresponding points.

A *k-subspace* of P, where k is a nonnegative integer, consists of a set Q of $k + 1$ linearly independent points, together with all other points dependent on Q, that is, expressible as linear combinations of members of Q. Thus a 0-subspace consists of a single point, a 1-subspace has exactly three points, and a 2-subspace has exactly seven. We refer to 1-subspaces and 2-subspaces as *lines* and *planes* of P respectively. Evidently P has only one d-subspace, and we identify this with P itself. It is often convenient to recognize a (-1)-subspace of P which has no points at all.

Let the rank of $\Delta(G)$ be r. Rejecting some trivial cases we suppose r to be at least 1. Choose a basis

$$\{f_1, f_2, \ldots, f_r\}$$

of $\Delta(G)$, and enumerate the edges of G from A_1 to A_n. We can represent the basis by a matrix M of r rows and n columns. The entry in the ith row and jth column is $f_i(A_j)$.

Let us take d to be $r - 1$. Let S denote the set of those points of P whose coordinate vectors appear as columns of M. We have now obtained our representation of G as a set of points in a finite projective geometry. We proceed to interpret the edges, cycles, and coboundaries of G in terms of S.

The edges of G are, of course, represented by the points of S. Moreover distinct edges of G are represented by distinct points of S; if it were not so there would be two edges having equal coefficients in every coboundary and therefore having the same two ends, contrary to the conditions imposed on G. Since M has r independent rows it must also have r independent columns. Hence S has r independent points. Accordingly S generates P in the sense that each point of P is a linear combination of points of S.

The orthogonality between cycles and coboundaries enables us to interpret the cycles of G as the linear relations between the points of S. Denoting the ith column of M by C_i we can say that a 1-chain f is a cycle if and only if

$$\sum_i f(A_i) \,.\, C_i = 0,$$

where the symbol 0 stands for a zero vector. This linear relation between the columns of M can be described equally well as a linear relation between the points of S; we can say that the subset of S corresponding to the nonzero coefficients $f(A_i)$ is dependent.

The nonzero coboundaries of G can be put into one-to-one correspondence with the $(d - 1)$-subspaces of P. For let Q be such a subspace. It defines a 1-chain q such that $q(A_i) = 0$ if and only if Q contains the point of S corresponding to A_i. The orthogonality of this chain to every cycle is easily demonstrated, with the above interpretation of the cycles. It is known moreover that P has exactly $2^r - 1$ $(d - 1)$-subspaces, and these must give rise between them to all the $2^r - 1$ nonzero coboundaries of G.

In the above argument it is assumed implicitly that M has no zero column, this assumption being equivalent to the condition that G shall have no loop.

The above representation of G by a set S of points in a geometry P is called the *direct* one in what follows. There is also a *dual* representation defined analogously with $\Gamma(G)$ replacing $\Delta(G)$. In this representation d is $s - 1$, where s is the rank of $\Gamma(G)$. Coboundaries are interpreted as linear relations between points of S, and cycles correspond to the $(d - 1)$-subspaces.

For the dual representation it is necessary for G to have no isthmus. This ensures that all the columns of M will be nonzero. For the correspondence between $E(G)$ and S to be one-to-one it is necessary and sufficient that there shall be no coboundary with exactly two nonzero coefficients. Actually no new major difficulty arises if the correspondence is not one-to-one, either in the direct or the dual representation. The ban on multiple joins could be relaxed.

4. Colorings

By an *n-coloring* of a graph G we mean a coloring of the vertices in n colors in such a way that no two vertices of the same color are joined by an edge. It is not necessary that all the n colors shall actually be used.

The famous 4-Color Conjecture asserts that every planar graph has a 4-coloring. In one approach to this conjecture we try to classify the graphs that have no 4-coloring, hoping to show eventually that these are all nonplanar. In what follows we refer to such graphs simply as "noncolorable."

The simplest noncolorable graph is of course the complete 5-graph, which has five vertices and ten edges joining them in pairs. But there are many others, enough to make the task of direct classification seem impossible. Usually we define some operation which replaces a given graph by another with fewer edges. We then try to classify those noncolorable graphs that are minimal with respect to this operation, i.e., cannot be reduced to smaller noncolorable graphs by repetitions of the operation.

We may consider for example the operation of deleting edges, and say that noncolorable graphs minimal with respect to this operation are *deletion-minimal*. Another possible operation is that of contracting edges. (To contract an edge we delete it and then identify its two ends. If this operation introduces loops we delete them,[1] and if it introduces multiple joins we reduce them to single edges.) If a noncolorable graph is minimal with respect to contraction we call it *contraction-minimal*. It is possible for a 4-colorable graph to give rise to a noncolorable one by edge-contraction, but not by edge-deletion.

The proposition known as Hadwiger's Conjecture asserts that if a graph has no n-coloring, and is connected, it can be transformed into a complete $(n + 1)$-graph by edge-contractions. No counterexample is known. For the case $n = 4$ Hadwiger's Conjecture asserts that the complete 5-graph is the only contraction-minimal noncolorable graph. Hadwiger's Conjecture for the case $n = 4$ implies the 4-Color Conjecture, for the complete 5-graph is nonplanar and the property of planarity is invariant under edge-contraction.

It is Hadwiger's Conjecture for $n = 4$ for which we seek a geometrical generalization.

Consider the 4-colorings of a graph G. As colors we take the four ordered pairs (0, 0), (0, 1), (1, 0), and (1, 1) of elements of $GF(2)$. Let the vertices be enumerated from v_1 to v_m. Then a 4-coloring is an assignment of an ordered

[1] Here I deviate into "Michigan" terminology, at the expense of some later complication.

pair (a_i, b_i) of elements of $GF(2)$ to each vertex v_i of G in such a way that whenever v_i and v_j are directly joined by an edge we have either $a_i + a_j = 1$ or $b_i + b_j = 1$. Accordingly a 4-coloring of G can be described as an ordered pair (f, g) of 0-chains on G such that no edge has a zero coefficient in both δf and δg.

If G is connected there are exactly two 0-chains with a given coboundary, and they differ only by an interchange of zero and nonzero coefficients. Hence the ordered pair (f, g) is uniquely determined by the unordered pair $(\delta f, \delta g)$ to within a permutation of the four colors among themselves. We therefore define a *coloring* of $\Delta(G)$ as an unordered pair (h_1, h_2) of coboundaries of G such that no edge of G has zero coefficient in both h_1 and h_2. Then $\Delta(G)$ has a coloring if and only if G has a 4-coloring.

If we reject the trivial case in which G is bipartite we can be sure that the zero boundary never appears as a member of a coloring of $\Delta(G)$.

We define a coloring of $\Gamma(G)$ analogously as an unordered pair of cycles of G not both zero on any edge. The zero cycle cannot appear as a member of a coloring except in the case of an Eulerian graph.

If G is trivalent, the colorings of $\Gamma(G)$ are associated with the "Tait colorings" of G. Such a Tait coloring is a coloring of the edges of G in three colors a, b, and c in such a way that no two edges of the same color have a common end. It is well known that the 4-Color Conjecture is equivalent to the proposition that every trivalent planar map, without an isthmus, has a Tait coloring. In a Tait coloring the edges of colors a and b define an Eulerian subgraph, corresponding to a cycle k_c, and the edges of colors b and c similarly give rise to a cycle k_a. Clearly (k_a, k_b) is a coloring of $\Gamma(G)$. Conversely if we are given a coloring (k_1, k_2) of $\Gamma(G)$ we can assign the color b to edges with unit coefficient in both k_1 and k_2 and colors a and c to the remaining edges according as their unit coefficients are in k_1 or k_2. The trivalency of G then ensures that one edge of each color is incident with each vertex. We thus have a Tait coloring.

The best-known example of a trivalent graph without an isthmus but with no Tait coloring is the Petersen graph. This is derived from the graph of edges and vertices of the regular dodecahedron by identifying diametrically opposite points.

Let us interpret the colorings of $\Delta(G)$ in terms of the direct representation of G. If G is not bipartite a coloring (h_1, h_2) of $\Delta(G)$ corresponds to a pair (Z_1, Z_2) of distinct $(d-1)$-subspaces of P whose intersection contains no point of S. We accordingly define a *color-space* in P as a $(d-2)$-subspace that does not meet S. Now any $(d-2)$-subspace can be represented as the intersection of two $(d-1)$-spaces. Accordingly we can say that the colorings of $\Delta(G)$ are represented, though not in a one-to-one manner, by the color-spaces in P. Thus $\Delta(G)$ has a coloring if and only if a color-space exists in the direct representation.

In the same way we can relate the colorings of $\Gamma(G)$ to the color-spaces in the dual representation of G.

In the following section we interpret the operations of deletion and contraction of edges in terms of geometrical representations.

5. Deletion and Contraction

Consider the direct representation of G by a set S of points in a projective space P.

Let H be the graph derived from G by deleting a set Q of edges. The Eulerian subgraphs of H are those Eulerian subgraphs of G that have no edges in Q. Hence the cycles of H are the restrictions to $E(H)$ of those cycles of G whose coefficients in Q are all zero. They are represented by the linear relations holding in the subset S_H of S corresponding to $E(H)$. It can be deduced from this result that we can obtain a direct representation of H by replacing S by S_H, and P by the subspace of P generated by S_H. This is the operation of "point-deletion." It corresponds to edge-deletion in G.

Next we consider the operation of contracting a set Q of edges in G. We suppose Q to be *closed* in the sense that the contraction leaves no loop in the resulting graph K. A necessary and sufficient condition for this is that no Eulerian subgraph of G shall have exactly one edge outside Q. The Eulerian subgraphs of K are obtained from those of G by contracting in them any edges belonging to Q. Consider a direct representation of K by a set $S(K)$ of points in a projective geometry $P(K)$. Let us identify each point of $S(K)$ with the point of S representing the same edge. Then the above result on Eulerian subgraphs shows that the linearly dependent sets in $S(K)$ are the nonnull intersections with $S(K)$ of the linearly dependent sets in S.

Consider the subset Q' of S corresponding to members of Q. The closure of Q defined above is equivalent to the closure of Q' in the following sense. Each point of S dependent on Q', that is expressible as a linear combination of points of Q', is itself a member of Q'. The closed set Q' generates a subspace Σ of P, of dimension h say, which contains all the points of Q but no point of $S - Q'$. By the theory of projective geometries there is a subspace $P(K)$ of P, of dimension $d - h - 1$, that has no common point with Σ. Any point X of $S - Q'$ generates with Σ a subspace of dimension $h + 1$ of P, and this meets $P(K)$ in exactly one point X'. We call X' the *projection* of X from Σ onto $P(K)$. We write $S(K)$ for the set of projections from Σ onto $P(K)$ of the points of $S - Q'$.

Each point of $S(K)$ corresponds to an edge of K in the following sense: it is the projection of the member of $S - Q'$ representing that edge in P. It may happen that some point Y of $S(K)$ is the projection of two or more points of

$S - Q'$. But these would correspond to edges of K with the same ends, and so all but one of them would be deleted in the construction of K. We can show that, with respect to this correspondence, the set $S(K)$ in the projective space $P(K)$ constitutes a direct representation of K. For, first, the correct linear relations hold between the points of $S(K)$. Second, $S(K)$ generates $P(K)$, for otherwise Σ and $S - Q'$ would together generate a proper subspace of P, whereas in fact S generates the whole of P. We say that the pair $PK)$, $((S(K))$ is derived from the pair (P, S) by *projection* from the closed subset Q' of S. This is the geometrical operation corresponding to edge-contraction in G.

It is not difficult to see that, when we delete a set Q of edges of G to form H, the coboundaries of H are the restrictions to $E(H)$ of the coboundaries of G. In order that H shall have a dual representation as a set S_H of points in a projective space P_H it is necessary and sufficient that Q shall be *isthmus-closed*, that is, its deletion must leave no isthmus in H. Perhaps it is an unnecessary refinement to insist that the correspondence between $V(H)$ and S_H shall be one-to-one, but this can be arranged by following the edge-deletion by the contraction of just enough edges to remove every coboundary having exactly two nonzero coefficients. With these auxiliary contractions deletion of an isthmus-closed set of edges becomes the graph-theoretical dual of the operation of contraction defined in Section 4, applied to a closed set of edges. Accordingly we find that the corresponding operation in the dual representation (P, S) of G is projection from a closed subset of S.

Let us refer to the dual operation of edge-deletion as *pure edge-contraction*. It is defined as is edge-contraction, but without deletion of loops and members of multiple joins. If we form a graph K from G by pure edge-contraction of a set Q of edges we find that the coboundaries of K are the restrictions to $E(K)$ of those coboundaries of G that have only zero coefficients in Q. (A loop has zero coefficient in every coboundary.) We deduce that pure edge-contraction corresponds to point-deletion in the dual representation of G.

6. Blocks

Consider a set S of points in a projective space P, and suppose S to generate the whole of P. It may happen that the pair (P, S) constitutes the direct representation of some graph. If so, we say it is *graphic*. If it constitutes the dual representation of a graph, we say it is *cographic*. The problem of deciding whether a given pair (P, S) is graphic, or cographic, can be solved by the methods of matroid theory, but we shall not discuss it here.

Letting d denote the dimension of P we define a *color-space* of (P, S) as a $(d - 2)$-subspace of P which does not contain any point of S. We have of course already made this definition in Section 4 for the graphic and cographic cases. If (P, S) has no color-space we call it a *2-block*. More generally (P, S)

is a k-block if every $(d-k)$-subspace of P contains a point of S.

A 2-block (P, S) is *tangential* if it cannot be transformed into another 2-block by projection from a nonnull closed subset of S. Since such projections reduce the number of members of S it is clear that any 2-block can be transformed into a tangential one by a sequence of them, and it is easy to show that any such sequence can be reduced to a single projection. Accordingly a tangential 2-block is characterized by the following properties. First, it is a 2-block. Second, if Q is any closed subset of S, there is a $(d-2)$-subspace of P that contains all the points of Q but no point of $S-Q$. Such $(d-2)$-subspaces correspond to the color-spaces in the projection from Q. We refer to them as the *tangents* of Q in (P, S).

It is noteworthy that a tangential 2-block (P, S) cannot be transformed into a 2-block (P', S') by point-deletion of a nonnull subset $Q = S - S'$ of S. For suppose such a transformation is possible. Choose $A \in Q$. Then the set $\{A\}$ has a tangent T in (P, S), and the intersection of T with P' must be a color-space of (P', S').

By the theory of the preceding sections the graphic tangential 2-blocks are the direct representations of the contraction-minimal noncolorable graphs. Dually it is found that at least one of the cographic tangential 2-blocks corresponds to a trivalent graph that has no Tait coloring, and cannot be transformed into another such trivalent graph by the dual operation of edge-contraction.

In [1] we determine the tangential 2-blocks up to $d = 5$. We find one 2-dimensional one in which S is the set of all seven points of the plane P. We call this the *Fano block*. By a theorem proved by Whitney [2] it is found to be neither graphic nor cographic. There is a 3-dimensional tangential 2-block which is graphic, but not cographic, and which corresponds to the complete 5-graph. In this block the ten points of S define a 3-dimensional Desargues configuration. We speak therefore of the *Desargues block*. Finally there is a 5-dimensional tangential 2-block that is cographic, but not graphic, and which is the dual representation of the Petersen graph.

The Hadwiger Conjecture for $n = 4$ puts a limitation on the number of contraction-minimal noncolorable graphs. A geometrical generalization of it should limit the number of tangential 2-blocks. For the time being a suitable generalization is that the only tangential 2-blocks are the Fano, Desargues, and Petersen blocks.

References

1. W. T. Tutte, On the Algebraic Theory of Graph Colorings, *J. Combinatorial Theory* **1** (1966), 15–50.
2. H. Whitney, The Abstract Properties of Linear Dependence, *Amer. J. Math.* **57** (1935), 507–533.

CONTRIBUTED PAPERS

REMARK ON A COMBINATORIAL THEOREM OF ERDŐS AND RADO

H. L. Abbott

UNIVERSITY OF ALBERTA
EDMONTON, ALBERTA

and

B. Gardner

MEMORIAL UNIVERSITY OF NEWFOUNDLAND
ST. JOHN'S, NEWFOUNDLAND

Erdős and Rado [2] proved that, to each pair of positive integers n and k, with $k \geq 3$, there corresponds a least positive integer $\phi(n, k)$ such that, if F is a family of more than $\phi(n, k)$ sets, each set with n elements, then some k of the sets have pairwise the same intersection. They also proved

$$(k-1)^n \leq \phi(n, k) \leq n!(k-1)^n\left\{1 - \sum_{t=0}^{n-1} \frac{t}{(t+1)!(k-1)^t}\right\}. \tag{1}$$

Abbott [1] proved that

$$\phi(n, k) \geq \begin{cases} \phi(2, k)^{n/2} & \text{if} \quad n \quad \text{is even} \\ (k-1)\phi(2, k)^{(n-1)/2}, & \text{if} \quad n \quad \text{is odd,} \end{cases} \tag{2}$$

and that

$$\phi(2, k) \geq (k-1)^2 + \left[\frac{k-1}{2}\right]. \tag{3}$$

The following theorem gives a slight improvement of (3) for the case where k is odd.

THEOREM 1.

$$\phi(2, k) \geq k(k-1) \qquad \textit{if} \qquad k \quad \textit{is odd.} \tag{4}$$

PROOF: Let $k = 2\lambda + 1$ and let

$$F_1 = \{(i, j) : 1 \leq i < j \leq k\}, \qquad F_2 = \{(i, j) : k + 1 \leq i < j \leq 2k\}.$$

Then

$$|F_1| = |F_2| = \binom{k}{2}.$$

Let $F = F_1 \cup F_2$. Then

$$|F| = |F_1| + |F_2| = \binom{k}{2} + \binom{k}{2} = k(k-1).$$

To complete the proof it is sufficient to show that no k members of F have pairwise the same intersection. Now each of $1, 2, \ldots, 2k$ appears in $k - 1$ members of F. Hence if k sets are to have pairwise the same intersection they must be pairwise disjoint, but this would contradict the fact that $2(\lambda + 1) > 2\lambda + 1 = k = |\bigcup F_1| = |\bigcup F_2|$. Hence no k sets have pairwise the same intersection. The proof of the theorem is now complete.

Erdős and Rado observed that $\phi(1, k) = k - 1$ for all values of k and that $\phi(2, 3) = 6$. Up to the present these were the only known values of $\phi(n, k)$ and the evaluation of $\phi(n, k)$ for larger values of n and k seems to be a very difficult problem. However, we have been able to evaluate $\phi(2, 4)$ and $\phi(3, 3)$. Since the proofs are somewhat long the details are not presented, but rather a sketch of the argument used is given.

THEOREM 2.

$$\phi(2, 4) = 10. \tag{5}$$

PROOF: That $\phi(2, 4) \geq 10$ follows from (3). To prove that $\phi(2, 4) \leq 10$, consider any family of eleven sets, each set with two elements. If we assume that no four of these sets have pairwise the same intersection, it can be shown that a contradiction follows.

Clearly, it can be assumed that no element appears in more than three sets. If there are n_1 elements which appear exactly once, n_2 elements which appear exactly twice, and n_3 elements which appear exactly three times, then

$$n_1 + 2n_2 + 3n_3 = 22. \tag{6}$$

By using (6) it can be shown that if two elements a and b appear exactly once they do not appear in the same set; in fact we must have (ac), (bc). It also follows from this argument that if a, b, c appear exactly once we must have (ad), (bd), (cd) and that there cannot be four or more elements which appear exactly once.

There are thus four cases to be considered.

CASE 1. Three elements a, b, c appear exactly once, in which case we have (ad), (bd), (cd).

CASE 2. Two elements a, b appear exactly once, in which case we have (ac), (bc).

CASE 3. One element appears exactly once.

CASE 4. No element appears exactly once.

It follows from (6) that in each case there are four sets with pairwise the same intersection.

COROLLARY.

$$\phi(n, 4) \leq n!\,3^n\left\{\frac{5}{9} - \frac{1}{9}\sum_{t=2}^{n-1}\frac{t}{(t+1)!\,3^{t-2}}\right\}. \tag{7}$$

PROOF: This follows from (5) by using the Erdős–Rado recurrence inequality

$$\phi(n, k) \leq (k-1)n\phi(n-1, k) - (k-1)(n-1). \tag{8}$$

THEOREM 3.

$$\phi(3, 3) = 20. \tag{9}$$

PROOF: That $\phi(3, 3) \geq 20$ follows from the fact that in the following family of sets no three sets have pairwise the same intersection:

$$\{(abc), (abd), (ace), (adf), (aef), (bcf), (bde), (bef), (cde), (cdf),$$
$$(mnp), (mnq), (mpr), (mqs), (mrs), (nps), (nqr), (nrs), (pqr), (pqs)\}.$$

To show that $\phi(3, 3) \leq 20$ consider an arbitrary family F of twenty-one sets each set with three elements. It must be shown that three of the sets have pairwise the same intersection. This portion of the proof makes use of the following lemmas.

LEMMA 1. *Let F be a family of six sets each set with two elements, no three members of F having pairwise the same intersection. Then*

$$F = \{(ab), (ac), (bc), (de), (df), (ef)\}.$$

LEMMA 2. *Let F be a family of five sets, each set with two elements, no three members of F having pairwise the same intersection. Then either*

$$F = \{(ab), (ac), (bc), (de), (df)\}$$

or

$$F = \{(ab), (ac), (bd), (ce), (de)\}.$$

To return to the proof of the theorem it is clear that if an element appears in more than six or if no element appears in more than four members of F there is no problem.

The importance of the lemmas now becomes clear. If an element a appears in six sets then these sets must be

$$(abc), \quad (abd), \quad (acd), \quad (aef), \quad (aeg), \quad (afg).$$

In addition, if an element a appears in exactly five sets we must have one of the following:

$$(abc), \quad (abd), \quad (acd), \quad (aef), \quad (aeg)$$

or

$$(abc), \quad (abd), \quad (acf), \quad (ade), \quad (aef).$$

Select a set (abc) in F and specify the number of sets containing a, b, and c. Let $F_1 = \{\mathrm{F} : \mathrm{F} \in F, \mathrm{F} \cap (abc) \neq \Phi\}$. The structure of F_1 is then largely determined. It can then be shown that either three members of F_1 are pairwise disjoint or one member of $F - F_1$ and two members of F_1 are pairwise disjoint, and the proof of the theorem is complete.

COROLLARY.

$$\phi(n, 3) \leq n!\,2^n\left\{\frac{5}{12} - \frac{1}{8}\sum_{t=3}^{n-1}\frac{t}{(t+1)!\,2^{t-3}}\right\}. \tag{10}$$

PROOF: This follows easily from (8) and (9).

Any result of this type will not improve the general upper bound by more than a constant factor. However, Theorem 3 together with the fact that for all positive integers a, b, and k, with $k \geq 3$,

$$\phi(a + b, k) \geq \phi(a, k)\phi(b, k), \tag{11}$$

which was proved by Abbott [1], yields a substantially better lower bound for $\phi(n, 3)$. It gives the following:

COROLLARY.

$$\phi(n, 3) \geq c(20)^{n/3}. \tag{12}$$

REFERENCES

1. H. L. ABBOTT, Some Remarks on a Combinatorial Theorem of Erdős and Rado, *Canad. Math. Bull.* **9** (2) (1966), 155–160.
2. P. ERDŐS and R. RADO, Intersection Theorems for Systems of Sets, *J. London Math. Soc.* **35** (1960), 85–90.

THE FOUR LEADING COEFFICIENTS OF THE CHROMATIC POLYNOMIALS $Q_n(u)$ AND $R_n(x)$ AND THE BIRKHOFF–LEWIS CONJECTURE

Ruth Bari

GEORGE WASHINGTON UNIVERSITY
WASHINGTON, D.C.

In their paper Chromatic Polynomials [1], Birkhoff and Lewis introduced a strong form of the 4-color conjecture.

If P_{n+3} is any map with $n + 3$ regions and only triple vertices, so that $P_n(0) = P_n(1) = P_n(2) = 0$, and $P_{n+3}(\lambda)$ is the chromatic polynomial associated with P_{n+3}, then Birkhoff and Lewis surmised that

$$(\lambda - 3)^n << \frac{P_{n+3}(\lambda)}{\lambda(\lambda - 1)(\lambda - 2)} << (\lambda - 2)^n \qquad \text{for} \qquad \lambda \geq 4.$$

The relation $S(y) << T(y)$ holds if, when $S(y)$ and $T(y)$ are expanded in powers of $z = y - 4$, the coefficients of $S(z)$ are nonnegative and not greater than the corresponding coefficients of $T(z)$.

If $P_{n+3}(\lambda)$ satisfies the Birkhoff–Lewis conjecture, then $P_{n+3}(4) \geq 24$.

Let $x = \lambda - 4$, and let

$$R_{n+3}(x) = \frac{P_{n+3}(\lambda)}{\lambda(\lambda - 1)(\lambda - 2)},$$

where $P_{n+3}(x)$ is expanded in powers of x. Here $R_{n+3}(x)$ will be referred to as the R-polynomial.

Birkhoff and Lewis also found that the computation of chromatic polynomials could be simplified by using the Q-polynomial $Q_n(u)$.

If P_n is a regular map, define

$$Q_n(u) = \frac{P_n(\lambda)}{\lambda(\lambda - 1)(\lambda - 2)(\lambda - 3)},$$

where $Q_n(u)$ is expanded in powers of $u = \lambda - 3$. Since $P_n(3) = 0$ for all regular maps which have at least one odd-sided region and $P_n(3) = 6$ only for the few maps containing no odd-sided regions (Heawood [2]), $Q_n(u)$ is a polynomial of degree $n-4$, or, if the map contains no odd-sided regions, a polynomial of degree $n-4$ plus a term u^{-1}.

In studying the 4-color conjecture, it is sufficient to consider only those regular major maps which have no proper 4-rings and no proper 5-rings except around a single region. We shall call these maps *5-regular maps.*

Using the method of Whitney [3], we shall find formulas for the four leading coefficients of $Q_n(u)$ and $R_n(x)$, and prove that the four coefficients of $R_n(x)$ satisfy the Birkhoff–Lewis conjecture for all 5-regular maps.

Let R_n be a 5-regular map, with n regions, and let

$$P_n(\lambda) = m_0 \lambda^n + m_1 \lambda^{n-1} + m_2 \lambda^{n-2} + m_3 \lambda^{n-3} + \cdots$$

be the chromatic polynomial associated with P_n.

Then Whitney's formulas yield

$$m_0 = 1, \qquad m_1 = -3(n-2),$$

$$m_2 = \binom{3n-6}{2} = \frac{n-2}{2}(9n-25),$$

$$m_3 = -\binom{3n-6}{3} + (3n-6-2)2(n-2) = \frac{n-2}{2}(3n-8)(3n-11),$$

so that

$$P_n(\lambda) = \lambda^n - 3(n-2)\lambda^{n-1} + \frac{n-2}{2}(9n-25)\lambda^{n-2} + \frac{n-2}{2}(3n-8)(3n-11)\lambda^{n-3} + \cdots.$$

Letting $u = \lambda - 3$, and expanding $P_n(\lambda)$ in powers of u, we get

$$P_n(u) = u^n + 6u^{n-1} + (n+7)u^{n-2} + 4(n-2)u^{n-3} + \cdots.$$

But then

$$Q_n(u) = \frac{P_n(u)}{u(u+1)(u+2)(u+3)} = u^{n-4} + Ou^{n-3} + (n-4)u^{n-4} - 2(n-5)u^{n-7} + \cdots.$$

Now, to derive the first four coefficients of $R_{n+3}(u)$ we first find $P_m(u)$, where $m = n+3$:

$$P_{n+3}(u) = u^{n-1} + Ou^{n-2} + (n-1)u^{n-3} - 2(n-2)u^{n-4} + \cdots.$$

Since the R-polynomial has divisors $\lambda(\lambda-1)(\lambda-2)$, we then find $uQ_{n+3}(u)$:

$$uQ_{n+3}(u) = u^n + Ou^{n-1} + (n-1)u^{n-2} - 2(n-2)u^{n-3} + \cdots.$$

Let $x = \lambda - 4 = u - 1$, and expand uQ_{n+3} in powers of x:

$$R_{n+3}(x) = x^n + nx^{n-1} + \frac{n-1}{2}(n+2)x^{n-2}$$

$$+ \frac{n-2}{6}(n^2 + 5n - 18)x^{n-3} + \cdots.$$

In order to prove that these four coefficients satisfy the Birkhoff–Lewis conjecture, we need only prove that they dominate the corresponding coefficients of $(x+1)^n$ and are dominated by those of $(x+2)^n$, or

$$1 \le 1 \le 1, \tag{1}$$

$$n \le n \le 2n, \tag{2}$$

$$\frac{n-1}{2}n \le \frac{n-1}{2}(n+2) \le \frac{n-1}{2}(4n), \tag{3}$$

$$\frac{n-2}{6}(n^2 - n) \le \frac{n-2}{6}(n^2 + 5n - 18) \le \frac{n-2}{6}8(n^2 - n). \tag{4}$$

Equations (1), (2), and (3) are obviously true for all natural numbers n, while (4) holds for all $n > 3$. Since every 5-regular map has at least 12 regions, $n > 3$ for all maps of this type, and the Birkhoff–Lewis conjecture is confirmed.

References

1. G. D. Birkhoff and D. C. Lewis, Chromatic Polynomials, *Trans. Amer. Math. Soc.* **60**, No. 3 (1946), 355–451.
2. P. J. Heawood, Map-Color Theorem, *Quart. J. Math.* **24** (1890), 332–338.
3. H. Whitney, The Coloring of Graphs, *Ann. Math.* (2) **33** (1952), 688–718.

SIMPLE PROOFS FOR THE RAMANUJAN CONGRUENCES $p(5m+4) \equiv 0 \pmod 5$ AND $p(7m+5) \equiv 0 \pmod 7$

J. M. Gandhi

DEPARTMENT OF MATHEMATICS
UNIVERSITY OF MANITOBA
WINNIPEG, MANITOBA

Let $p(n)$ denote the unrestricted partition of a number n. Then it is known that

$$p(5m + 4) \equiv 0 \pmod 5 \tag{1}$$

$$p(7m + 5) \equiv 0 \pmod 7. \tag{2}$$

Gandhi [2] generalized those congruences and proved the generalized congruences by an elementary method. However his proof makes use of some of his theorems of [1]. Although the central idea is the same it will probably be worthwhile to publish the following simple proof which does not make use of any theorems other than the standard ones.

We recall the following well-known results [3],

$$\prod_{n=1}^{\infty}(1 - x^n)^{-1} = \sum_{n=1}^{\infty} p(n)x^n, \tag{3}$$

$$\prod_{n=1}^{\infty}(1 - x^n)^3 = \sum_{j=0}^{\infty}(-1)^j(2j + 1)x^{j(j+1)/2}, \tag{4}$$

$$\prod_{n=1}^{\infty}(1 - x^n) = \sum_{\beta=0}^{\infty}(-1)^{\beta}x^{\beta(3\beta\pm 1)/2}. \tag{5}$$

Now

$$\prod_{n=1}^{\infty}(1 - x^n)^{-1} = \left\{\prod_{n=1}^{\infty}(1 - x^n)^{-1}\right\}^5 \prod_{n=1}^{\infty}(1 - x^n)^4 \tag{6}$$

since

$$\left\{\prod_{n=1}^{\infty}(1 - x^n)^{-1}\right\}^5 \equiv \prod_{n=1}^{\infty}(1 - x^{5n})^{-1}$$

$$\equiv \sum_{n=1}^{\infty} p(n)x^{5n} \pmod 5.$$

Hence using (3), (4), and (5), we have from (6)

$$\sum_{n=0}^{\infty} p(n)x^n = \sum_{n=0}^{\infty} p(n)x^{5n} \sum_{\beta=0}^{\infty} \sum_{j=0}^{\infty} (-1)^{\beta} x^{\beta(3\beta \pm 1)/2}$$

$$\times (2j + 1)(-1)^j x^{j(j+1)/2} \pmod 5 \tag{7}$$

For $j \equiv 2 \pmod 5$, $2j + 1 \equiv 0 \pmod 5$. Now $p(n)$ must be divisible by 5 for all values of n for which x^n does not occur in the right member of (7), and such values of n are given by

$$n \not\equiv \frac{j(j+1)}{2} + \frac{\beta(3\beta \pm 1)}{2} \pmod 5 \qquad \text{with} \qquad j \neq 2.$$

Now the residues for $j(j + 1)/2 \pmod 5$ with $j \neq 2$ are 0 and 1 while the residues for $\beta(3\beta \pm 1)/2 \pmod 5$ are 0, 1, and 2. Hence the residues for $j(j + 1)/2 + \beta(3\beta \pm 1)/2 \pmod 5$ are 0, 1, 2, and 3. Since 4 is not the residue, hence

$$p(5m + 4) \equiv 0 \pmod 5.$$

Similarly congruence (2) can be proved.

Acknowledgment

I am thankful to Professor N. S. Mendelsohn for his kind encouragement.

References

1. J. M. Gandhi, Congruences for $p_r(n)$ and Ramanujan's τ Function, *Amer. Math. Monthly* **70** (1963), 265–274.
2. J. M. Gandhi, Generalization of Ramanujan's Congruences $p(5m + 4) \equiv 0 \pmod 5$ and $p(7m + 5) \equiv 0 \pmod 7$, *Monatsh. Math.* **69** (1965), 389–392.
3. G. H. Hardy and E. M. Wright, *Introduction to the Theory of Numbers*, Chapt. 9, Oxford Univ. Press, London, 1960.

CONNECTED EXTREMAL EDGE GRAPHS HAVING SYMMETRIC AUTOMORPHISM GROUP

*Allan Gewirtz**
and
Louis V. Quintas

MATHEMATICS DEPARTMENT
PACE COLLEGE
NEW YORK, NEW YORK

1. Introduction

By a *graph* we mean a finite undirected graph (as defined in [2, p. 2]) without loops and without multiple edges. The *automorphism group* of a graph consists of those permutations of the vertex set of the graph which preserve adjacency relations (cf. [2, p. 239]). In [3] the least and greatest number of edges realizable by a graph having n vertices and automorphism group isomorphic to S_m, the *symmetric group of degree m*, was determined for all admissible n. In this paper this is done for the case where the graphs are required to be connected. In obtaining this result the following lemmas play the central role.

Lemma 1. *If* $m \geq 3$, *there exists exactly one connected graph having* $m + 2$ *vertices and automorphism group* S_m, *namely, the graph obtained from the complete* $(m + 1)$*-point by adjoining a pendant edge at one of its vertices* (cf. Section 2).

Lemma 2. *If* $m \geq 2$, *there exists exactly one connected graph having* $m + 1$ *vertices and automorphism group* S_m, *namely, the m-star* (cf. proof of Theorem 2 in [3]).

* Present address: Department of Mathematics, Brooklyn College of The City University of New York.

THEOREM. (a) *Let* $\tilde{e}(S_m, n)$ *denote the least integer for which there exists a connected graph having* $\tilde{e}(S_m, n)$ *edges, n vertices, and automorphism group* S_m $(m \geq 2)$. *Then,*

(i) $\tilde{e}(S_m, n)$ *is not defined for* $n < m$,
(ii) $\tilde{e}(S_2, n) = n - 1$, $n = 2, 3, \ldots$, *and*
(iii) *if* $m \geq 3$, *then*

$$\tilde{e}(S_m, n) = \begin{cases} m(m-1)/2, & n = m \\ m, & n = m+1 \\ (m(m+1)+2)/2, & n = m+2 \\ n-1, & n = m+3, m+4, \ldots. \end{cases}$$

(b) *Let* $\tilde{E}(S_m, n)$ *denote the greatest integer for which there exists a connected graph having* $\tilde{E}(S_m, n)$ *edges, n vertices, and automorphism group* S_m $(m \geq 2)$. *Then, except for the case* $(n = m + 1, m \geq 3)$,

$$\tilde{E}(S_m, n) = E(S_m, n)$$

where $E(S_m, n)$ *is the greatest number of edges realizable by a (not necessarily connected) graph having n vertices and automorphism group* S_m; *if* $m \geq 3$, *then*

$$\tilde{E}(S_m, m+1) = m = \tilde{e}(S_m, m+1)$$

(cf. Section 3).

REMARK. $E(S_m, n)$ was determined for all n in [3].

2. OUTLINE OF THE PROOF OF LEMMA 1

Let K be a graph with $m + 2$ vertices and automorphism group G isomorphic to S_m. Since $m \geq 3$ by hypothesis, we note that all graphs considered in this proof have five or more vertices. The organization of the proof is separated into the following cases.

CASE 1. G leaves a vertex of K fixed. This will imply that G leaves exactly two vertices of K fixed and that K is the graph in the statement of the lemma.

CASE 2. G leaves no vertex of K fixed:

(a) G is transitive and primitive;
(b) G is transitive and imprimitive;
(c) G is intransitive.

For Case 1 if G leaves a vertex fixed, applying [1, Theorem II, p. 209] we deduce that if $m + 1 \neq 6$, G fixes exactly two points, and if $m + 1 = 6$, G fixes exactly two points or G is doubly transitive on the six vertices.

If G fixes exactly two points we deduce that the section graph on the remaining m points is either the complete m-point or the null m-point. That it is not the null m-point is clear when we attempt to adjoin one of the fixed points. We thus obtain, by adjoining this fixed point to the complete m-point, the complete $(m + 1)$-point to which the other fixed point may be adjoined in only one way, that is, to the first fixed point. This yields the graph in the statement of the lemma. If G is doubly transitive on the six vertices then K is the complete 7-point or the 6-star, neither of which has $G \approx S_5$ as its automorphism group.

For Case 2(a), from various theorems in [4] we deduce by group theoretic techniques the nonexistence of a graph K with automorphism group G with the stated properties. Using [1] and [4] for Case 2(b) we deduce that for $m \geq 5$, $G_\alpha \subset A_m \subset G$. $m = 3$ is impossible by group considerations, and if $m = 4$, K must be a connected regular graph with six vertices. One can verify that such a graph does not have automorphism group S_4. Returning to $m \geq 5$ we deduce that the orbits of A_m partition the $m + 2$ vertices of K into a complete block system of G consisting of the two blocks

$$\Delta_\alpha = \{\alpha_1, \ldots, \alpha_{(m+2)/2}\}, \qquad \Delta_\beta = \{\beta_1, \ldots, \beta_{(m+2)/2}\}.$$

From this we deduce that K contains as a subgraph the complete bipartite graph on the vertex sets Δ_α and Δ_β.

Since Δ_α and Δ_β are blocks, the section graphs on Δ_α and Δ_β are isomorphic graphs. Denote these by $S(\Delta_\alpha)$ and $S(\Delta_\beta)$, respectively.

Let $\overline{K}$ be the complementary graph of K. Then,

$$\overline{K} = \overline{S(\Delta_\alpha)} + \overline{S(\Delta_\beta)}.$$

By [3, Lemma 4] with $m \geq 3$ we have, since $\overline{K}$ is not connected, that $\overline{K}$ consists of m isomorphic asymmetric graphs (by an *asymmetric graph* we mean a graph with an identity automorphism group) plus possibly some other type of asymmetric graph W. But the structure of $\overline{K}$ is such that all components come in even numbers. Thus, there cannot be an asymmetric graph W as part of $\overline{K}$ since this would give rise to automorphisms other than those in G. Therefore, $\overline{K}$ consists precisely of m isomorphic asymmetric graphs. Since asymmetric graphs have $v = 1$ or $v \geq 6$ vertices, it is impossible to have m such graphs on a total of $m + 2$ vertices.

Thus we have shown that if $G \approx S_m$ is transitive and imprimitive, there are no graphs on $m + 2$ vertices with such a group. This completes the discussion of the transitive fixed-point free cases, which are parts (a) and (b) of Case 2.

We now consider Case 2(c), where we assume G is intransitive and has no fixed points.

In a manner similar to that used in 2(b) we deduce for Case 2(c) that there are no graphs K with such an automorphism group G.

We thank Charles C. Sims for his suggestions on how this lemma could be proved. A complete proof will appear elsewhere.

3. Proof of Theorem

Part (a) There are no graphs having $n < m$ vertices and automorphism group S_m. This is clear since there are not enough vertices to generate the necessary number of permutations. Thus,

$$\tilde{e}(S_m, n) \qquad \text{is not defined for} \quad n < m. \tag{1}$$

If $n = m$, there are exactly two graphs having group S_m, the complete m-point and the null m-point. Thus,

$$\tilde{e}(S_m, m) = m(m-1)/2. \tag{2}$$

Since we are only considering connected graphs, we have

$$\tilde{e}(S_m, n) \geq n - 1. \tag{3}$$

Thus, in those cases where there exist trees with group S_m this lower bound for $\tilde{e}(S_m, n)$ will be realized.

If $m = 2$, then the simple arc on n vertices has group S_2. Thus,

$$\tilde{e}(S_2, n) = n - 1 \qquad (n = 2, 3, \ldots). \tag{4}$$

If $n = m + 1$, then the m-star realizes the bound (3). Thus,

$$\tilde{e}(S_m, m + 1) = m \qquad (m = 3, 4, \ldots). \tag{5}$$

If $n = m + 3, m + 4, \ldots$, then the tree depicted in Fig. 1 has $n = m + k$ ($k = 3, 4, \ldots$) vertices and group S_m. Thus,

$$\tilde{e}(S_m, n) = n - 1 \qquad (n = m + 3, m + 4, \ldots). \tag{6}$$

If $n = m + 2$ and $m \geq 3$, there exists a unique connected graph having $m + 2$ vertices and group S_m (see Lemma 1, Section 1). Since this graph has $(m(m + 1) + 2)/2$ edges, we have

$$\tilde{e}(S_m, m + 2) = (m(m + 1) + 2)/2. \tag{7}$$

The statements (1), (2), and (4)–(7) taken together are the assertions of Part (a) of the theorem.

Part (b) In [3] minimum edge (not necessarily connected) graphs having group S_m were defined and used to prove Theorem 2 of that paper. In all but two cases these minimum edge graphs contained a singleton graph as a component and no vertex of degree $n - 1$. The complementary graph of such

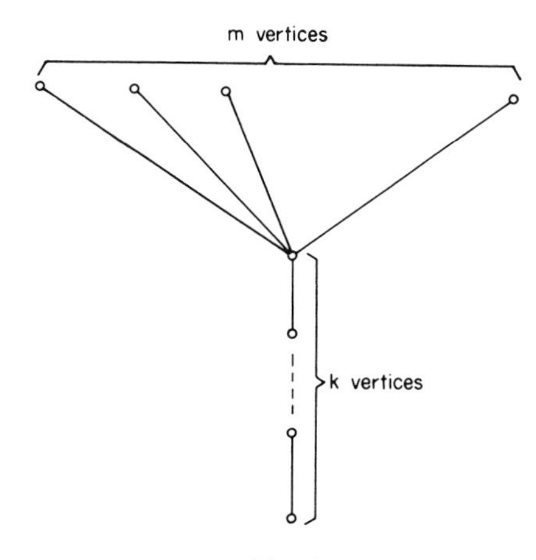

FIG. 1.

a graph is connected, has group S_m, and a maximum number of edges. Thus, $\tilde{E}(S_m, n)$ for these cases is the same number obtained in [3]. The two exceptional cases are $(n = m + 1, m \geq 3)$ and $(n = m + 3, m \geq 3)$. In the latter case the graph has no vertex of degree $n - 1$ and does have two nonadjacent vertices of degree 1 which are not adjacent to a common vertex. This implies that the complementary graph is connected. In the former case there exists only one connected graph having $m + 1$ vertices, the m-star (see Lemma 2, Section 1). Therefore, we have

$$\tilde{E}(S_m, m + 1) = m = \tilde{e}(S_m, m + 1) \qquad (m \geq 3).$$

REFERENCES

1. W. BURNSIDE, *Theory of Groups of Finite Order*, Dover, New York, 1955.
2. O. ORE, *Theory of Graphs*, Colloq. Pub. Vol. 38, Amer, Math. Soc., Providence, Rhode Island 1962.
3. L. V. QUINTAS, The Least Number of Edges for Graphs Having Symmetric Automorphism Group, *J. Combinatorial Theory* **5** (1968).
4. H. WIELANDT, *Finite Permutation Groups*, Academic Press, New York, 1964.

BOUNDS ON THE CHROMATIC AND ACHROMATIC NUMBERS OF COMPLEMENTARY GRAPHS

*Ram Prakash Gupta**, †

DEPARTMENT OF STATISTICS
UNIVERSITY OF NORTH CAROLINA
CHAPEL HILL, NORTH CAROLINA

1. Introduction

The graphs considered below are assumed to be nonnull, finite, undirected, which are to have no loops and no multiple edges.

A *graph* G consists of a set $V(G)$ of *vertices* together with a set $E(G)$ of unordered pairs $[u, v]$ of distinct vertices $u, v \in V(G)$. The elements of $E(G)$ are called *edges* of the graph G. If $[u, v]$ is an edge of G, the vertices u and v are said to be adjacent. The *order* of a graph G is the number of its vertices. For any $S \subseteq V(G)$, the *subgraph* G' of G, induced by S, is defined as follows:

Definition. $V(G') = S$ and for any $u, v \in S$, $[u, v] \in E(G')$ if and only if $[u, v] \in E(G)$.

Two graphs G and $\bar{G}$ are called *complementary* if they have the same set of vertices and any two vertices are adjacent in one of G or $\bar{G}$ but not in both, i.e., $V(G) = V(\bar{G})$ and $[u, v] \in E(G)$ if and only if $[u, v] \notin E(\bar{G})$.

Consider a graph G and let $\alpha_1, \alpha_2, \ldots, \alpha_k$ represent k distinct colors. Any function f which assigns to each vertex v of G a unique color $f(v) \in \{\alpha_1, \alpha_2, \ldots, \alpha_k\}$ is called *coloring* or, more specifically, a *k-coloring* of G. If $f(v) = \alpha$, we say that v is colored α or that v is an α-vertex. For any $S \subseteq V(G)$, $f(S)$

* This research was supported by the National Science Foundation Grant No. GP-5790.

† The author's present address is the Department of Mathematics, Ohio State University, Columbus, Ohio.

denotes the set of colors $\{f(v)/v \in S\}$. Any k-coloring f of G induces a decomposition of $V(G)$,

$$V(G) = V_1 \cup V_2 \cup \cdots \cup V_k, \qquad V_i \cap V_j = \emptyset, \qquad i \neq j, \tag{1}$$

where V_i is precisely the set of all α_i-vertices. Conversely, any decomposition (1) of $V(G)$ induces a k-coloring f of G such that $f(v) = \alpha_i$ whenever $v \in V_i$, $1 \leq i \leq k$. Thus, there is a natural one-to-one correspondence between k-colorings of G and decompositions of $V(G)$ into k mutually disjoint sets.

In the present paper, we shall consider k-colorings f of a graph G, with (1) as the induced decomposition of $V(G)$, which satisfy one or both of the following two conditions:

(R) For any two vertices u and v of G, $[u, v] \in E(G)$ implies $f(u) \neq f(v)$ or, equivalently, for each i, $1 \leq i \leq k$, $u, v \in V_i$ implies u and v are not adjacent.

(C) For any two colors α_i and α_j, $1 \leq i < j \leq k$, there is an edge $[u, v] \in E(G)$ with $f(u) = \alpha_i$ and $f(v) = \alpha_j$ or, equivalently, for each i and j, $1 \leq i < j \leq k$, there exist vertices $u \in V_i$, $v \in V_j$ such that u and v are adjacent.

A k-coloring f of G is called *regular*, *pseudocomplete*, or *complete* according as it satisfies the condition (R), (C), or both (R) and (C), respectively. The *chromatic number* of G, denoted by $\chi(G)$, is the minimum number k for which a regular k-coloring of G exists. The *pseudoachromatic number* of G, denoted by $\psi_s(G)$, is defined to be the maximum number k for which a pseudocomplete k-coloring of G exists. Finally, the *achromatic number* of G, denoted by $\psi(G)$, is the maximum number k for which a complete k-coloring of G exists. From the definitions, it is obvious that for any graph $\psi(G) \leq \psi_s(G)$. Also, if $\chi(G) = k$ and f is any regular k-coloring of G, then it is easily seen that f must also be pseudocomplete, so that clearly $\chi(G) \leq \psi(G)$. Hence, for any graph G, we always have

$$\chi(G) \leq \psi(G) \leq \psi_s(G). \tag{2}$$

Further, there exist graphs G for which the strict inequalities $\chi(G) < \psi(G)$ and/or $\psi(G) < \psi_s(G)$ may hold. For instance, if G is the graph of Fig. 1, then it is

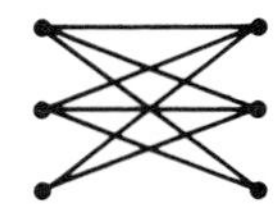

FIG. 1.

easily verified that $\chi(G) = 2$, $\psi(G) = 3$, and $\psi_s(G) = 4$. From the above observations, it is evident that the concept of pseudoachromatic number of a graph as defined here is a proper generalization of the concept of achromatic number of a graph, introduced by Hedetniemi [1].

Let G and $\bar{G}$ be complementary graphs defined on a set of p vertices. Then it is known [2] that

$$2\sqrt{p} \leq \chi(G) + \chi(\bar{G}) \leq p + 1. \tag{3}$$

In the present paper, we determine the following upper bounds:

$$\chi(G) + \psi_s(\bar{G}) \leq p + 1, \tag{4}$$

$$\psi_s(G) + \psi_s(\bar{G}) \leq \{\tfrac{4}{3}p\}. \tag{5}$$

The bounds (4) and (5) are shown to be exact. As a corollary to (4), we obtain the bound

$$\chi(G) + \psi(\bar{G}) \leq p + 1, \tag{6}$$

conjectured by Hedetniemi [1], and also the upper bound in (3) due to Nordhaus and Gaddum [2]. From (5), we obtain the (exact) upper bound

$$\psi(G) + \psi(\bar{G}) \leq \{\tfrac{4}{3}p\} \tag{7}$$

which answers a question by Hedetniemi [1] in the negative.

2. Bounds

In the following $|A|$ denotes, as usual, the number of elements in the set A; $[x]$ denotes the integral part of the number x and $\{x\}$ is the smallest integer greater than or equal to x.

We first prove the following.

THEOREM 2. *If G and $\bar{G}$ are complementary graphs of order p, then*

$$\chi(G) + \psi_s(\bar{G}) \leq p + 1, \tag{8}$$

$$\chi(G) + \psi(\bar{G}) \leq p + 1, \tag{9}$$

$$\chi(G) + \chi(\bar{G}) \leq p + 1. \tag{10}$$

For any $p \geq 1$, the bounds (8), (9), *and* (10) *are attainable.*

PROOF: We shall first prove (8). The bounds (9) and (10) then follow immediately from (8) and (2).

Let G and $\bar{G}$ be complementary graphs of order p, and let $\psi_s(\bar{G}) = k$. Obviously, $1 \leq k \leq p$. If $k = 1$, then since clearly $\chi(G) \leq p$, we have the inequality (8). We may therefore assume that $k > 1$. Now, consider any pseudocomplete k-coloring of $\bar{G}$ and let

$$V(G) = V(\bar{G}) = V_1 \cup V_2 \cup \cdots \cup V_k, \qquad V_i \cap V_j = \varnothing, \qquad i \neq j, \tag{11}$$

be the induced decomposition of $V(G)$. Let G_r denote the subgraph of G induced by the set of vertices $V(G_r) = V_1 \cup V_2 \cup \cdots \cup V_r$ and let $|V(G_r)| = p_r$, $r = 1, 2, \ldots, k$. We assert that $\chi(G_r) \le p_r - r + 1$.

To prove the assertion, it is evidently sufficient to show that G_r possesses a $(p_r - r + 1)$-coloring f_r which is regular. Now, for $r = 1$, we may define f_1 by assigning p_1 distinct colors to the vertices of G_1 so that any two vertices are colored differently. Let us assume as induction hypothesis that we have already defined a (fixed) regular $(p_r - r + 1)$-coloring f_r of G_r for some r, $1 \le r < k$. Now consider the graph G_{r+1}. Since the decomposition (11) is induced by a pseudocomplete coloring of $\bar{G}$, by definition it is clear that there exist vertices $v_1 \in V_1$, $v_2 \in V_2$, $\ldots$, $v_r \in V_r$ such that for some vertices $u_i \in V_{r+1}$ $(1 \le i \le r)$, which need not all be distinct, $[v_i, u_i] \in E(\bar{G})$, so that clearly $[v_i, u_i] \notin E(G_{r+1})$. Now, we define a coloring f_{r+1} of G_{r+1} by using the coloring f_r of G_r. If each of the colors $f_r(v_1), f_r(v_2), \ldots, f_r(v_r)$ is also assigned by f_r to some vertex in $V(G_r) - \{v_1, v_2, \ldots, v_r\}$, then evidently the number of colors actually used by f_r does not exceed $|V(G_r) - \{v_1, v_2, \ldots, v_r\}| = p_r - r$. In this case, we define f_{r+1} as follows: assign $|V_{r+1}| = p_{r+1} - p_r$ new colors to the vertices in V_{r+1} and let $f_{r+1}(v) = f_r(v)$ for the rest of the vertices $v \in V(G_{r+1})$. And, if there is a color α among $f_r(v_1), f_r(v_2), \ldots, f_r(v_r)$ which is not assigned by f_r to any vertex in $V(G_r) - \{v_1, v_2, \ldots, v_r\}$ so that all the α-vertices $v_{i_1}, v_{i_2}, \ldots, v_{i_t}$ $(1 \le t \le r)$, say, are among $v_1, v_2, \ldots, v_r$, then, we define f_{r+1} as follows: assign $p_{r+1} - p_r$ new colors to the vertices in V_{r+1}; delete the color α by putting $f_{r+1}(v_{i_j}) = f_{r+1}(u_{i_j})$, $1 \le j \le t$, where $u_{i_j} \in V_{r+1}$ is such that $[v_{i_j}, u_{i_j}] \notin E(G_{r+1})$; for all other vertices $v \in V(G_{r+1})$, let $f_{r+1}(v) = f_r(v)$. In either case, clearly f_{r+1} is a $(p_{r+1} - \overline{r + 1} + 1)$-coloring of G_{r+1}, and it is easily verified that f_{r+1} is regular. The assertion is now proved by finite induction. In particular, since $G_k = G$, we have $\chi(G) \le p - k + 1$ from which we obtain (8) immediately.

To see that the bounds (8)–(10) are attainable, it is sufficient to let G be a complete graph of order p, i.e., G has p vertices each pair of which is adjacent in G so that no two vertices are adjacent in $\bar{G}$. Then, clearly $\chi(G) = p$ $\chi(\bar{G}) = \psi(\bar{G}) = \psi_s(\bar{G}) = 1$ and the equalities in (8)–(10) hold. This completes the proof of the theorem.

We shall now prove the following.

THEOREM 2. *If G and $\bar{G}$ are complementary graphs of order p, then*

$$\psi_s(G) + \psi_s(\bar{G}) \le \{\tfrac{4}{3}p\}, \tag{12}$$

$$\psi(G) + \psi_s(\bar{G}) \le \{\tfrac{4}{3}p\}, \tag{13}$$

$$\psi(G) + \psi(\bar{G}) \le \{\tfrac{4}{3}p\}. \tag{14}$$

For any $p \ge 1$, the bounds (12)–(14) *are attainable.*

PROOF: We shall first prove (12). The bounds (13) and (14) then follow immediately from (12) and (2).

Let G and $\bar{G}$ be complementary graphs of order p, and let $\psi_s(G) = k$. Obviously, $1 \leq k \leq p$. If $k = 1$, then since clearly $\psi_s(\bar{G}) \leq p$, we have the inequality (12). We may therefore assume that $k > 1$. Now, consider any pseudocomplete k-coloring of G and let

$$V(G) = V(\bar{G}) = V_1 \cup V_2 \cup \cdots \cup V_k, \qquad V_i \cap V_j = \varnothing, \qquad i = j, \quad (15)$$

be the induced decomposition of $V(\bar{G})$. Clearly ,$V_i \neq \varnothing$ for $i = 1, 2, \ldots, k$. Let the number of sets among $V_1, V_2, \ldots, V_k$ which consist of exactly one vertex each be r where $r \geq 0$. To be definite, we may assume that $V_i = \{v_i\}$, $i = 1, 2, \ldots, r$ (if $r > 0$). Then, the remaining sets $V_{r+1}, \ldots, V_k$ consist of at least two vertices each. Since the sets V_i are mutually disjoint, we have evidently

$$k \leq r + \left[\frac{p-r}{2}\right]. \quad (16)$$

Now, let $\psi_s(\bar{G}) = k'$ and let f be any pseudocomplete k'-coloring of $\bar{G}$. Since the decomposition (15) is induced by a pseudocomplete coloring of G, it is observed that each pair of vertices in $\{v_1, v_2, \ldots, v_r\}$ must be adjacent in G so that no two vertices in $\{v_1, v_2, \ldots, v_r\}$ are adjacent in $\bar{G}$. Therefore, since f is a pseudocomplete coloring of $\bar{G}$, it is easily seen that there can be at most one color in $\{f(v_1), f(v_2), \ldots, f(v_r)\}$ which is not assigned by f to any vertex in $V(\bar{G}) - \{v_1, v_2, \ldots, v_r\}$. Hence, we have evidently

$$k' \leq |V(\bar{G}) - \{v_1, v_2, \ldots, v_r\}| + 1$$

or

$$k' \leq p - r + 1. \quad (17)$$

We shall next prove the following inequality

$$k' \leq p - \left[\frac{k}{2}\right]. \quad (18)$$

To this end, consider the sets of colors $f(V_1), f(V_2), \ldots, f(V_k)$. If for each index i, $1 \leq i \leq k$, we have either $|f(V_i)| < |V_i|$ or $f(V_i) \cap f(V_j) \neq \varnothing$ for some $j, j \neq i$, then it is easily seen (by induction on k) that

$$k' = \left|\bigcup_{i=1}^{k} f(V_i)\right| \leq \left|\bigcup_{i=1}^{k} V_i\right| - \left\{\frac{k}{2}\right\} = p - \left\{\frac{k}{2}\right\}$$

and hence, *a fortiori*, we have the inequality (18). We may therefore assume that $f(V_1)$, say, is such that $|f(V_1)| = |V_1| = p_1$ and $f(V_1) \cap f(V_j) = \varnothing$ for

$2 \le j \le k$. If possible, let there be another set $f(V_2)$, say, such that $|f(V_2)| = |V_2|$ and $f(V_2) \cap f(V_j) = \varnothing$ for $j \neq 2$. Now, since there exist vertices $v_1 \in V_1$, $v_2 \in V_2$ such that $[v_1, v_2] \in E(G)$ or $[v_1, v_2] \notin E(\bar{G})$, it is seen that there can be no edge $[v, u] \in E(\bar{G})$ with $f(v) = f(v_1)$ and $f(u) = f(v_2)$. This, however, contradicts the fact that f is a pseudocomplete coloring of $\bar{G}$. Hence, we must have for every index i, $2 \le i \le k$, either $|f(V_i)| < |V_i|$ or $f(V_i) \cap f(V_j) \neq \varnothing$ for some j, $j \neq i$. Hence, as above, we have

$$k' - p_1 = \left| \bigcup_{i=2}^{k} f(V_i) \right| \le \left| \bigcup_{i=2}^{k} V_i \right| - \left\{ \frac{k-1}{2} \right\} = p - p_1 - \left[\frac{k}{2} \right]$$

whence we obtain (18) immediately.

Now, we shall derive (12), from (16), (17), and (18). If $r \ge [p/3] + 1$, then from (16) and (17), we obtain

$$\psi_s(G) + \psi_s(\bar{G}) = k + k' \le p + \left[\frac{p-r}{2} \right] + 1 \le \{\tfrac{4}{3}p\}.$$

(The last inequality is obtained by substituting $r = [p/3] + 1$ and some elementary simplification.) If $r \le [p/3] + 1$, then from (16) and (18) we obtain similarly $\psi_s(G) + \psi_s(\bar{G}) \le \{\tfrac{4}{3}p\}$. This completes the proof of (12).

It now remains to show that for any $p \ge 1$, the bounds (12)–(14) are attainable. For $p \le 3$, this is obvious. In general, it is evidently sufficient if we show that the bound (14) is attainable for all p of the form $3r + 1$, $r = 1, 2, \ldots$. To this end, we construct below examples of graphs G_r of order $3r + 1$ successively for $r = 1, 2, \ldots$, and define $(2r + 1)$-colorings f_r and $\bar{f}_r$ of G_r and $\bar{G}_r$, respectively. For $r = 1$, we define $G_1, f_1, \bar{f}_1$ as follows:

$$V(G_1) = \{v_0, v_1, v_2, v_3\},$$

$$E(G_1) = \{[v_0, v_2], [v_1, v_3], [v_2, v_3]\};$$

$$f_1(v_0) = f_1(v_1) = \alpha_1, \qquad f_1(v_2) = \alpha_2, \qquad f_1(v_3) = \alpha_3;$$

$$\bar{f}_1(v_0) = \alpha_1, \qquad \bar{f}_1(v_1) = \alpha_2, \qquad \bar{f}_1(v_2) = \bar{f}_1(v_3) = \alpha_3.$$

Suppose that we have already defined $G_r, f_r, \bar{f}_r$ for some $r \ge 1$. Now, we define $G_{r+1}, f_{r+1}, \bar{f}_{r+1}$ as follows:

$$V(G_{r+1}) = V(G_r) \cup \{v_{3r+1}, v_{3r+2}, v_{3r+3}\},$$

$$\begin{aligned} E(G_{r+1}) = E(G_r) &\cup \{[v_{3r+1}, v_{3i-1} \mid i = 1, 2, \ldots, r\} \\ &\cup \{[v_{3r+2}, v_{3i}] \mid i = 0, 1, \ldots, r+1\} \\ &\cup \{[v_{3r+3}, v_{3i-1}], [v_{3r+3}, v_{3i}] \mid i = 1, 2, \ldots, r\} \\ &\cup \{[v_{3r+3}, v_1]\}; \end{aligned}$$

$$f_{r+1}(v_{3r+1}) = f_{r+1}(v_{3r+2}) = \alpha_{2r+2}, \qquad f_{r+1}(v_{3r+3}) = \alpha_{2r+3},$$

$$f_{r+1}(v_i) = f_r(v_i) \qquad \text{for} \qquad i = 0, 1, \ldots, 3r;$$

$$\bar{f}_{r+1}(v_{3r+1}) = \alpha_{2r+2}, \qquad \bar{f}_{r+1}(v_{3r+2}) = \bar{f}_{r+1}(v_{3r+3}) = \alpha_{2r+3},$$

$$\bar{f}_{r+1}(v_i) = \bar{f}_r(v_i) \qquad \text{for} \qquad i = 0, 1, \ldots, 3r.$$

It is easily checked that f_r and $\bar{f}_r$ are complete $(2r + 1)$-colorings of G_r and $\bar{G}_r$, respectively. Hence, by definition, we have $\psi(G_r) + \psi(\bar{G}_r) \geq 2(2r + 1) = \{\frac{4}{3}(3r + 1)\}$. But, from (14), the equality $\psi(G_r) + \psi(\bar{G}_r) = \{\frac{4}{3}(3r + 1)\}$ must hold.

This completes the proof of the theorem.

REMARK. The research work presented in this note was motivated by the suggestions made by Hedetniemi [1]. He conjectured that for any complementary graphs G and $\bar{G}$ of order p, we have: (a) $\psi(G) + \psi(\bar{G}) \leq p + 1$, and further, asked the question: (b) $\psi(G) + \psi(\bar{G}) \leq p + 2$? Clearly, Theorem 1 yields a stronger result than (a) and Theorem 2 answers the question (b) in the negative.

REFERENCES

1. S. T. HEDETNIEMI, Homomorphism of Graphs and Automata, Univ. of Michigan, Tech. Rept., pp. 24–25. Ann Arbor, Michigan (1966).
2. E. A. NORDHAUS and J. W. GADDUM, On Complementary Graphs, *Amer. Math. Monthly* **63** (1956), 175–177.

A PROBLEM OF ZARANKIEWICZ

Richard K. Guy

DEPARTMENT OF MATHEMATICS
UNIVERSITY OF CALGARY
CALGARY, ALBERTA

and

Stefan Znám

SLOVENSKÁ AKADÉMIE VIED KABINET MATEMATIKY
BRATISLAVA, CZECHOSLOVAKIA

1. Introduction

Zarankiewicz [12] and others [1–10] have asked for the least integer $k = k_{i,j}(m, n)$, $2 \leq i \leq m$, $2 \leq j \leq n$, such that an $m \times n$ matrix containing k ones and $mn - k$ zeros, must contain an $i \times j$ submatrix all of whose elements are 1, however the elements are distributed. The case $i = j$, $m = n$ has attracted particular attention where we write $k = k_j(n)$.

Hartman *et al.* [5] showed that

$$(\tfrac{3}{4}2^{1/3} - \varepsilon)n^{4/3} < k_2(n) < (2 + \varepsilon)n^{3/2}, \tag{1}$$

Kővári *et al.* [7, see 11 for graph-theoretic connections] showed that

$$k_j(n) < jn + [(j - 1)^{1/j} n^{2 - 1/j}], \tag{2}$$

where brackets denote "integer part," and Hylteń-Cavallius [6] observed that their method yields

$$k_{i,j}(m, n) < (i - 1)n + [(j - 1)^{1/i} n^{1 - 1/i} m], \tag{3}$$

and for $i = 2$ improved this to

$$k_{2,j}(m, n) \leq 1 + [\tfrac{1}{2}n + \{(j - 1)nm(m - 1) + \tfrac{1}{4}n^2\}^{1/2}]. \tag{4}$$

Reiman [8] also gave this result in the case $j = 2$, showed that equality occurred infinitely often, and noted the connection with finite projective and affine planes. We will be concerned with the cases $j \geq i \geq 3$.

One of us (Znám [13]) improved (3) in the case $i = j$, $m = n$ to

$$k_j(n) \leq 1 + [\tfrac{1}{2}n(j-1) + (j-1)^{1/j}n^{2-(1/j)}], \tag{5}$$

noted that it was not as good as (4) for $j = 2$, and later [14] made the further improvements

$$k_j(n) \leq 1 + [\tfrac{1}{2}n(j-1) + (j-1)^{1/j}n(n - \tfrac{3}{8}(j-1))^{1-(1/j)}], \tag{6}$$

$$k_j(n) \leq 1 + [n(j-1)/e + (j-1)^{1/j}n^{2-(1/j)}], \tag{7}$$

where $e = (2f-1)/(f-1)$, $(j-1)f^j = n$, and, using the methods of [2, 13, 14], gave (unpublished) the further improvements

$$k_j(n) \leq 1 + \left[\frac{1}{2}n(j-1) + n(j-1)^{1/j} \times \left(n^2 - (j-1)n + \frac{1}{4}(j-1)^{2-(2/j)}n^{(2/j)}\right)^{(j-1)/2j}\right], \tag{8}$$

$$k_3(n) \leq 1 + n + \left[n\left(2n^2 - \frac{11}{2}n + \frac{9}{2}\right)^{1/3}\right], \tag{9}$$

$$k_4(n) \leq 1 + \left[\frac{3}{2}n + n\left(3n^3 - \frac{397}{24}n^2 + \frac{89}{3}n - \frac{257}{16}\right)^{1/4}\right]. \tag{10}$$

Guy [4] showed that (7) is sharper than (6) for $j = 2$ and 3, but for $j \geq 4$, only when n is in the range

$$4 \leq j \leq n \leq \frac{3^j(j-1)^{j+1}}{2^j(j-3)^j}.$$

Similar methods shows that for $j = 3$, (9) is better than (8) for $n \geq 3[3]$, which in turn is better than (7) for $n \geq 39[41]$; while (9) is better than (7) for $n \geq 5[8]$. The values in brackets are the least n for which the formulas give an actual improvement for k, which is necessarily an integer. For $j = 4$, (10) is better than (8) for $n \geq 4[4]$, which in turn is better than (6) for $n \geq 4[4]$, which in turn is better than (7) for $n \geq 1226[1255]$; while (8) is better than (7) for $n \geq 25[29]$ and (10) is better than (6) for $n \geq 4[4]$, and better than (7) for $n \geq 4[6]$. For $n \geq j \geq 4$, it can be shown that (8) is better than (7) for $n \geq (j-1)(2j-1)^j/(2j-4)^j$.

Theorem 2 improves on (5)–(8), and Theorems 3–9 give specific improvements, e.g., on (9) and (10), in the cases $3 \leq i \leq 9$.

2. General Theorems

We make this paper self-contained by giving a new and simpler proof of the main result of the methods developed in [2, 13, 14].

THEOREM 1. *If an $m \times n$ matrix contains more than nu ones, and it can be shown that*

$$n\binom{u}{i} \geq (j-1)\binom{m}{i}, \tag{11}$$

then there is an $i \times j$ submatrix consisting entirely of ones.

NOTE 1. u is not restricted to integer values; assume $u \geq i$ and define

$$\binom{u}{i} \quad \text{as} \quad u(u-1)\cdots(u-i+1)/i!.$$

NOTE 2. It follows from the definition of $k_{i,j}(m, n)$ that under the conditions of Theorem 1,

$$k_{i,j}(m, n) \leq 1 + [nu]. \tag{12}$$

PROOF OF THEOREM 1: Suppose column l contains c_l ones, so that

$$\sum_{l=1}^{n} c_l > nu. \tag{13}$$

The number of sets of i ones in the same column, totaled over all columns, is

$$\sum_{l=1}^{n} \binom{c_l}{i}$$

and if this exceeds

$$(j-1)\binom{m}{i},$$

then by the pigeon-hole principle there are at least j columns which each contain corresponding sets of i ones, forming an $i \times j$ submatrix of ones. It remains to show

$$\sum_{l=1}^{n} \binom{c_l}{i} > n\binom{u}{i}, \tag{14}$$

since this, with (11), proves the theorem.

Consider the function $y = x(x-1)\cdots(x-i+1)$. For $x \geq i$, y and its first two derivatives are positive so the function is convex, and the average of its values, taken over $x = c_l$, $1 \leq l \leq n$, is at least equal to (greater than unless c_l is constant) the value of y at the average point

$$\bar{x} = \frac{1}{n}\sum_{l=1}^{n} c_l.$$

Divide by $i!$ and

$$\frac{1}{n}\sum_{l=1}^{n} \binom{c_l}{i} = \binom{\bar{x}}{i},$$

which is greater than

$$\binom{u}{i}$$

by (13). Multiply by n, and formula (14) and Theorem 1 are proved.

THEOREM 2. *If* $3 \leq i \leq m$, $3 \leq j \leq n$, *and* $n \ll m^i$, *then* (12) *holds*,

$$k_{i,j}(m, n) \leq 1 + [nu], \qquad \text{where} \qquad u = v + \tfrac{1}{2}(i-1),$$

$$v = z + (i^2-1)/24z + (i^2-1)(i^2-9)/1920z^3 + (i^2-1)(i^2-4)(i^2-25)/41472z^5$$

and

$$nz^i = (j-1)m(m-1)\cdots(m-i+1).$$

NOTE. The last term in v has been artificially enlarged, to ensure inequality; the series naturally continues

$$(i^2-1)(i^2-25)(13i^2-61)/580608z^5 + (i^2-1)(i^2-49)(1967i^4-29470i^2+87983)/1393459200z^7 + \cdots.$$

PROOF OF THEOREM 2: This follows from Theorem 1, Note 2, if we show that u satisfies (11), which may be rewritten $u(u-1)\cdots(u-i+1) \geq z^i$. But the left member is greater than

$$v^i - i(i^2-1)v^{i-2}/24 + i(i^2-1)(i-2)(i-3)(5i+7)v^{i-4}/5760$$
$$- i(i^2-1)(i-2)(i-3)(i-4)(i-5)(35i^2+112i+93)v^{i-6}/2903040$$
$$= z^i + i(i^2-1)(i^2-25)(i^2+5)z^{i-6}/580608 + \cdots > z^i,$$

and the theorem is proved

3. SPECIAL CASES

Theorem i, $3 \leq i \leq 9$, gives a sharper result in these special cases. The method of proof is the same as that of Theorem 2, and the calculations are omitted. The symbols u, v, w, and z depend on i and are connected by the relations $u = v + \frac{1}{2}(i-1)$, $w = v^2$, $nz^i = (j-1)m(m-1)\cdots(m-i+1)$.

THEOREM 3. *If* $3 \leq m$, $2 \leq j \leq n$, *then* $k_{3,j}(m, n) \leq 1 + [nu]$, *where* $v = z + 1/3z$ $[= u - \frac{1}{2}(i-1)]$.

PROOF: $u(u-1)(u-2) = z^3 + 1/27z^3 > z^3 = (j-1)m(m-1)(m-2)/n$.

THEOREM 4. *If* $4 \leq m$, $2 \leq j \leq n$, *then* $k_{4,j}(m, n) \leq 1 + [nu]$, *where* $(v^2 =)\ w = (5/4) + (1 + z^4)^{1/2}$.

PROOF: $u(u-1)(u-2)(u-3) = z^4 = (j-1)m(m-1)(m-2)(m-3)/n$.

THEOREM 5. *If* $5 \leq m$, $2 \leq j \leq n$, *and* $300n < m^5(j-1)$, *then* $k_{5,j}(m, n) \leq 1 + [nu]$, *where* $v = z + 1/z + 1/5z^3 - 6/25z^7 - 6/25z^9$.

PROOF: $u(u-1)(u-2)(u-3)(u-4) = v(v^2-1)(v^2-4) > z^5 + 1/5z^5 - 36/25z^7 - \cdots > z^5$ provided z is sufficiently large, e.g., $z > 3.1$, $m^5(j-1) > 300n$.

THEOREM 6. *If* $6 \leq m$, $2 \leq j \leq n$, *then* $k_{6,j}(m, n) \leq 1 + [nu]$, *where* $w = t + 35/12$, $t = \alpha + \beta$, $\alpha^3 = z^6 + 160/27 - 21952/729z^6$, $\beta = 28/9z^2$.

PROOF:

$$\begin{aligned} u(u-1)\cdots(u-5) &= (v^2-(1/4))(v^2-(9/4))(v^2-(25/4)) \\ &= w^3 - 35w^2/4 + 259w/16 - 225/64 \\ &= t^3 - 28t/3 - 160/27 = z^6 + t(3\alpha\beta - 28/3) > z^6. \end{aligned}$$

THEOREM 7. *If* $7 \leq m$, $2 \leq j \leq n$, *then* $k_{7,j} \leq 1 + [nu]$, *where* $v = z + 2/z + 1/z^3 + 8/7z^5$.

PROOF: $u(u-1)\cdots(u-6) = z^7 + 62/z^3 + \cdots > z^7$.

THEOREM 8. *If* $8 \leq m$, $2 \leq j \leq n$, *then* $k_{8,j} \leq 1 + [nu]$, *where* $(v^2 =)$ $w = z^2 + 21/4 + 21/2z^2 + 16/z^4 + 231/8z^6$.

PROOF: $u(u-1)\cdots(u-7) = z^8 + 6899/4z^4 + 10752/z^6 + 1644657/32z^8 + \cdots > z^8$.

NOTE. The theorem gives a sharp estimate provided $n \ll m^4$. A result which is sharp without this restriction may be obtained from an exact solution of the quartic $w^4 - 21w^3 + (987/8)w^2 - (3229/16)w + (11025/256) = z^8$.

THEOREM 9. *If* $9 \leq m$, $2 \leq j \leq n$, *then* $k_{9,j} \leq 1 + [nu]$, *where* $v = z + 10/3z + 3/z^3 + 620/81z^5 + 4735/243z^7$.

PROOF:

$$u(u-1)\cdots(u-8) = z^9 + 3635419/729z^3 + 1184283550/19683z^5 + \cdots > z^9.$$

4. Exact Results

Exact results have been given in [4] for $2 \leq i, j \leq 3$, and small m and n. We extend these to the case $i = 2$, $j = 4$ by a theorem of Čulík [3], of which Theorems 10 and 11 are special cases, and by the methods of [4], which yield Theorems 12 and 13, and the other results of this section. We write $k = k_{2,4}$.

THEOREM 10. *If* $m \geq 2$ *and*

$$n \geq 3\binom{m}{2},$$

then

$$k(m, n) = 1 + n + 3\binom{m}{2}.$$

THEOREM 11. *If* $n \geq 4$ *and*

$$m \geq \binom{n}{4},$$

then

$$k(m, n) = 1 + 3m + \binom{n}{4}.$$

THEOREM 12. *If* $m \geq 2$ *and*

$$\binom{m}{2} + g \leq n \leq 3\binom{m}{2} + 1,$$

then

$$k(m, n) = 1 + \left[3\left(n + \binom{m}{2}\right)\Big/2\right],$$

where $g = 0$, *or the nearest odd integer to* $(m - 1)/3$, *according as m is odd or even.*

THEOREM 13. *If* $n \geq 4$ *and*

$$\frac{1}{5}\binom{n}{4} + h \leq m \leq \binom{n}{4} + 1,$$

then

$$k(m, n) = 1 + \left[15\left(m + \binom{n}{4}\right)\Big/4\right],$$

where h depends on the residue class to which n belongs, modulo 30, *and is of order at most* $n^3/60$.

Other values of $k = k_{2,4}$ are as follows:

$$
\begin{aligned}
k(m, 6) &= 4m + 2 && (2 \le m \le 11).\\
k(m, 7) &= 4m + 4 && (3 \le m \le 23).\\
k(m, 8) &= 5m + 2 && (2 \le m \le 5),\\
k(m, 8) &= 5m + 1 && (6 \le m \le 8),\\
k(m, 8) &= 4m + 9 && (8 \le m \le 38).\\
k(4, n) &= 3n + 2 && (4 \le n \le 5),\\
k(4, n) &= 2n + 6 && (6 \le n \le 8).\\
k(5, 9) &= 28. && \\
k(6, n) &= [(7n + 38)/3] && (8 \le n \le 12),\\
k(6, n) &= 2n + 15 && (13 \le n \le 17).\\
k(7, n) &= [(9n + 74)/4] && (8 \le n \le 17),\\
k(7, 18) &= 58, && \\
k(7, 19) &= 59, && \\
k(7, 20) &= 61. &&
\end{aligned}
$$

REFERENCES

1. W. G. BROWN, On Graphs Which Do Not Contain a Thomsen Graph, *Canadian Math. Bull.* **9** (1966), 281–285.
2. K. ČULÍK, Poznámka k Problému K. Zarankiewicze, *Práce Brnen. Základny Ceskoslov. Akad. Vied.* **26** (1955), 341–348.
3. K. ČULÍK, Teilweise Lösung eines verallgemeinerten Problem von K. Zarankiewicz, *Ann. Polon. Math.* **3** (1956), 165–168.
4. R. K. GUY, A Problem of Zarankiewicz, in *Theory of Graphs* (P. Erdős and G. Katona, eds.), pp. 119–150, Academic Press, New York, and Publ. House Hungar. Acad. Sci., Budapest, 1968.
5. S. HARTMAN, J. MYCIELSKI, and C. RYLL-NARDZEWSKI, *Colloq. Math.* **3** (1954), 84–85.
6. C. HYLTEŃ-CAVALLIUS, On a Combinatorial Problem, *Colloq. Math.* **6** (1958), 59–65.
7. T. KŐVÁRI, V. T. SÓS and P. TURÁN, On a Problem of Zarankiewicz, *Colloq. Math.* **3** (1954), 50–57.
8. I. REIMAN, Über ein Problem von K. Zarankiewicz, *Acta Math. Acad. Sci. Hungar.* **9** (1958), 269–279.
9. W. SIERPIŃSKI, Sur un Problème Concernant un Réseau à 36 Points, *Ann. Polon. Math.* **24** (1951), 173–174.
10. W. SIERPIŃSKI, *Problems in the Theory of Numbers* (translated by A. Sharma), p. 16, Macmillan (Pergamon), 1964.
11. P. TURÁN, On the Theory of Graphs, *Colloq. Math.* **3** (1954), 19–30.
12. K. ZARANKIEWICZ, Problem P. 101, *Colloq. Math.* **2** (1951), 301.
13. Š. ZNÁM, On a Combinatorical Problem of K. Zarankiewicz, *Colloq. Math.* **11** (1963), 81–84.
14. Š. ZNÁM, Two Improvements of a Result Concerning a Problem of K. Zarankiewicz, *Colloq. Math.* **13** (1965), 255–258.

INFINITE GRAPHS AND MATROIDS

D. A. Higgs

DEPARTMENT OF PURE MATHEMATICS
UNIVERSITY OF WATERLOO
WATERLOO, ONTARIO

1. INTRODUCTION

Bean [1] has considered the problem of extending finite matroid theory to the nonfinitary case, and in particular the extent to which this theory may be carried over to infinite graphs when as circuits we take, not only the polygons, but also the two-way infinite arcs of the graph considered. Here we characterize those graphs whose polygons and two-way infinite arcs give rise to matroids in the sense of [3]. It will be helpful if we first describe the relevant definitions and results of [3], to which we refer for the proofs of these results.

2. SPACES AND MATROIDS

The notion of a *space* on a set E used here is nearly the same as that of a Fréchet (V)-space, as described for example in Chapter I of Sierpinski's "General Topology" [6]. For our purposes, a space on E is most conveniently specified by means of the *derived set* operator ∂, which maps the set of subsets of E into itself and which satisfies the conditions: $\partial A \subseteq \partial B$ whenever $A \subseteq B \subseteq E$; $x \in \partial(A \backslash x)$ whenever $x \in \partial A$ and $A \subseteq E$. This concept of a space is in fact slightly more general than that of a Fréchet (V)-space, the latter being obtained when one further insists that $\partial \varnothing = \varnothing$. The class of spaces is very wide, containing for instance all topological spaces. Indeed, any set E, on which a closure operator $A \mapsto \bar{A}$ is defined, becomes a space if we put $\partial A = \{x \in E;$ $x \in \overline{A \backslash x}\}$. The spaces which arise in this way are characterized by the condition $\partial(A \cup \partial A) \subseteq A \cup \partial A$ for all $A \subseteq E$, and we call a space satisfying this condition a *transitive* space. (The closure operator associated with a transitive space may be recovered from ∂ by means of the equation $\bar{A} = A \cup \partial A$.)

The examples of spaces most relevant here are obtained as follows. Let $E(G)$ denote the edge-set of a graph G and, for $A \subseteq E(G)$, let ∂A be the set of x in $E(G)$ for which there exists a polygon of G with edge-set P such that $x \in P \subseteq A \cup x$. Then $(E(G), \partial)$ is a transitive space (indeed a matroid, according to the definition below) in which the minimal dependent sets are the edge-sets of the polygons of G. Our task is to consider what happens when we admit two-way infinite arcs, as well as polygons, in the definition of ∂. We still obtain a space, but we shall see [(7) below] that this space will only be transitive when a certain type of subgraph is excluded from G.

This last circumstance suggests that there is some point in studying spaces which are not necessarily transitive, a suggestion which is reinforced by the fact that the duality originally defined by Whitney [8] for the class of finite matroids actually extends to the class of all spaces (but not to the class of transitive spaces). This fact was discovered by Sierpinski [5] (see also [6, p. 15]), who was apparently unaware of Whitney's work, or at least of its relation to his own. (Incidentally, in order to stay within the class of Fréchet (V)-spaces, Sierpinski had to impose the (slight) restriction of applying duality only to those Fréchet (V)-spaces which were "dense-in-themselves.")

The *dual* of a space (E, ∂) is defined to be the space (E, ∂^*) with $\partial^* A = E \backslash \partial(E \backslash A)$ for all $A \subseteq E$. (One readily verifies that (E, ∂^*) is a space. Note that $\partial^{**} = \partial$.) If P is any property of spaces, we say that a space (E, ∂) is *dually* P if and only if (E, ∂^*) is P. It may be verified that a space (E, ∂) is dually transitive if and only if $A \cap \partial A \subseteq \partial(A \cap \partial A)$ for all $A \subseteq E$. We define a *matroid* to be a space (E, ∂) which is both transitive and dually transitive. Certainly for the case of E finite, this notion of matroid is coextensive with the usual one.

A pleasant feature of spaces is that the relations established by Tutte [7], between the reductions and contractions of a finite matroid and its dual, carry over to spaces in general. If (E, ∂) is a space and $S \subseteq E$, define the *subspace* of (E, ∂) on S to be the space $(S, \partial \cdot S)$ with $(\partial \cdot S)A = S \cap \partial A$ for all $A \subseteq S$. Write $\partial \times S$ for $(\partial^* \cdot S)^*$ and refer to a space of the form $(T, (\partial \cdot S) \times T)$, where $T \subseteq S \subseteq E$, as a *minor* of (E, ∂). Then the results 3.331 through 3.36 of [7] remain valid for arbitrary spaces.

A subset A of a space (E, ∂) is said to be *dense* (or *spanning*) if $A \cup \partial A = E$; *discrete* (or *independent*) if $A \cap \partial A = \varnothing$; and a *base* if both dense and discrete. Then A is dense in (E, ∂) if and only if $E \backslash A$ is discrete in (E, ∂^*); and A is a base of (E, ∂) if and only if $E \backslash A$ is a base of (E, ∂^*).

We say that a space (E, ∂) is *finitely transitive* if $x \in \partial(A \cup y)$ and $y \in \partial A$ implies $x \in \partial A$ for all $A \subseteq E$ and distinct $x, y \in E \backslash A$; and we say that (E, ∂) is *exchange* if $x \in \partial(A \cup y)$ implies $x \in \partial A$ or $y \in \partial(A \cup x)$ for all $A \subseteq E$ and distinct $x, y \in E \backslash A$. A transitive space is finitely transitive, and a space is finitely transitive if and only if it is dually exchange (thus a dually transitive space is exchange). A space (E, ∂) is said to be B_1 if each discrete set in it is contained in a base of it; (E, ∂) is a *B-matroid* if it is transitive and each of its subspaces is B_1. The class of B-matroids, like the class of matroids, is closed under the taking of duals—from which it follows that B-matroids are indeed matroids. The classes of transitive spaces, matroids, finitely transitive spaces, exchange spaces, and B-matroids are each closed under the taking of minors.

Besides the two above-mentioned works of Sierpinski, we might also mention, in connection with general spaces, Hammer's papers on extended topology (see [2]) and the paper [4] of Schmidt, who discusses a number of the concepts arising in matroid theory.

3. Graph Terminology

Graphs are undirected, possibly with loops and "multiple" edges. A monovalent vertex of a graph G is called an *end* of G. Let G be a connected nonnull graph with e ends and its remaining vertices divalent. Then G is a *polygon* if it is finite and $e = 0$; G is a *two-way infinite arc* if it is infinite and $e = 0$; G is a *ray* if $e = 1$; and G is a *finite arc* if $e = 2$. A graph is *rayed*, or *rayless*, according as it contains or does not contain a ray. Let H, K, and L be subgraphs of some graph. Then H and K are *connected via* L if H and K have a vertex in common or if there is a finite arc in L with one end in H and the other in K. If a graph H is obtained from a graph G by replacing the edges of G with finite arcs then H is a *subdivision* of G. We will not distinguish between subgraphs having no isolated vertices and their corresponding edge-sets.

4. The Space $(E(G), \partial)$ Associated with a Graph G

Let G be any graph. For each $A \subseteq E(G)$, define ∂A to be the set of x in $E(G)$ such that $x \in P \subseteq A \cup x$ for some polygon or two-way infinite arc P in G. Then $(E(G), \partial)$ is a space, which we call the space *associated with* G. Note that a subset A of $E(G)$ is discrete in $(E(G), \partial)$ if and only if A contains no polygons or two-way infinite arcs.

(1) *The space $(E(H), \partial)$ associated with a subgraph H of a graph G is the same as the subspace on $E(H)$ of the space $(E(G), \partial)$ associated with G.*

This is an immediate consequence of the definitions.

(2) *Let G be a graph. Then every minor of $(E(G), \partial)$ is dually transitive (therefore also exchange) and finitely transitive.*

PROOF: It is sufficient to consider only the space $(E(G), \partial)$ itself since the properties in question are preserved under the taking of minors.

To show that $(E(G), \partial)$ is dually transitive, suppose that $x \in A \cap \partial A$; we wish to deduce that $x \in \partial(A \cap \partial A)$. Since $x \in \partial A$, there is a polygon or two-way infinite arc P such that $x \in P \subseteq A \cup x$, where $A \cup x = A$ since $x \in A$. Then clearly $P \subseteq \partial P \subseteq \partial A$ and hence $x \in P \subseteq A \cap \partial A$, so that $x \in \partial(A \cap \partial A)$ as required.

To show that $(E(G), \partial)$ is finitely transitive, we first remark that, as shown in Corollary (2-16) of [1], the polygons and two-way infinite arcs of any graph satisfy Whitney's postulate (C_2) for circuits, namely that if P_1 and P_2 are circuits and $y \in P_1 \cap P_2$, $x \in P_1 \backslash P_2$, then there is a circuit $P_3 \subseteq P_1 \cup P_2$ such that $x \in P_3$, $y \notin P_3$. Now suppose that $x \in \partial(A \cup y)$ and $y \in \partial A$, where x and y are distinct and not in A. We wish to conclude that $x \in \partial A$. Let P_1 and P_2 be polygons or two-way infinite arcs such that $x \in P_1 \subseteq A \cup y \cup x$ and $y \in P_2 \subseteq A \cup y$. Then $x \notin P_2$; also we may clearly suppose that $y \in P_1$. By (C_2), there exists P_3 such that $P_3 \subseteq P_1 \cup P_2$, $x \in P_3$, and $y \notin P_3$. It follows that $x \in P_3 \subseteq A \cup x$, and we have $x \in \partial A$ as required.

The next result extends Lemma (2.62) of [1] and is the main fact needed in the characterization of those graphs whose associated space is matroidal.

(3) *Let G be a graph and let $A \subseteq D \subseteq E(G)$ be such that, for each x in $E(G) \backslash D$, there is a polygon or two-way infinite arc P, having only finitely many vertices in common with each rayless component of A, for which $x \in P \subseteq D \cup x$. Then there exists a set B which is minimal with respect to the requirement that it be dense in $(E(G), \partial)$ and such that $A \subseteq B \subseteq D$.*

PROOF: Let $\mathscr{M}$ be maximal such that

(i) $\mathscr{M}$ is a set of rays in D containing a ray from each rayed component of A,

(ii) the only polygons in $A \cup (\bigcup \mathscr{M})$ are those already in A, and

(iii) no two rays of $\mathscr{M}$ are connected via A.

Next let B be maximal such that

(iv) $A \cup (\bigcup \mathscr{M}) \subseteq B \subseteq D$,

(v) the only polygons in B are those already in A, and

(vi) no two rays of $\mathscr{M}$ are connected via B.

[Any set $\mathscr{M}$ of rays in A, exactly one from each rayed component of A, will satisfy (i), (ii), and (iii); and if $\mathscr{M}$ satisfies (i), (ii), and (iii) then $B = A \cup (\bigcup \mathscr{M})$ will satisfy (iv), (v), and (vi). Zorn's lemma then guarantees the existence of a maximal $\mathscr{M}$ and of a maximal corresponding B.]

We have

(a) *If vertices u and v of G are connected via D then either they are connected via B or each is connected via B to a ray of $\mathcal{M}$.*

(b) *If a ray R in D has only finitely many vertices in common with each rayless component of A then R is connected via A to a ray of $\mathcal{M}$.*

Proof of (a): Suppose that (a) fails for some pair of vertices of G. Then there is a minimal finite arc C in D whose ends u and v fail to satisfy (a). We cannot have $C \subseteq B$; let x in $C \backslash B$ have ends u' and v', where u, u', v', v occur in that order in C. By the maximality of B, either $B \cup x$ fails to satisfy (v), or $B \cup x$ fails to satisfy (vi) [$B \cup x$ necessarily satisfies (iv)]. In the former case, u' and v' are connected via B; and in the latter, each is connected via B to a ray of $\mathcal{M}$. Also, u and u' satisfy (a) by the minimality of C, and so likewise do v' and v. It follows that u and v satisfy (a) after all, contrary to what was supposed. Hence (a) holds.

Proof of (b): Suppose that R is a ray for which (b) fails. Then R has only finitely many vertices in common with each component of A (indeed, none at all with the rayed components of A) and we may modify R according to the following description (in which the ordering of the vertices of R implicitly referred to is the natural one, beginning at the end). Let u_1 be the first vertex of R in A, let A_1 be the component of A containing u_1, and let v_1 be the last vertex of R in A_1. Replace the finite arc in R with ends u_1 and v_1 by a finite arc in A_1 with the same ends. Now do the same for the ray in R whose end is the vertex of R immediately following v_1, and continue in this way all along R (the process terminates if a ray in R is reached having no vertices in A). The ray R' which results in this way from R is like R in that it is in D and is not connected via A to any ray in $\mathcal{M}$—and in addition it is such that $A \cup (\bigcup \mathcal{M}) \cup R'$ contains no polygons not already in A. This however is contrary to the maximality of $\mathcal{M}$. Thus (b) holds.

We can now show that if A and D satisfy the hypotheses of (3) then B is dense in $(E(G), \partial)$. For let x be in $E(G) \backslash B$ and let u and v be the ends of x. We distinguish three cases. In the first case, u and v are connected via B. If this is so then there is a polygon Q such that $x \in Q \subseteq B \cup x$, and we have $x \in \partial B$. In the second case, u and v are not connected via B but they are connected via D. If this happens then by (a) each of u and v is connected via B to a ray of $\mathcal{M}$, so that each is the end of a ray in B (the two rays have no vertices in common since u and v are not connected via B). Hence there is a two-way infinite arc Q such that $x \in Q \subseteq B \cup x$, and we again have $x \in \partial B$. In the last case, u and v are not connected via D. Then in particular x is not in D and there must exist a two-way infinite arc P having only finitely many vertices in common with each rayless component of A such that $x \in P \subseteq D \cup x$. From (b) we see that the ray in $P\backslash x$ with end u is connected via A to a ray of $\mathscr{M}$. Since A and $P\backslash x$ are subsets of D, u is connected via D to a ray of $\mathscr{M}$. By virtue of (a), this implies that u is in fact connected via B to a ray of $\mathscr{M}$ and u is therefore the end of a ray in B. Applying the same reasoning to v, we obtain the situation already arrived at in the second case. Thus we have $x \in \partial B$ whenever $x \in E(G)\backslash B$, that is, B is dense in $(E(G), \partial)$.

It remains for us to prove that B is a *minimal* set dense in $(E(G), \partial)$ and such that $A \subseteq B \subseteq D$. This will be done if we show that

(c) *The only polygons and two-way infinite arcs in B are those already in A.*

Condition (v) states that (c) is true as far as polygons are concerned. Suppose that P is a two-way infinite arc in B but not in A. By (v), each ray in B satisfies the hypothesis of (b) above and is therefore connected via A to a ray of $\mathscr{M}$. Together with (vi), this implies that there exists a (unique) ray R of $\mathscr{M}$ to which each ray in P is connected via A. Let x be in $P\backslash A$, let P_1 and P_2 be the two rays into which $P\backslash x$ falls, and let B_1 and B_2 be the two components of $B\backslash x$ containing P_1 and P_2, respectively [B_1 and B_2 are distinct by (v)]. Since P_1 and P_2 are both connected via A to R, x must be in R. It follows that $P \subseteq A \cup R$. Now one of B_1 and B_2, say B_2, contains a ray in R. Then $P_1\backslash R$ is a ray in A and by (i), the component A_1 of A which contains $P_1\backslash R$ must also contain a ray S of $\mathscr{M}$. But then S is connected via A to R, yet, being contained in B_1, S is distinct from R—and this is contrary to (iii). This completes the proof of (3).

It is convenient to have available the following dualized form of (3).

(4) *Let G be a graph and let $X \subseteq S \subseteq E(G)$. Then X is contained in a base Y of the subspace $(S, \partial^* \cdot S)$ of $(E(G), \partial^*)$ provided that the following condition holds: for each x in X there exists a polygon or two-way infinite arc P, having only finitely many vertices in common with each rayless component of $E(G)\backslash S$, such that $X \cap P = x$.*

Proof: Assume that this condition does hold. Then $A = E(G)\backslash S$ and $D = E(G)\backslash X$ satisfy the hypotheses of (3). Hence there exists a minimal set B such that $A \subseteq B \subseteq D$ and B is dense in $(E(G), \partial)$. Then $Y = E(G)\backslash B$ is a maximal set such that $X \subseteq Y \subseteq S$ and Y is discrete in $(E(G), \partial^*)$. It follows that Y is a maximal discrete set in the subspace $(S, \partial^* \cdot S)$ of $(E(G), \partial^*)$. Since $(E(G), \partial)$ is finitely transitive by (2), $(E(G), \partial^*)$, and hence also $(S, \partial^* \cdot S)$, is exchange. Now a maximal discrete set in an exchange space is actually a base of that space. Thus Y is a base of $(S, \partial^* \cdot S)$.

(5) *Let G be a graph. Then*

(i) *every minor of* $(E(G), \partial)$ *is* B_1, *and*
(ii) *every subspace of* $(E(G), \partial)$ *is dually* B_1.

(Note that since B_1 is the property that every discrete set is contained in a base, dually B_1 is the property that every dense set contains a base.)

PROOF OF (i): By (1) we need only show that every minor of $(E(G), \partial)$ of the form $(S, \partial \times S)$ is B_1—equivalently, that every subspace $(S, \partial^* \cdot S)$ of $(E(G), \partial^*)$ is dually B_1. We make use of the following results.

(d) *A base of a dense subspace of a transitive space is a base of the whole space.*

(e) *If every subspace of a transitive space has a base then every subspace of the space is dually* B_1.

[Result (d) is a direct consequence of transitivity. The conclusion of (e) states that if D is a dense set in a subspace $(S, \partial \cdot S)$ of the given transitive space then D contains a base of $(S, \partial \cdot S)$. By the hypothesis (e), the subspace $(D, \partial \cdot D)$ has a base, B say. Also $(S, \partial \cdot S)$, being a subspace of a transitive space, is itself transitive. Hence B is a base of $(S, \partial \cdot S)$ by (d).]

Returning to the proof of (i), we see from (e) and the transitivity of $(E(G), \partial^*)$ [$(E(G), \partial)$ is dually transitive from (2)] that (i) will be obtained if we show that every subspace $(S, \partial^* \cdot S)$ of $(E(G), \partial^*)$ has a base Y. This last fact however follows immediately from (4) on taking $X = \varnothing$.

PROOF OF (ii): By (1) we need only show that $(E(G), \partial)$ is dually B_1—equivalently, that in $(E(G), \partial)$ every dense set D contains a base B. Let D be dense in $(E(G), \partial)$ and put $A = \varnothing$: then the hypotheses of (3) are fulfilled. Hence there exists a minimal set B such that $B \subseteq D$ and B is dense in $(E(G), \partial)$. Thus B is a minimal dense set in $(E(G), \partial)$. Now a minimal dense set in a finitely transitive space is in fact a base of that space. Hence B is a base of $(E(G), \partial)$, in view of (2).

Condition (i) of (5) shows that if $(E(G), \partial)$ is transitive then it is a B-matroid. On looking at (4), we see that what prevents the transitive space $(E(G), \partial^*)$ from being a B-matroid is the presence in G of some two-way infinite arc having infinitely many vertices in common with some rayless and connected subgraph of G. We now analyze this situation further.

5. THE BEAN GRAPH

This is the graph shown in Fig. 1 of [1] and in Fig. 1 here.

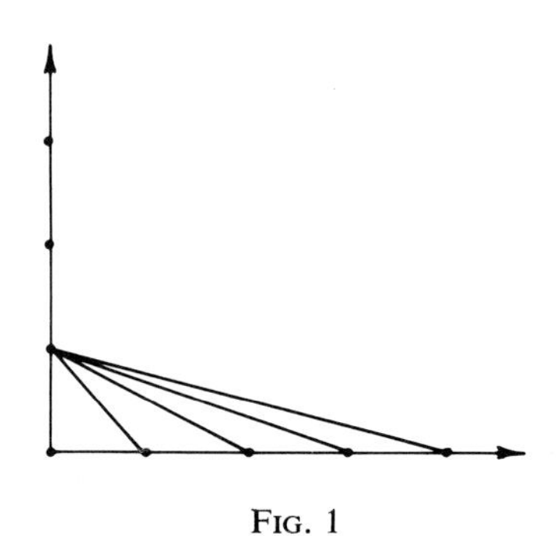

FIG. 1

(6) *If a graph G contains a two-way infinite arc P having infinitely many vertices in common with a rayless and connected subgraph C of G then G has a subdivision of the Bean graph as subgraph.*

PROOF: Replacing C if necessary by a maximal tree contained in C, we may without loss of generality take C to be a tree. We may further suppose that C is the union of finite arcs having their ends in P, for the union of the arcs in C of this type itself satisfies the hypotheses on C. Now since C is rayless, infinite, and connected, it must have an infinite-valent vertex, v say. For each edge x in C which is incident with v, let C_x be a minimal finite arc in C containing x and having v as one end and a vertex of P as the other end. Then it is easy to see that $P \cup \left(\bigcup_x C_x\right)$ contains a subdivision of the Bean graph.

The main theorem of this paper can now be proved.

(7) *Let G be a graph. Then the following conditions are equivalent:*

(i) *$(E(G), \partial)$ is a matroid;*
(ii) *$(E(G), \partial)$ is a B-matroid;*
(iii) *G has no subdivision of the Bean graph as subgraph.*

PROOF: The equivalence of (i) and (ii) is a consequence of condition (i) of (5) and the fact that any B-matroid is a matroid. To see that (i) implies (iii), suppose that G contains a subdivision H of the Bean graph and let A be the set of edges in H which are drawn in full in Fig. 2. Then the edge marked x is not in $A \cup \partial A$ but is in $\partial(A \cup \partial A)$. Thus $(E(G), \partial)$ is not transitive and is therefore not a matroid. Finally suppose that (iii) holds. We then claim that every subspace $(S, \partial^* \cdot S)$ of $(E(G), \partial^*)$ is B_1. For let X be discrete in $(S, \partial^* \cdot S)$: then X is discrete in $(E(G), \partial^*)$, and therefore $E(G)\backslash X$ is dense in $(E(G), \partial)$. Hence for each x in X there exists a polygon or two-way infinite arc P such that $x \in P \subseteq (E(G)\backslash X) \cup x$, or, what is the same thing, such that $X \cap P = x$.

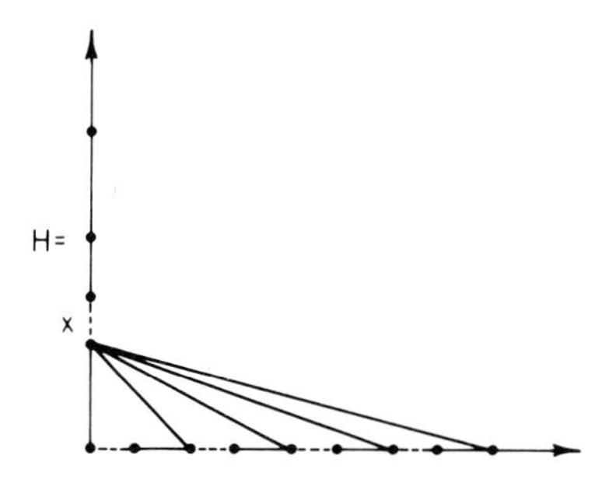

FIG. 2. The broken lines represent single edges and the full lines represent finite arcs.

From (iii) and (6), it is also the case that P has only finitely many vertices in common with each rayless component of $E(G)\backslash S$. Hence the condition of (4) holds, and X is contained in some base Y of $(S, \partial^* \cdot S)$, as required for our claim. Since $(E(G), \partial^*)$ is in any case transitive, we have shown that $(E(G), \partial^*)$ is a B-matroid. The fact that the class of B-matroids is closed under the taking of duals now gives (ii).

ACKNOWLEDGEMENTS

I would like to express my thanks to Dr. D. W. T. Bean for several stimulating discussions, and to Dr. C. St. J. A. Nash-Williams for his helpful suggestions concerning the presentation of this paper (and for pointing out an error in one of the arguments).

REFERENCES

1. D. W. T. BEAN, Infinite Exchange Systems, Ph.D. thesis, McMaster University, Hamilton, Ontario, 1967.
2. G. GASTL and P. C. HAMMER, Extended Topology: Neighborhoods and Convergents, *Proc. Coll. Convexity, Copenhagen, 1965* (1967), pp. 104–116.
3. D. A. HIGGS, Matroids and Duality. (to appear in *Colloq. Math.*).
4. J. SCHMIDT, Mehrstufige Austauchstrukturen, *Z. Math. Logik Grundlagen Math.* **2** (1956), 233–249.
5. W. SIERPINSKI, Sur les Espaces (V) de M. Fréchet Denses en Soi, *Fund. Math.* **33** (1945), 174–176.
6. W. SIERPINSKI, General Topology, Univ. of Toronto Mathematical Expositions No. 7. Toronto, 1952.
7. W. T. TUTTE, Lectures on Matroids, *J. Res. Nat. Bur. Standards Sect.* B **69** (1965), 1–47.
8. H. WHITNEY, On the Abstract Properties of Linear Dependence, *Amer. J. Math.* **57** (1935), 509–533.

A SURVEY OF THICKNESS

Arthur M. Hobbs

FACULTY OF MATHEMATICS
UNIVERSITY OF WATERLOO
WATERLOO, ONTARIO

1. Introduction

Many parts of graph theory have grown from the needs of other branches of science and engineering; for example, the original questions which led to the invention of trees in graphs came from chemistry. We are interested in the questions which arose from problems in the design of printed and integrated circuits and which led to the property of graphs called "thickness." Two wires on the same surface of the board of a printed circuit cannot cross unless a junction of the two is intended. To prevent a junction, one of the two wires must pass through the board to the other side; this kind of passage is costly and should be minimized. Essentially the same considerations hold when integrated circuits are designed, although more than two surfaces are commonly used in such structures. We are drawn naturally to the question: "What is the smallest number of surfaces required in order to draw a given network on these surfaces, without having lines drawn on the same surface cross?"

We need the following definitions.

A *graph* is a finite set of points, called *vertices*, together with a finite set of line segments, called *edges*, such that each end point of an edge is a vertex, no point of an edge other than an end point may be a vertex, and two edges may meet only at a vertex. A *subgraph* H of a graph G is a subset of the edges and vertices of G such that the end points of every edge in H are also in H. An edge is *incident on*, or *with*, each of its end points, and its end points are *joined* by the edge. The *valency* of a vertex is the number of edges incident on it.

The *genus* of a graph G is the smallest integer k such that G can be drawn on a surface of genus k with no two distinct points of G mapped to the same point of the surface. A graph is *planar* if its genus is zero. The *thickness* $\theta(G)$ of a graph G is defined [27] as the minimum number of planar subgraphs of G whose union is G. A graph is *biplanar* if its thickness is 2. A *thickness-minimal*, or *t-minimal*, graph is a graph of thickness t which contains no proper subgraph of thickness t.

A *path* P in a graph G is a set of vertices and edges $\{v_0, e_1, v_1, \ldots, e_n, v_n\}$ of G such that edge e_i is incident with vertices v_{i-1} and v_i for all $i = 1, 2, \ldots, n$, and $v_i = v_j$ implies that $i = j$; v_0 is called the first vertex and v_n the last vertex of the path P. A path is said to *connect* any two vertices in it. A graph G is *connected* if for every two vertices a and b in G there exists a path in G connecting a and b. A vertex is *removed* from a graph G by taking the vertex and all edges incident on it from G. The *connectivity* of a graph G is the smallest number of vertices of G whose removal from G results in a graph without edges or a graph which is not connected. A set $\{P_i\}$ of paths all having the same first and last vertices v_0 and v_n are called *vertex-disjoint* if the only vertices shared by any two of them are v_0 and v_n. A theorem of Menger [10, 22] states that the connectivity of a graph G is equal to the minimum over all pairs of vertices of G of the maximum number of vertex-disjoint paths joining each pair. A connected graph is *separable* if its connectivity is one.

A *circuit* C in a graph G is a set of vertices and edges $\{v_0, e_1, v_1, \ldots, v_{n-1}, e_n, v_0\}$ of G such that edge e_i is incident with vertices v_{i-1} and v_i for all $i = 1, 2, \ldots, n-1$; e_n is incident with v_{n-1} and v_0; and $v_i = v_j$ implies that $i = j$. The *girth* of a graph is the smallest number of edges in any circuit of G.

A graph is *complete* if every pair of vertices is joined by an edge; a complete graph with n vertices is denoted by K_n. A graph G is *bipartite* if the vertices of G can be partitioned into two sets A and B such that every edge of G joins a vertex in A with a vertex in B. A bipartite graph is *complete* if every vertex in A is joined to every vertex in B; such a graph is denoted $K_{m,n}$, where m is the number of vertices in the set A and n is the number of vertices in the set B. Given an integer $n > 0$, let the n-tuples $(x_1, \ldots, x_n)$ with $x_i = 0$ or 1 represent the n^2 vertices of a graph Q_n. Let two vertices of Q_n be joined by an edge iff the representing n-tuples of the two vertices differ in exactly one location. Then Q_n is called the *n-dimensional cubic* graph.

We denote the greatest integer less than or equal to a real number x by $[x]$. We let $\{x\} = -[-x]$.

The earliest work relevant to the thickness of graphs was the determination, in a paper published in 1930 by Kuratowski [20], of a method for determining the planarity or nonplanarity of a graph without actually drawing it in a plane. He showed that a graph is nonplanar if and only if it contains a subgraph homeomorphic to one of the two graphs K_5 or $K_{3,3}$. Although this was only one of many different characterizations of the planarity of graphs discovered at about that time, in the history of thickness Kuratowski's result stands out because it characterizes planarity in terms of graphs which are minimally nonplanar. However, this fact was not recognized as being particularly important for many years.

In 1961, Kodres [19] as an elementary problem asked for a proof that $K_{7,7}$ could not "be factored into two planar factors." This proof was supplied by Marsh [21] in December 1961, as well as by David Sachs and Kodres. H. Eves, the editor of the elementary problems section of the *American Journal of Mathematics*, noted then that $K_{4m-1,4m-1}$ was at least $m+1$-planar.

At about the same time Selfridge, working with printed circuit designs, noticed that K_9 did not seem to be biplanar and mentioned this conjecture to Harary [13, 14]. No easy proof that $\theta(K_9) = 3$ appeared then (and no easy one is known even now). Therefore, Harary published the conjecture in the *Bulletin of the American Mathematical Society* [13], commenting that the conjecture was easily shown to be true for K_n, $n \geq 11$. He showed that the comment was correct early the next year [14]. Using an enumeration of cases with many cases eliminated because of symmetry, Battle, *et al.* [1], Tutte [26], Picard [23], and probably others, have now shown that the thickness of K_9 is 3. For the English-speaking reader, the proof by Tutte [26] is the most complete published version.

Tutte considered the problem further and in 1963 [27] generalized the concept, inventing the term "thickness of a graph." This happy choice of words was almost immediately adopted by all who were concerned with the problem.

The history of the thickness of graphs serves very well as an example of the way some developments are called forth by the current state of knowledge. As mentioned before, the idea of splitting graphs into planar subgraphs was conceived in 1961. Tutte, without knowledge of the 1961 work, reinvented the concept and studied it in detail. Finally, Shirakawa, *et al.* [25] again conceived the property and considered the problem of the thickness of K_n, obtaining a partial solution which had been anticipated by Beineke [5]. Thus, within a space of six years the same property was independently invented three times.

In recent years, the study of the thickness of graphs has proceeded along three paths:

(1) find the thickness of particular graphs and find examples of graphs with given thickness and other properties;

(2) find theorems about the thickness of graphs and about the relation of the thickness to other properties of graphs, such as to connectivity; and

(3) generalize the concept of thickness.

2. Examples

Beineke has done a major part of the work of determining the thickness of particular graphs. Much of this work has been summarized before by Harary [15] and Beineke [6–8], but we include it here for completeness. One of the first results was that

$$\theta(K_{m,n}) = \tfrac{1}{2}m$$

if m is even and $n > \frac{1}{2}(m-2)^2$, and that

$$\theta(K_{m,n}) = \tfrac{1}{2}(m+1)$$

if m is odd and $n > (m-1)(m-2)$. By an ingenious argument, Beineke *et al.* [2] extended this result to the following:

Theorem 1. *If $m \leq n$,*

$$\theta(K_{m,n}) = \{mn/2(m+n-2)\},$$

except, possibly, when m and n are both odd and there exists an integer k such that

$$n = [2k(m-2)/(m-2k)].$$

Beineke [6] has also found that

$$\theta(K_{4t-3,\,4t+1}) = t$$

for all $t \geq 4$.

Beineke and Harary [3, 5, 7] have determined the thickness of K_n for five out of every six integers n, as follows:

Theorem 2. *If $n \equiv -1, 0, 1, 2, 3 \pmod 6$ and $n \neq 9$, then*

$$\theta(K_n) = [(n+7)/6]. \tag{1}$$

Using the result that $\theta(K_9) = 3$ we have $\theta(K_{10}) = 3$, so that the equality in formula (1) fails for the two cases $n = 9$ and 10. However, we know that K_4 is planar, Beineke [7] has shown that $\theta(K_{28}) = 5$, and Hobbs and Grossman [17] have shown $\theta(K_{22}) = 4$; thus the equality of formula (1) holds for these cases of $n \equiv 4 \pmod 6$. In fact we conjecture that the formula (1) is valid for all n except $n = 9$, 10, and 16, with the conjectured [15] value $\theta(K_{16}) = 4$.

A third class of graphs whose thickness has been determined is that of the cubic graphs Q_n. Kleinert [18] proved:

Theorem 3. *If Q_n is the n-dimensional cube, then*

$$\theta(Q_n) = 1 + [n/4].$$

It is a popular activity (particularly in Hungary) to ask for graphs with a given property which are critical in the sense that no proper subgraph has the same property. When speaking of thickness we call such graphs *thickness-minimal*, or *t-minimal*. Tutte [27] pointed out that the homeomorphs of K_5 and $K_{3,3}$ are the only 2-minimal graphs, and that K_9 is a 3-minimal graph. Beineke proved [6]:

THEOREM 4. *For every* $t \geq 2$, $K_{2t-1, 4t^2-10t+7}$ *is t-minimal.*

He further showed that $K_{7,7}$ is 3-minimal and claimed that $K_{8r-5, 8r-5}$ is $2r$-minimal. Hobbs and Grossman [16] extended this last result to the following:

THEOREM 5. *For all* $t \geq 2$, $K_{4t-5, 4t-5}$ *is t-minimal.*

Thus the results of research to date on determining which graphs are t-minimal are:

All homeomorphs of K_5 *and* $K_{3,3}$ *are* 2-*minimal and these graphs constitute the complete set of* 2-*minimal graphs.*

K_9, $K_{7,7}$, *and* $K_{5,13}$ *are* 3-*minimal, but there are an infinite number of non-isomorphic* 3-*minimal graphs.*

If $\theta(K_{16}) = 4$, K_{16} *is* 4-*minimal* [16].

$K_{2t-1, 4t^2-10t+7}$ *and* $K_{4t-5, 4t-5}$ *are t-minimal for all* $t \geq 2$.

In addition, we conjecture that K_{6t-7} is t-minimal for all $t \geq 5$.

3. THEOREMS

Some graph-theoretical properties, such as connectivity, can be greatly affected by the addition of a vertex to a graph if the vertex is isolated or joined to the graph by a single edge. The thickness of the graph, however, is clearly not affected by such an operation. To find a relation between thickness and these other properties, we find it useful to discuss t-minimal graphs.

To prepare for the study of t-minimal graphs, Tutte [27] first proved the basic theorem that the removal of a single edge or a single vertex from a graph will not reduce the thickness of the graph by more than one. In addition, he showed that, given $k \leq t$, every graph of thickness t contains a k-minimal subgraph. Next he proved:

THEOREM 6. *Every t-minimal graph is connected and nonseparable.*

Thus, if a graph is t-minimal it is at least 2-connected.

Tutte [27] next considered the local structure of a t-minimal graph, proving:

THEOREM 7. *If G is a t-minimal graph, $t > 0$, then $m \geq t$, where m is the smallest integer occurring as the valency of a vertex in G.*

Furthermore, if u is the largest integer occurring as the valency of a vertex in the graph G, then:

THEOREM 8. *Let G be a graph of thickness $t \geq 2$. Then $u \geq 2t - 1$. Moreover, if $u = 2t - 1$, there are more than $2t - 2$ vertices of valency u in G.*

Theorem 7 pointed up one fact in particular: for any $t > 2$, there cannot be homeomorphic t-minimal graphs which are not isomorphic. But it seemed likely that, since two nonhomeomorphic 2-minimal graphs existed, there were probably several nonisomorphic t-minimal graphs for each $t > 2$. In the same remarkable paper [27], Tutte more than verified this conjecture with the following theorem.

THEOREM 9. *For each integer $t \geq 2$ there exist infinitely many nonisomorphic t-minimal graphs of upper valency $2t - 1$, smallest number of edges in any path between any two vertices of valency $2t - 1$ greater than any specified integer M, and girth greater than any specified integer N.*

This theorem has shaken our hope that there is any easy characterization of t-minimal graphs for $t > 2$. Where $t = 2$, of course, the characterization is simple in terms of just the two graphs K_5 and $K_{3,3}$. But since no two t-minimal graphs are homeomorphic for $t > 2$, we are faced with an infinite class of t-minimal graphs, no one of which can be generated from another in any obvious manner. Here the matter stands at present.

Continuing with local structures, Hobbs and Grossman [17] generalized Theorem 7 to cut sets, which are minimal sets of edges whose removal disconnects a connected graph:

THEOREM 10. *If G is a t-minimal graph, $t > 0$, then $c \geq t$, where c is the smallest number of edges found in any cut set of G.*

This result suggested the conjecture that the connectivity of a t-minimal graph is bounded from below by t. The conjecture seemed reasonable since the edge connectivity was so bounded, but unfortunately they found:

THEOREM 11. *For every $t \geq 2$ there exist infinitely many nonisomorphic t-minimal graphs having connectivity exactly 2 and girth greater than any specified integer N.*

Thus the connectivity of a graph cannot be used to find an upper bound for its thickness. This disappointing result was lightened somewhat by the following upper bound for connectivity.

THEOREM 12. *The connectivity of a graph of thickness t does not exceed* $6t - 1$.

Hobbs and Grossman were not able to show that the bound given in Theorem 12 was the best possible. However, it is possible to demonstrate a graph of thickness t with connectivity $6t - 2$.

THEOREM 13. *For every $t > 1$, let $H(t)$ be the graph with $6t$ vertices in which each vertex has valency $6t - 2$. Then $\theta(H(t)) = t$ and the connectivity of $H(t)$ is exactly $6t - 2$.*

PROOF: Beineke and Harary [5] proved that $\theta(H(t)) = t$. Further, since each vertex has valency $6t - 2$, the connectivity of $H(t)$ cannot exceed this value.

CASE 1. Suppose vertices a and b in $H(t)$ are not joined by an edge in $H(t)$. Then, since there are only $6t - 2$ vertices in $H(t)$ other than a and b, both a and b must be joined by edges to all of these vertices. There are thus $6t - 2$ vertex-disjoint paths joining the two vertices.

CASE 2. Suppose vertices a and b are joined by an edge e in $H(t)$. Then there exist vertices $c \neq d$ distinct from a and b in $H(t)$ such that a is not joined to c and b is not joined to d by an edge. Since all of these four vertices have valency $6t - 2$, there must be an edge e' joining a with d, an edge f joining c with d, and an edge f' joining b with c. Furthermore, both a and b are joined by an edge to each of the remaining $6t - 4$ vertices in $H(t)$; let this set of $6t - 4$ vertices be called A. Then the edge e, the path consisting of the edges e', f, f', and their end points, and the remaining two-edge paths, each passing through one of the vertices in A, constitute $6t - 2$ vertex-disjoint paths from a to b.

Since between every pair of vertices in $H(t)$ there are $6t - 2$ vertex-disjoint paths, by Menger's theorem [22] the connectivity of $H(t)$ is $6t - 2$.

Thus for each $t \geq 2$ we have a graph which falls short of the upper limit to connectivity, given by Theorem 12, by only one. When $t = 1$, the graph of the icosahedron is seen to have thickness 1 and connectivity 5, thus showing the bound is the best possible in this case.

4. GENERALIZATIONS

There are two generalizations of thickness which have attracted attention: the "coarseness" of a graph and the "n-thickness" of a graph. The work so far on both of these topics has been to find the "coarseness" and "n-thickness" of specific classes of graphs, particularly of the complete graphs and the complete bipartite graphs.

DEFINITION (BEINEKE AND HARARY [4]). The *n-thickness* of a graph G is the minimum number of subgraphs of genus at most n whose union is G. The n-thickness of a graph G is denoted by $\theta_n(G)$. Ringel [24], and Beineke and Harary [4] have found

$$\theta_1(K_n) = \left[\frac{n+4}{6}\right].$$

In addition, Beineke and Harary [4] showed that

$$\theta_2(K_n) = \left[\frac{n+3}{6}\right].$$

Considering that it has not yet proved possible to determine the ordinary thickness of all complete graphs, it appears easier to find the n-thickness of graphs for $n > 0$ than for $n = 0$. This situation is similar to that of the chromatic number, which has been determined for surfaces of every genus except zero. Although the n-thickness for $n \geq 3$ has not been learned for any class of graphs as yet, it is reasonable to believe that further results will not be long in coming.

The other primary generalization of thickness is the "coarseness" of graphs, whose definition is an interesting contrast to the definition of thickness.

DEFINITION. The *coarseness* $c(G)$ of a graph G is the *maximum* number of edge-disjoint *nonplanar* subgraphs into which G may be decomposed.

Although the concept was apparently originated by Erdös, the major part of the work on this subject has been done by Guy and Beineke [11, 12]. Among the results so far found are the following:

For $n = 3r$,

$$c(K_n) = \binom{r}{2} + [r/5], \qquad r \geq 10,$$

$$c(K_n) = \binom{r}{2}, \qquad r \leq 5.$$

For $n = 3r + 1$,

$$\binom{r}{2} + 2[r/3] \leq c(K_n) \leq \binom{r}{2} + [2r/3], \qquad r \neq 3, 4,$$

$$c(K_{10}) = 4, \qquad 7 \leq c(K_{13}) \leq 8.$$

For $n = 3r + 2$,

$$c(K_n) = \binom{r}{2} + [(14r + 1)/15].$$

It will be noticed that these formulas give exact values for the coarseness of most complete graphs and give a choice of two possible values for the coarseness of each of the remaining complete graphs.

In addition, Beineke [8] has found the coarseness of the complete bipartite graph $K_{n,n}$ to be

$$\left[\left[\frac{n}{3}\right]\frac{n}{3}\right],$$

and he also obtained the lower bound

$$\min\left(\left[\left[\frac{m}{3}\right]\frac{n}{3}\right], \left[\left[\frac{n}{3}\right]\frac{m}{3}\right]\right)$$

for the coarseness of $K_{m,n}$.

REFERENCES

1. J. BATTLE, F. HARARY, and Y. KODAMA, Every Planar Graph with Nine Points Has a Nonplanar Complement, *Bull. Amer. Math. Soc.* **68** (1962), 569–571.
2. L. W. BEINEKE, F. HARARY, and J. W. MOON, On the Thickness of the Complete Bipartite Graph, *Proc. Cambridge Philos. Soc.* **60** (1964), 1–5.
3. L. W. BEINEKE and F. HARARY, On the Thickness of the Complete Graph, *Bull. Amer. Math. Soc.* **70** (1964), 618–620.
4. L. W. BEINEKE and F. HARARY, Inequalities Involving the Genus of a Graph and Its Thickness, *Proc. Glasgow Math. Assoc.* **7** (1965), 19–21.

5. L. W. BEINEKE and F. HARARY, The Thickness of the Complete Graph, *Canad. J. Math.* **17** (1965), 850–859.
6. L. W. BEINEKE, Complete Bipartite Graphs: Decomposition into Planar Subgraphs, in *A Seminar on Graph Theory* (F. Harary, ed.), Chapt. 7, pp. 42–53, Holt, New York, 1967.
7. L. W. BEINEKE, The Decomposition of Complete Graphs into Planar Subgraphs, in *Graph Theory and Theoretical Physics* (F. Harary, ed.), Chapt. 4, pp. 139–153, Academic Press, New York, 1967.
8. L. W. BEINEKE, Topological Aspects of Complete Graphs, in *Theory of Graphs* (*Proc. Colloq. Tihany*, 1966) (P. Erdös and G. Katona, eds.), pp. 19–26, Academic Press, New York, and Publ. House Acad. Sci. Hungar., Budapest, 1968.
9. G. A. DIRAC, and S. SCHUSTER, A Theorem of Kuratowski, *Nederl. Akad. Wetensch. Proc. Ser. A* **57** (1954), 343–348.
10. G. A. DIRAC, Short Proof of Menger's Graph Theorem, *Mathematika* **13** (1966), 42–44.
11. R. K. GUY, A Coarseness Conjecture of Erdös, *J. Combinatorial Theory* **3** (1967), 38–42.
12. R. K. GUY and L. W. BEINEKE, The Coarseness of the Complete Graph, *Canad. J. Math.* **20** (1968), 888–894.
13. F. HARARY, Research Problem No. 28, *Bull. Amer. Math. Soc.* **67** (1961), 542.
14. F. HARARY, A Complementary Problem on Nonplanar Graphs, *Math. Mag.* **35** (1962), 301–303.
15. F. HARARY, Recent Results in Topological Graph Theory, *Acta Math. Acad. Sci. Hungar.* **15** (1964), 405–412.
16. A. M. HOBBS and J. W. GROSSMAN, A Class of Thickness-Minimal Graphs, *J. Res. Nat. Bur. Standards Sect. B* **72B** (1968), 145–153.
17. A. M. HOBBS and J. W. GROSSMAN, Thickness and Connectivity in Graphs, *J. Res. Nat. Bur. Standards Sect. B* **72B** (1968), 239–244.
18. M. KLEINERT, Die Dicke des n-dimensionalen Würfel-Graphen, *J. Combinatorial Theory* **3** (1967), 10–15.
19. U. R. KODRES, Problem No. E 1465, *Amer. Math. Monthly* **68** (1961), 379.
20. C. KURATOWSKI, Sur le Problème des Courbes Gauches en Topologie, *Fund. Math.* **15** (1930), 271–283.
21. D. C. B. MARSH, Solution to Problem No. E 1465, *Amer. Math. Monthly*, **68** (1961), 1009.
22. K. MENGER, Zur allgemeinen Kurventheorie, *Fund. Math.* **10** (1927), 96–115.
23. C. PICARD, Graphes Complementaires et Graphes Planaires, *Rev. Francaise Informat. Recherche Operationelle* **8** (1964), 329–343.
24. G. RINGEL, Die toroidal Dicke des vollständigen Graphen, *Math. Z.* **87** (1965), 19–26.
25. I. SHIRAKAWA, H. TAKAHASHI, and H. OZAKI, On the Decomposition of a Complete Graph into Planar Subgraphs, *J. Franklin Inst.* **283** (1967), 379–388.
26. W. T. TUTTE, The Nonbiplanar Character of the Complete 9-Graph, *Canad. Math. Bull.* **6** (1963), 319–330.
27. W. T. TUTTE, The Thickness of a Graph, *Nederl. Akad. Wetensch. Proc. Ser. A* **66** (1963), 567–577.

LINE-SYMMETRIC TOURNAMENTS

Michel Jean

COLLÈGE MILITAIRE ROYAL
SAINT-JEAN, QUÉBEC*

1. Introduction

A tournament[1] T is called *line-symmetric* if, for every pair of lines (u_1, u_2) and (v_1, v_2) of T, there is an automorphism f of T such that $f(u_1) = v_1$ and $f(u_2) = v_2$; it is called *transitive* if, whenever (u, v) and (v, w) are lines of T, (u, w) is also a line of T.

Theorem. *Let T be a tournament with more than four points. The subtournaments of T obtained by deleting any pair of points are isomorphic iff T is either transitive or line-symmetric.*

The condition is obviously sufficient. For the proof of the necessity of the condition we will need to investigate a new class of tournaments, which we call regular.

2. Regular Tournaments

Let $T = \langle V, L \rangle$ be a tournament; V is the set of points and L is the set of lines of T; uLv means that (u, v) is a line of T. Let $|V| = p$.[2] For every $u \in V$, $uL = \{v \in V : uLv\}$ is the set of points of T adjacent from u and $s_u = |uL|$ is

* This research was sponsored in part by DRB grant No. 9540–08. The paper was written while the author was a fellow of the Summer Research Institute (1968) of the Canadian Mathematical Congress at the University of Montreal.

[1] Terms which are not defined and notations which are not explained are those of [1].

[2] If X, Y are sets then $X \times Y$ denotes the set of ordered pairs (x, y), $x \in X$ and $y \in Y$; $X \sim Y$ is the set of elements of X which do not belong to Y and $|X|$ is the cardinality of X. Every set considered is finite.

the outdegree of u. The integral-valued function s defined on V is called the score function of T. For every k, $0 \le k \le p-1$, let $\delta_k = |\{u \in V : s_u = k\}|$; the function δ is called the density function of T.

The score and density functions of the subtournament T^u obtained from T by removing the point u and the lines adjacent to u, are denoted, respectively, by s^u and δ^u.

For $u, v \in V, u \ne v$,

$$s_v{}^u = \begin{cases} s_v - 1 & \text{if} \quad vLu \\ s_v & \text{if} \quad uLv. \end{cases}$$

For $u \in V$ and $0 \le k \le n-2$

$$\begin{aligned} \delta_k{}^u &= |\{v \in V : s_v = k \quad \text{and} \quad uLv\}| \\ &\quad + |\{v \in V : s_v = k+1 \quad \text{and} \quad vLu\}| \\ &= |\{v \in V : s_v = k \quad \text{and} \quad uLv\}| \\ &\quad + |\{v \in V : s_v = k+1\} \sim (\{v \in V : s_v = k+1 \quad \text{and} \quad uLv\} \cup \{u\})|. \end{aligned}$$

If we let $\lambda_k{}^u = |\{v \in V : s_v = k \text{ and } uLv\}|$, then

$$\delta_k{}^u = \begin{cases} \lambda_k{}^u + \delta_{k+1} - \lambda_{k+1}^u & \text{if} \quad s_u \ne k+1 \\ \lambda_k{}^u + \delta_{k+1} - \lambda_{k+1}^u - 1 & \text{if} \quad s_u = k+1. \end{cases} \tag{1}$$

Let

$$m = \min\{s_u : u \in V\}, \qquad M = \max\{s_u : u \in V\}.$$

For $k < m$ and for $k > M$ and for all $u \in V$, $\delta_k = \lambda_k{}^u = 0$.

The tournament T is called *regular* if for all $u, v \in V$, $\delta^u = \delta^v$. An obvious necessary condition for a tournament to be regular is given by the following lemmas.

Lemma 1. *If the subtournaments T^u, $u \in V$, are isomorphic, then T is regular.*

In the following three lemmas it is assumed implicitly that T is a regular tournament.

Lemma 2. *If $u, v \in V$, then*

$$\lambda_k{}^v = \begin{cases} \lambda_k{}^u & \text{if} \quad s_u,\ s_v \le k, \text{ or } k < s_u, s_v \\ \lambda_k{}^u + 1 & \text{if} \quad s_u \le k < s_v. \end{cases}$$

Proof: If $k < m$, then $\lambda_k{}^v = \lambda_k{}^u = 0$ and the lemma is true. The proof is by induction on k, starting with $k = m$. Without loss of generality we may assume that $s_u = m$ and hence, using (1), $\delta_{m-1}^u = \delta_m - \lambda_m{}^u - 1$.

If $s_v = m$, then $\delta_{m-1}^v = \delta_m - \lambda_m{}^v - 1$ and $\lambda_m{}^v = \lambda_m{}^u$.

If $m < s_v$, then $\delta_{m-1}^v = \delta_m - \lambda_m{}^v$ and $\lambda_m{}^v = \lambda_m{}^u + 1$.

Let us assume that the lemma is true for some $k \geq m$. Then $\delta_k{}^u = \lambda_k{}^u + \delta_{k+1} - \lambda_{k+1}^u$.

To prove the lemma for $k+1$, we distinguish three cases to evaluate $\lambda_k{}^v$ from the induction hypothesis and to evaluate $\delta_k{}^v$ from (1).

CASE 1. $k+1 < s_v$. Then $\lambda_k{}^v = \lambda_{k+1}^u$ and $\delta_k{}^v = \lambda_k{}^v + \delta_{k+1} - \lambda_{k+1}^v$. Therefore $\lambda_{k+1}^v = \lambda_{k+1}^u + 1$.

CASE 2. $k+1 = s_v$. Then $\lambda_k{}^v = \lambda_k{}^u + 1$ and $\delta_k{}^v = \lambda_k{}^v + \delta_{k+1} - \lambda_{k+1}^v - 1$. Therefore $\lambda_{k+1}^v = \lambda_{k+1}^u$.

CASE 3. $s_v < k+1$. Then $\lambda_k{}^v = \lambda_k{}^u$ and $\delta_k{}^v = \lambda_k{}^v + \delta_{k+1} - \lambda_{k+1}^v$. Therefore $\lambda_{k+1}^v = \lambda_{k+1}^u$.

LEMMA 3. *If* $m \leq k \leq M$, *then* $\delta_k \neq 0$.

PROOF: On the contrary, let us assume that $\delta_k = 0$ for some k, $m \leq k \leq M$. Let $u, v \in V$ such that $s_u = m$ and $s_v = M$. By Lemma 2, $\lambda_k{}^v = \lambda_k{}^u + 1$, but from the definition of λ, $\lambda_k{}^v = \lambda_k{}^u = 0$.

LEMMA 4. *If* $s_u = k$, *then* $\lambda_k{}^u = (\delta_k - 1)/2$.

PROOF: Let $X = \{v \in V : s_v = k\}$. Then $|X| = \delta_k$ and for $v \in X$, $\lambda_k{}^v = \lambda_k{}^u$ by Lemma 2. Therefore $|L \cap (X \times X)| = \sum_{v \in X} \lambda_k{}^v = \delta_k \lambda_k{}^u$. Also $|L \cap (X \times X)|$ is the number of lines in the subtournament determined by X and is equal to $\delta_k(\delta_k - 1)/2$. Hence $\delta_k(\delta_k - 1)/2 = \delta_k \lambda_k{}^u$ and since $\delta_k \neq 0$ (Lemma 3), $\lambda_k{}^u = (\delta_k - 1)/2$.

Lemmas 2–4 may be improved in the following way.

PROPOSITION 5. If T is a regular tournament, then there is an integer d, called the degree of T, such that

$$\text{(a)}\quad \delta_k = d \quad \text{if} \quad m \leq k \leq M,$$

$$\text{(b)}\quad \lambda_k{}^u = \begin{cases} \dfrac{d-1}{2} & \text{if} \quad s_u \leq k \leq M, \\[2ex] \dfrac{d+1}{2} & \text{if} \quad m \leq k < s_u, \end{cases}$$

$$\text{(c)}\quad \delta_k{}^u = \begin{cases} \dfrac{d-1}{2} & \text{if} \quad k = m-1 \quad \text{or} \quad M, \\[1ex] d & \text{if} \quad m \leq k < M, \end{cases}$$

PROOF: Let k, l be integers such that $m \leq k < l \leq M$. We will show that $\delta_k = \delta_l$. Let $X = \{u \in V : s_u = k\}$ and $Y = \{v \in V : s_v = l\}$.[3] Then $|X| = \delta_k$, $|Y| = \delta_l$, and by the definition of a tournament, $|L \cap (X \times Y \cup Y \times X)| = \delta_k \cdot \delta_l$. By Lemmas 2 and 4, for any $u \in X$, $v \in Y$:

$$\lambda_k^{\ v} = \lambda_k^{\ u} + 1 = \frac{\delta_k + 1}{2}, \qquad \lambda_l^{\ v} = \lambda_l^{\ u} = \frac{\delta_l - 1}{2}.$$

Therefore

$$|L \cap (X \times Y)| = \sum_{u \in X} \lambda_l^{\ u} = \delta_k \cdot \frac{\delta_l - 1}{2},$$

$$|L \cap (Y \times X)| = \sum_{v \in Y} \lambda_k^{\ v} = \delta_l \cdot \frac{\delta_k + 1}{2},$$

$$\delta_k \cdot \delta_l = \delta_k \frac{\delta_l - 1}{2} + \delta_l \cdot \frac{\delta_k + 1}{2}$$

$$= \delta_k \delta_l + \frac{\delta_l - \delta_k}{2}.$$

and hence $\delta_k = \delta_l$.

Parts (b) and (c) then follow from part (a), Lemmas 2 and 4, and Eq. (1).

The integers p, d, m, and M are related by the following equations which may be verified easily:

$$p = (M - m + 1)d, \qquad p(p-1) = (M + m)(M - m + 1)d,$$

$$m = (M - m + 1)\frac{d - 1}{2},$$

which yield

$$m = \frac{p(d-1)}{2d} \qquad \text{and} \qquad M + m = p - 1.$$

If the degree of T is 1, then $m = 0$, $M = p - 1$, and the map which sends the integer k, $0 \leq k \leq p - 1$, on the unique $u \in V$ such that $s_u = k$ is an isomorphism of the natural ordering of the integers $0, 1, \ldots, p - 1$ onto T; in that case T is transitive. If T is regular and if some T^u is transitive, then all subtournaments T^v, $v \in V$ are also transitive, and T itself is transitive provided $p \geq 4$. If T is regular but not transitive, then its degree d is greater than 1 and by Proposition 5(c), $d = 2\delta_{m'}^u + 1$ where $u \in V$ and $m' = \min\{s_v^{\ u} : v \in A \sim \{u\}\} = m - 1$. We have thus shown:

[3] See footnote 2.

COROLLARY 6. If T is regular and $p \geq 4$, then the degree of T is determined by any one of its subtournaments T^u.

3. PROOF OF THE THEOREM

Let us assume that $p \geq 5$ and that all subtournaments obtained from T by deleting any two points are isomorphic. Let $u, v \in V$ such that uLv; T^{uv} will denote the subtournament obtained from T by deleting u and v.

The subtournaments obtained from T^u and T^v by deleting a point are isomorphic and hence they are both regular by Lemma 1. Then T^{uv} is a tournament common to T^u and T^v, hence by Corollary 6, T^u and T^v have the same degree d'. Let $m' = (p-1)(d'-1)/2d'$ and $M' = p - 2 - m'$; then by Proposition 5(a), $\delta_k{}^u = \delta_k{}^v = d'$ for $m' \leq k \leq M'$, and since u, v were chosen arbitrarily, T itself is regular. Let d be the degree of T, $m = p(d-1)/2d$, and $M = p - 1 - m$.

By Proposition 5(c):

$$\delta_k{}^u = \delta_k{}^v = \begin{cases} \dfrac{d-1}{2} & \text{if} \quad k = m-1 \quad \text{or} \quad M, \\ d & \text{if} \quad m \leq k < M. \end{cases}$$

Therefore either $d = d' = 1$ and T is transitive or $m = M$ and $d' = (d-1)/2$. Let us assume the latter case. Then $m = M = (p-1)/2$, $d = p$, $m' = (p-3)/2$, $M' = (p-1)/2$, and $d' = (p-1)/2$.

Let

$$\begin{aligned} X &= \{w \in V : wLu \text{ and } wLv\}, \\ Y_1 &= \{w \in V : wLu \text{ and } vLw\}, \\ Y_2 &= \{w \in V : uLw \text{ and } wLv\}, \\ Z &= \{w \in V : uLw \text{ and } vLw\}. \end{aligned}$$

To complete the proof we will characterize the sets X, Y_1, Y_2, and Z in terms of T^{uv}; it will follow that the tournament T is uniquely determined by T^{uv}.

$$w \in X \quad \text{iff} \quad |wL \sim \{u, v\}| = \frac{p-1}{2} - 2 = \frac{p-5}{2} = m' - 1,$$

$$w \in Y_1 \cup Y_2 \quad \text{iff} \quad |wL \sim \{u, v\}| = \frac{p-1}{2} - 1 = \frac{p-3}{2} = m',$$

$$w \in Z \quad \text{iff} \quad |wL \sim \{u, v\}| = \frac{p-1}{2} = M'.$$

Therefore

$$X = \left\{ w \in V \sim \{u, v\} : |wL \sim \{u, v\}| = \frac{p-5}{2} \right\}, \tag{2}$$

$$Z = \left\{ w \in V \sim \{u, v\} : |wL \sim \{u, v\}| = \frac{p-1}{2} \right\}, \tag{3}$$

and by Proposition 5(c), applied to the regular tournament T^v:

$$|Z| = \frac{d'-1}{2} = \frac{p-3}{4}, \qquad |Y_1 \cup Y_2| = d' = \frac{p-1}{2}.$$

Also

$$w \in Y_1 \cup Z \qquad \text{iff} \quad |wL \sim \{v\}| = \frac{p-1}{2} = M',$$

and therefore by Proposition 5(a), applied to the regular tournament T^v

$$|Y_1 \cup Z| = d' = \frac{p-1}{2}, \qquad |Y_1| = \frac{p-1}{2} - \frac{p-3}{4} = \frac{p+1}{4},$$

$$|Y_2| = \frac{p-1}{2} - \frac{p+1}{4} = \frac{p-3}{4}.$$

Let $w_1, w_2 \in V$, $w_1 \neq w_2$. By Proposition 5(c), applied to the regular tournament T^{w_2}

$$|\{w \in V : \{w_1, w_2\} \subseteq wL\}| = \frac{d'-1}{2} = \frac{p-3}{4}.$$

Therefore $\{wL : w \in V\}$ is a symmetric block design [2] and $|w_1 L \cap w_2 L| = (p-3)/4$.

If w_1 and w_2 belong to Y_1 or if they belong to Y_2, then

$$|w_1 L \cap w_2 L \sim \{u, v\}| = \frac{p-3}{4} - 1 = \frac{p-7}{4}.$$

Hence

$$Y_1 = \left\{ w \in V \sim \{u, v\} : |wL \sim \{u, v\}| = \frac{p-3}{2} \quad \text{and} \right.$$

$$\left| \left\{ w' \in V \sim \{u, v\} : \quad \text{either} \quad w' = w \quad \text{or} \quad w' \neq w \quad \text{and} \right. \right.$$

$$\left. \left. |w'L \cap wL \sim \{u, v\}| = \frac{p-7}{4} \right\} \right| = \frac{p+1}{4} \right\}, \tag{4}$$

$$Y_2 = \left\{ w \in V \sim \{u, v\} : |wL \sim \{u, v\}| = \frac{p-3}{2} \quad \text{and} \right.$$

$$\left| \left\{ w' \in V \sim \{u, v\} : \quad \text{either} \quad w' = w \quad \text{or} \quad w' \neq w \quad \text{and} \right. \right.$$

$$\left. \left. |w'L \cap wL \sim \{u, v\}| = \frac{p-7}{4} \right\} \right| = \frac{p-3}{4} \right\}. \tag{5}$$

Let u', $v' \in V$ such that $u'Lv'$ and define similarly the sets X', Y_1', Y_2', and Z'; these sets satisfy equations analogous to (2)–(5). Let f be an isomorphism of T^{uv} onto $T^{u'v'}$. Then $f(X) = X'$, $f(Y_1) = Y_1'$, $f(Y_2) = Y_2'$, and $f(Z) = Z'$. The extension $\bar{f}$ of f to V defined by $\bar{f}(u) = u'$ and $\bar{f}(v) = v'$ is an automorphism of T.

References

1. F. Harary and L. Moser, The Theory of Round Robin Tournaments, *Amer. Math. Monthly* **73** (1966), 231–246.
2. H. J. Ryser, *Combinatorial Mathematics*, Wiley, New York, 1963.

ON THE RAMSEY NUMBER $N(4, 4; 3)$*

J. G. Kalbfleisch

UNIVERSITY OF WATERLOO
WATERLOO, ONTARIO

1. Introduction

Let S be a set of n vertices in general position in r-space. Each subset of S containing r vertices defines a hyperplane or *r-face* of S. Here 2-faces and 3-faces will be called *edges* and *faces*, respectively. The r-faces of S are colored, some red and the others blue. Let p, q, r be integers with p, $q \geq r \geq 1$. A $(p, q; r)$-coloring of S is an assignment of colors to the r-faces of S in such a way that no p-subset of S has all its r-faces red, and no q-subset of S has all its r-faces blue. The Ramsey number $N(p, q; r)$ is the least integer such that for $n \geq N(p, q; r)$ there exists no $(p, q; r)$-coloring on n vertices. The existence of $N(p, q; r)$ is guaranteed by a theorem of Ramsey [11]. Another statement and proof of Ramsey's theorem will be found in Ryser [12].

It follows from symmetry that $N(p, q; r) = N(q, p; r)$. There will be a blue r-face unless all r-faces are colored red; and then there is a p-subset of S with all its r-faces red if $n \geq p$. Thus $N(p, r; r) = p$. Finally, the pigeon-hole principle gives $N(p, q; 1) = (p - 1) + (q - 1) + 1$.

For $r = 2$, the n points of S and its 2-faces may be thought of as a complete graph on n vertices. A $(p, q; 2)$-coloring is one in which no complete subgraph on p vertices has all its edges red, and no complete subgraph on q vertices has all its edges blue. A few of the Ramsey numbers $N(p, q; 2)$ are known (see [3, 4, 6, 8, 10]). Those of importance here are

$$N(3, 4; 2) = N(4, 3; 2) = 9; \qquad N(4, 4; 2) = 18.$$

Good upper and lower bounds are available for a few others (see [2, 5–7, 13]).

* This paper was presented at the Third Waterloo Conference on Combinatorics, May 20, 1968. The results of Section 5 appeared in the author's doctoral thesis. The remaining work was sponsored by grants from the National Research Council of Canada and the Department of University Affairs.

Very little is known about $N(p, q; r)$ for $r > 2$. In this paper new bounds are obtained for $N(4, 4; 3)$. The upper bound is reduced from 19 to 17 by exploiting a relationship between (4, 4; 3)- and (4, 4; 2)-colorings. A method for constructing face-colorings and checking them for the desired properties is given and used to obtain a (4, 4; 3)-coloring on 11 vertices. It then follows that $12 \leq N(4, 4; 3) \leq 17$. The possibility of further improving these bounds is discussed.

2. Induced Edge-Colorings

Suppose there exists a (4, 4; 3)-coloring on $n + 1$ vertices $V_0, V_1, \ldots, V_n$. Each of the

$$\binom{n+1}{3}$$

faces is colored red or blue, and there exists no tetrahedron whose four faces are of one color. Each face containing V_0 is incident with two vertices in $V_1, V_2, \ldots, V_n$. *The edge-coloring induced by* V_0 *on* $V_1, V_2, \ldots, V_n$ is obtained by coloring edge $V_i V_j$ the same color as face $V_0 V_i V_j$ for each pair i, j from $1, 2, \ldots, n$.

If edges $V_i V_j$, $V_j V_k$, and $V_i V_k$ are red, then faces $V_0 V_i V_j$, $V_0 V_j V_k$, $V_0 V_i V_k$ are red, and $V_0 V_i V_j V_k$ will be a red tetrahedron unless $V_i V_j V_k$ is a blue face. Thus if $V_i V_j V_k$ is a red triangle in the edge-coloring, face $V_i V_j V_k$ is blue in the face-coloring. It follows that the induced edge-coloring can contain no set of four vertices interjoined by red edges only, for then the face-coloring would contain a red or blue tetrahedron. Similarly the edge-coloring contains no set of four vertices interjoined by blue edges only. Thus

Lemma 1. *The edge-coloring induced by* V_0 *on* $V_1, V_2, \ldots, V_n$ *is a* (4, 4; 2)-*coloring.*

Since there must exist a (4, 4; 2)-coloring on n vertices, it follows that $n \leq N(4, 4; 2) - 1 = 17$. It is therefore impossible to construct a (4, 4; 3)-coloring on more than 18 vertices, and $N(4, 4; 3) \leq 19$. In Section 4, Lemma 1 is used to reduce this bound still further.

The above arguments may be generalized to give

$$N(p, q; r) \leq N(x, y; r - 1) + 1; \qquad x = N(p - 1, q; r), \qquad y = N(p, q - 1; r).$$

This result forms the basis of the proof of Ramsey's theorem given in [1].

3. Some Results for Edge-Colorings

Lemma 1 of the last section establishes a link between face-colorings and edge-colorings. In exploiting this link, some structural theorems for edge-colorings will be required.

Lemma 2. *Let V be any vertex in a* $(4, 4; 2)$*-coloring on n vertices. Let V be joined by red edges to x vertices determining a complete subgraph $\mathscr{X}$, and by blue edges to y vertices determining a complete subgraph $\mathscr{Y}$. Then $x + y + 1 = n$, $x \le 8$, and $y \le 8$. Furthermore $\mathscr{X}$ is* $(3, 4; 2)$*-colored*, and *$\mathscr{Y}$ is* $(4, 3; 2)$*-colored.*

Proof: A red triangle in $\mathscr{X}$ together with V would give four vertices interjoined by red edges only. This is impossible in a $(4, 4; 2)$-coloring. Since the whole graph contains no set of four vertices interjoined by blue edges only, neither does a subgraph $\mathscr{X}$. Thus $\mathscr{X}$ is $(3, 4; 2)$-colored, and similarly $\mathscr{Y}$ is $(4, 3; 2)$-colored. But $N(3, 4; 2) = N(4, 3; 2) = 9$, and thus $x \le 8$ and $y \le 8$.

Note that Lemma 2 implies $n = x + y + 1 \le 17$, and if $n = 17$, then $x = y = 8$. It is proved in [9] that there exists a unique $(4, 4; 2)$-coloring on 17 vertices. If V is any vertex in this coloring, the red edges interjoining vertices of $\mathscr{X}$ have the configuration shown in Fig. 3.

For $n = 16$, $x = 7$ and $y = 8$ (V is a vertex of type A), or $x = 8$ and $y = 7$ (V is a vertex of type B). If there are a vertices of type A and b of type B, then $a + b = 16$. The number of red edges is $(7a + 8b)/2 = 56 + b/2$, so that a and b are both even. In [6] it is shown that $a \ge 2$ and $b \ge 2$, so that there exist vertices of both types in a $(4, 4; 2)$-coloring on 16 vertices. Lemma 3 discusses possible coloring schemes for $\mathscr{X}$ and $\mathscr{Y}$ in Lemma 2. A proof of Lemma 3 will be found in [6], for example.

Lemma 3 (i) *There are three nonisomorphic* $(3, 4; 2)$*-colorings on eight vertices with* 10, 11, *and* 12 *red edges, respectively.* [The red edges in these colorings are given in Figs. 1, 2, and 3, respectively. The blue edges are given in Figures 4, 5, and 6 (obtained by taking the complements of the graphs in Figs. 1, 2, and 3, and rearranging).]

(ii) *There are nine nonisomorphic* $(4, 3; 2)$*-colorings on seven vertices with* 15, 14, 13, 12, *or* 11 *red edges* (not illustrated).

The corresponding results for $(4, 3; 2)$-colorings on eight vertices and $(3, 4; 2)$-colorings on seven vertices are obtained by interchanging red and

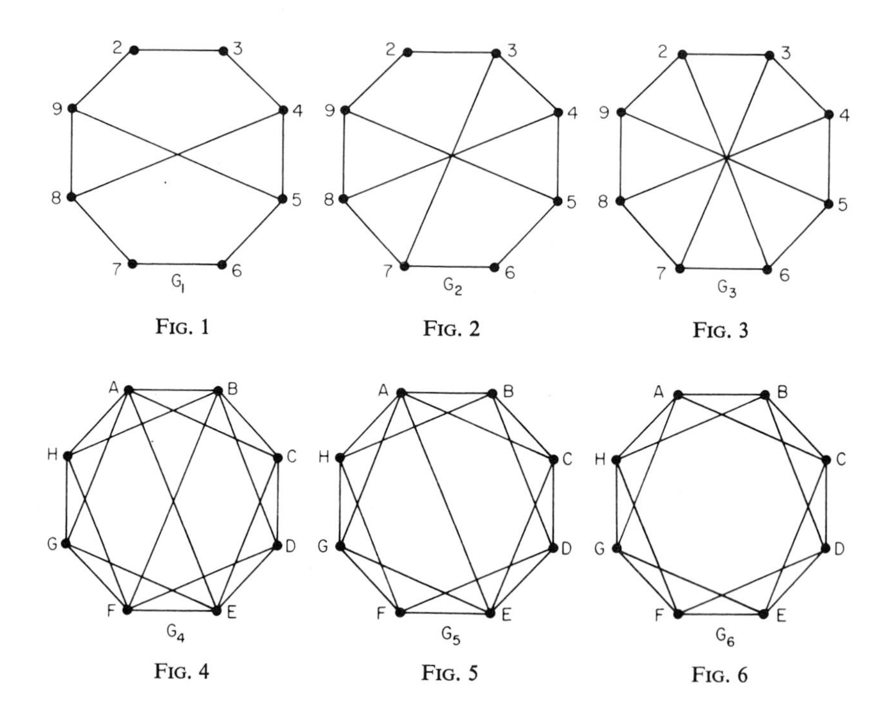

FIG. 1 FIG. 2 FIG. 3

FIG. 4 FIG. 5 FIG. 6

blue in Lemma 3. Let $G_1, G_2, \ldots, G_6$ denote the graphs illustrated in Figs. 1–6, respectively.

LEMMA 4. *G_4 contains no subgraph isomorphic to G_3.*

PROOF: Suppose that by removing six edges from G_4 it is possible to obtain a graph isomorphic to G_3. Now G_3 contains no triangles, and every complete subgraph on four vertices contains at least one edge. Thus the six edges must be removed from G_4 in such a way that no triangle remains, no empty four is introduced, and every vertex becomes 3-valent.

If AE and BG are both removed from G_4, G_6 is obtained, and four more edges must be removed to give G_3. Suppose that GA is removed. G must be 3-valent, so GH, GF, and GE remain. Since no triangle can remain, FH and FE must be removed, but then F is only 2-valent. Thus GA must remain, and by symmetry so must $HB, AC, \ldots, FH$. If AB and BC are removed, B is only 2-valent. Thus the four edges to be removed must be AB, CD, EF, and GH by symmetry. But then $ABFE$ is an empty four.

Thus either AE or BG must remain, say AE. If BF is removed, giving G_5, the argument above implies that neither GA nor GE may be removed, and AGE is a triangle. Thus both AE and BG must remain, and six other edges must be removed from G_6. If HB is removed, HA, HF, and HG remain, and AG and FG must be removed from triangles AGH and FGH. But this leaves G 2-valent. Thus HB, AC, FD, and ED all remain. But now AG must be removed from triangle AEG, and similarly BD, CE, and FH are removed. No more edges incident with C, D, H, or G may be removed, and thus AB and FE must be removed. Then $CEFH$ is an empty four. This completes the proof of Lemma 4.

LEMMA 5. *G_4 contains no subgraph isomorphic to G_2.*

PROOF: Suppose that by removing seven edges from G_4 it is possible to obtain a graph isomorphic to G_2. As in Lemma 4, suppose that AE and BG are removed. Then G_6 is obtained, and G_6 must contain a subgraph isomorphic to G_2. Thus the complement of G_2 contains a subgraph isomorphic to the complement of G_6. But the complements of G_2 and G_6 are G_5 and G_3, so G_5 contains a subgraph isomorphic to G_3. Since G_5 is isomorphic to a subgraph of G_4, it follows that G_4 contains a subgraph isomorphic to G_3, contradicting Lemma 4.

It follows that either AE and BG both remain, or AE is removed and BG remains. These two cases may be eliminated by arguments similar to those in Lemma 4. However the proofs are somewhat complicated by the lack of symmetry in G_2 and will not be given here.

Note that G_4 does contain a subgraph isomorphic to G_1. This fact somewhat complicates the proof of Theorem 2 in the next section.

4. A NEW UPPER BOUND

In this section the results of Section 3 are used with Lemma 1 to prove the nonexistence of (4, 4; 3)-colorings on 17 or 18 vertices.

THEOREM 1. $N(4, 4; 3) \leq 18$.

PROOF: Suppose there exists a (4, 4; 3)-coloring on 18 vertices V_0, $V_1, \ldots, V_{17}$. Apply Lemma 2 to the (4, 4; 2)-coloring induced by V_0 on $V_1, \ldots, V_{17}$. Then V_1 is joined by red edges to $x = 8$ vertices, and the red edges interjoining vertices of $\mathscr{X}$ have the configuration in Fig. 3. Any triangle which is red in the edge-coloring must be blue in the face-coloring. Thus faces

$$V_1V_2V_3,\quad V_1V_3V_4,\quad V_1V_4V_5,\quad V_1V_5V_6,\quad V_1V_6V_7,\quad V_1V_7V_8,$$
$$V_1V_8V_9,\quad V_1V_9V_2,\quad V_1V_2V_6,\quad V_1V_3V_7,\quad V_1V_4V_8,\quad V_1V_5V_9$$

are blue.

Now consider the (4, 4; 2)-coloring induced by V_1 on $V_0, V_2, V_3, \ldots, V_{17}$. Faces $V_0V_1V_2, V_0V_1V_3, \ldots, V_0V_1V_9$ are red, and thus $V_0V_2, V_0V_3, \ldots, V_0V_9$ are red edges in this induced coloring. Also from the above paragraph $V_2V_3, V_3V_4, \ldots, V_8V_9, V_9V_2, V_2V_6, V_3V_7, V_4V_8$, and V_5V_9 are blue edges. Thus the blue edges interjoining vertices of $\mathscr{X}$ in the coloring induced by V_1 contain a subgraph isomorphic to G_3. But $\mathscr{X}$ is (3, 4; 2)-colored, and the blue edges interjoining points of $\mathscr{X}$ are given by G_6, the complement of G_3. Thus G_6 contains a subgraph isomorphic to G_3. But G_6 is a subgraph of G_4 and this contradicts Lemma 4. Thus there can exist no (4, 4; 3)-coloring on 18 vertices, and Theorem 1 follows.

THEOREM 2. $N(4, 4; 3) \leq 17$.

PROOF: Suppose there exists a (4, 4; 3)-coloring on 17 vertices $V_0, V_1, \ldots, V_{16}$. Consider the (4, 4; 2)-coloring induced by V_0 on $V_1, \ldots, V_{16}$. By the comments following Lemma 2, each vertex is incident with seven or eight red edges in this coloring, and vertices of both valences exist. Let V_1 be a vertex incident with eight red edges joining it to $V_2, V_3, \ldots, V_9$ which span the complete subgraph $\mathscr{X}$ in the edge-coloring. By Lemma 2, $\mathscr{X}$ is (3, 4; 2)-colored. By Lemma 3, (i), the red edges interjoining vertices of $\mathscr{X}$ have one of the configurations in Figs. 1–3.

First suppose that for some vertex V_1, the red edges in $\mathscr{X}$ are as in Fig. 2 or Fig. 3. Any triangle whose edges are red in this edge-coloring must be blue in the face-coloring. Thus faces

$$V_1V_2V_3,\quad V_1V_3V_4,\quad V_1V_4V_5,\quad V_1V_5V_6,\quad V_1V_6V_7,\quad V_1V_7V_8,$$
$$V_1V_8V_9,\quad V_1V_9V_2,\quad V_1V_3V_7,\quad V_1V_4V_8,\quad V_1V_5V_9,$$

and possibly $V_1V_2V_6$ are blue in the face-coloring. Now consider the (4, 4; 2)-coloring induced by V_1 on $V_0, V_2, V_3, \ldots, V_{16}$. As in the proof of Theorem 1, $V_0V_2, V_0V_3, \ldots, V_0V_9$ are red edges, and $V_2V_3, V_3V_4, \ldots, V_8V_9, V_9V_2, V_3V_7, V_4V_8, V_5V_9$, and possibly V_2V_6 are blue edges. But $V_2, \ldots, V_9$ are the vertices of $\mathscr{X}$ in this edge-coloring, and thus the blue edges interjoining vertices of $\mathscr{X}$ contain a subgraph isomorphic to G_2 or G_3. But $\mathscr{X}$ is (3, 4; 2)-colored and the blue edges interjoining vertices of $\mathscr{X}$ form configuration G_4, G_5, or G_6 by Lemmas 2 and 3. Since G_5 and G_6 are subgraphs of G_4, and G_2 is a subgraph of G_3, this means that G_4 must contain a subgraph isomorphic to G_2—which is impossible to Lemma 5.

It follows that for every vertex V_1 of type B (incident with eight red edges), the red edges in $\mathscr{X}$ are as in Fig. 1. Interchanging red and blue in the above argument gives the result that for each vertex V_1 of type A (incident with eight blue edges), the blue edges in $\mathscr{Y}$ are as in Fig. 1. This case cannot be eliminated by the above arguments because G_4 does contain a subgraph isomorphic to G_1. Instead it is shown that there exists no (4, 4; 2)-coloring on 16 vertices with the desired properties.

Let V_i be any vertex of type B in this edge-coloring. Then V_i is joined by red edges to eight vertices spanning the complete subgraph $\mathscr{X}_i$, and by blue edges to seven vertices spanning $\mathscr{Y}_i$. Let $\mathscr{X}_i$ contain x_i vertices of type A, while $\mathscr{Y}_i$ contains y_i vertices of type A. Then $x_i + y_i = a = 16\text{-}b$. The number of red incidences with vertices in $\mathscr{X}_i$ is

$$7x_i + 8(8 - x_i) = 64 - x_i$$

and thus there are

$$64 - x_i - 2.10 - 8 = 36 - x_i$$

red edges from $\mathscr{X}_i$ to $\mathscr{Y}_i$ (ten red edges lie within $\mathscr{X}_i$).

But the number of red incidences with vertices of $\mathscr{Y}_i$ is

$$7y_i + 8(7 - y_i) = 56 - y_i = 40 + b + x_i$$

and thus there are

$$[(40 + b + x_i) - (36 - x_i)]/2 = (4 + b + 2x_i)/2$$

red edges lying within $\mathscr{Y}_i$.

But $\mathscr{Y}_i$ is (4, 3; 2)-colored by Lemma 2. By Lemma 3(ii), $\mathscr{Y}_i$ contains at least 11 red edges, and thus

$$4 + b + 2x_i \geq 22.$$

Upon rearrangement this gives

$$x_i \geq (18 - b)/2.$$

Summing x_i over all vertices V_i of type B gives the number of red edges from vertices of type A to vertices of type B. Thus there are at least $b(18 - b)/2$ red edges from vertices of type A to vertices of type B.

Now interchanging red and blue, there are at least $a(18 - a)/2$ blue edges from B's to A's, and in all there are at least

$$a(18 - a)/2 + b(18 - b)/2$$

edges joining A's to B's. But the number of such edges is ab, so that

$$a(18 - a)/2 + b(18 - b)/2 \leq ab.$$

This implies $a + b \geq 18$, contradicting $a + b = 16$. Thus there exists no (4, 4; 2)-coloring of the type required, and Theorem 2 is proved.

5. A Lower Bound

In this section a (4, 4; 3)-coloring on 11 vertices is constructed; that is, a coloring scheme for the

$$\binom{11}{3}$$

faces is given such that there is no tetrahedron of one color. This proves the following:

THEOREM 3. $N(4, 4; 3) \geq 12$.

Name the vertices $1, 2, \ldots, n$, $(n = 11)$. The "shape" of face ijk $(i < j < k)$ is defined to be $(j - i, k - j, n + i - k)$. It is specified by three positive integers a, b, c with $a + b + c = n$. A face with shape (a, b, c) is said to be congruent to faces with shapes (a, b, c), (b, c, a), or (c, a, b). However faces with shapes (a, b, c) and (a, c, b) are not congruent unless $b = c$. A coloring in which congruent faces are assigned the same color will be constructed. This method might be called a generalization of the constructions given in [5].

A tetrahedron with vertices i, j, k, m $(i < j < k < m)$ has faces ijk, jkm, ikm, and ijm. These faces have shapes

$$(a, b, c + d), \qquad (b, c, d + a), \qquad (a + b, c, d), \qquad (a, b + c, d)$$

where $a = j - i$, $b = k - j$, $c = m - k$, and $d = n + i - m$. The faces of a tetrahedron with vertices $i + s, j + s, k + s, m + s$ (addition modulo n) will be congruent to those above. Thus for the type of coloring under consideration, a tetrahedron may be represented by four positive integers a, b, c, d with $a + b + c + d = n$ and $a \leq b, c, d$; that is, by a partition (ordered) of n into four positive parts, the first of the parts being the least.

Now take $n = 11$. Corresponding to partition 1–1–1–8 there is a tetrahedron with face shapes $(1, 1, 9)$, $(1, 1, 9)$, $(1, 8, 2)$, $(8, 1, 2) \equiv (1, 2, 8)$. The tetrahedron corresponding to partition 1–1–2–7 has face shapes $(1, 1, 9)$, $(1, 2, 8)$, $(2. 7, 2) \equiv (2, 2, 7)$, and $(7, 1, 3) \equiv (1, 3, 7)$. In this way one can prepare a list of tetrahedrons together with their face shapes. Color a face red if it has shape

$$(1, 1, 9), \quad (1, 2, 8), \quad (1, 3, 7), \quad (1, 6, 4), \quad (2, 6, 3),$$
$$(2, 4, 5), \quad (2, 5, 4), \quad \text{or} \quad (3, 3, 5)$$

or is congruent to a face with one of these shapes. Otherwise color it blue. It is easily verified that every tetrahedron has both a red and a blue face, and thus a (4, 4; 3)-coloring has been constructed. Theorem 3 follows.

6. Discussion

The construction in Section 5 fails to yield a (4, 4; 3)-coloring on n vertices for $n = 12$ or 13, and it seems unlikely to work for larger n. Various other methods of construction have been attempted without success, so that at present the best lower bound for $N(4, 4; 3)$ is 12. A search program written by Mr. Lee James, a mathematics undergraduate at the University of Waterloo, discovered many (4, 4; 3)-colorings on 11 vertices, but failed to find any on 12 vertices. Although the search was not complete, this does indicate that (4, 4; 3)-colorings on 12 vertices are relatively few (if indeed any exist), and it leads one to speculate that the value of $N(4, 4; 3)$ is closer to the lower limit than to the upper one.[1]

Any improvement in the upper bound would appear to be quite difficult. The number of cases which must be considered in applying the methods of Section 4 to colorings on 16 vertices is almost prohibitive because of the large numbers of (4, 4; 2)-colorings on 15 vertices. It would seem that some fresh ideas are required if further progress is to be made.

References

1. P. Erdős and G. Szekeres, A Combinatorial Problem In Geometry, *Compositio Math.* **2** (1935), 463–470.
2. J. E. Graver and J. Yackel, An Upper Bound for Ramsey Numbers, *Bull. Amer. Math. Soc.* **72** (1966), 1076–1079.
3. J. E. Graver and J. Yackel, Some Graph Theoretic Results Associated with Ramsey's Theorem, *J. Combinatorial Theory* **4** (1968), 125–175.
4. R. E. Greenwood and A. M. Gleason, Combinatorial Relations and Chromatic Graphs, *Canad. J. Math.* **7** (1955), 1–7.
5. J. G. Kalbfleisch, Construction of Special Edge-Chromatic Graphs, *Canad. Math. Bull.* **8** (1965), 575–584.
6. J. G. Kalbfleisch, Chromatic Graphs and Ramsey's Theorem, Ph.D. thesis, Univ. of Waterloo, Waterloo, Ontario, 1966.
7. J. G. Kalbfleisch, Upper Bounds for Some Ramsey Numbers, *J. Combinatorial Theory* **2** (1967), 35–42.
8. J. G. Kalbfleisch, On An Unknown Ramsey Number, *Michigan Math. J.* **13** (1966), 385–392.
9. J. G. Kalbfleisch, A Uniqueness Theorem for Edge-Chromatic Graphs, *Pacific J. Math.* **21** (1967), 503–509.

[1] Note added in proof: A. Sobczyk proved that $N(4, 4; 3) \geq 11$ in his recent paper, Graph-Colouring and Combinatorial Numbers, *Canadian J. Math.* **20** (1968), 520–534.

10. G. KÉRY, Ramsey egy gráfélmeleti tételéröl, *Mat. Lapok*, **15** (1964), 204–224.
11. F. P. RAMSEY, On a Problem of Formal Logic, *Proc. Lond. Math. Soc.* [2] **30** (1930), 264–286.
12. H. J. RYSER, *Combinatorial Mathematics*, Chapt. 4, pp. 38–43, Wiley, New York, 1963.
13. K. WALKER, Dichromatic Graphs and Ramsey Numbers, *J. Combinatorial Theory* **5** (1968), 238–243.

SYLVESTER MATROIDS

U. S. R. Murty

UNIVERSITY OF WATERLOO
WATERLOO, ONTARIO

1. Introduction

The motivation for the problem considered in this paper is a theorem of Gallai which was originally formulated by J. J. Sylvester towards the end of the last century (see [1, p. 3]). Gallai's theorem states that *if a set of points in n-dimensional real projective space, not all points on a line, is such that on the line joining any two points of the set there is always a third point of the set, then the number of points cannot be finite.* We shall find it convenient to state the above theorem in terms of linear dependence of vectors.

Let $R_1, R_2, \ldots, R_m$ denote the set of rows of a matrix of rank n, $n \geq 3$, over a field F. Let us assume that the null-vector is not one of the rows and that no two rows are proportional. Let us assume further that given any two rows R_i and R_j there exists a third row R_k such that R_k is a linear combination of R_i and R_j. The theorem of Gallai would then imply that if F is the real number field then $m \geq N$ for all integral N.

One consequence of a theorem we are going to prove would be that

$$m \geq 2^n - 1 \tag{1}$$

regardless of which field F is.

A result of great generality such as (1) becomes possible because it owes nothing to the structure of the field under consideration but hinges merely on the abstract properties of linear dependence.

A study of abstract properties of linear dependence was started by Whitney [5]. Abstracting a set of axioms from the properties of linear dependence, Whitney defined mathematical systems called matroids. Matroids were eminently pursued by Tutte [4].

2. Matroids

Let E be a finite set. A class M of nonnull subsets of E is called a matroid on E if it satisfies the following two axioms.

Axiom 1. No member of M is contained in another.

Axiom 2. If X and Y are members of M and a and b are members of E such that $a \in X \cap Y$ and $b \in X - Y$, then there exists $Z \in M$ such that

$$b \in Z \subseteq (X \cup Y) - \{a\}.$$

We refer to the members of E and M as the *cells* and *circuits* of M, respectively.

One natural source for obtaining examples of matroids are matrices. Let E be the set of rows of a matrix over a field F. Let a subset A of E be a member of M if the members of A are linearly dependent over F and members of any proper subset of A are linearly independent over F. Then M is a matroid on E. A matroid is said to be *real* if it can be viewed as the matroid associated in this fashion with a matrix over the reals.

Let M be a matroid defined on E. A maximal subset of E which contains no circuit as a subset is called a *basis* of the matroid. It can be proved that different bases of the matroid have the same cardinality. The cardinality of a basis of the matroid is called the *rank* of the matroid.

A matroid is called a *Sylvester matroid* if any two cells of the matroid are contained in at least one circuit of size 3 (i.e., containing three cells). The following is an example of a Sylvester matroid.

Example. Let E denote the set of rows of the $(2^n - 1) \times n$ matrix, $n \geq 2$, whose rows are the $2^n - 1$ distinct nonnull n-tuples of elements from $GF(2)$. Let M be the matroid defined on E as described earlier. Then M is a Sylvester matroid.

Gallai's theorem asserts the nonexistence of real Sylvester matroids of rank greater than or equal to 3. A look at the Sylvester matroids in general gives us an interesting theorem which we prove in the next section.

3. Number of Cells of a Sylvester Matroid

In this section we shall prove the following.

THEOREM 1. *If a Sylvester matroid M defined on E is of rank n, $n \geq 2$, then*

$$|E| \geq 2^n - 1. \tag{2}$$

Let $B = \{x_1, x_2, \ldots, x_n\} \subseteq E$ be a basis of the matroid. If $X \subseteq B$, $e \in \bar{B}$, and $X + \{e\} \in M$, then we say that e is a *lid* of X. Every element of $e \in \bar{B}$ is a lid of a unique subset of B. This, in fact, is true for all matroids. We shall prove that each subset X of B with at least two elements has a lid, which need not be true for all matroids. The proof of this is by induction on the size of the set X. If $|X| = 2$, then by the definition of a Sylvester matroid X has a lid in $\bar{B}$. Suppose that all subsets X of B with $|X| < p > 2$ have a lid in $\bar{B}$. Let Y be a p-subset of B. Let $y \in Y$. Write $Y = X + \{y\}$. By the induction hypothesis there exists an element $\rho \in \bar{B}$ such that $C_1 = X + \{\rho\} \in M$. Any by the definition of a Sylvester matroid there exists a $t \in E$ such that $C_2 = \{y, t, \rho\} \in M$. Therefore by Axiom 2 there will exist a circuit $C_3 = \{y\} + Z + \{t\}$, Z being some subset of X. The subset Z cannot be a proper subset of X, for if it were a proper subset of X, then using Axiom 2 with reference to C_2 and C_3 we obtain the contradiction that ρ is a lid of two different subsets of B. Therefore t is a lid of Y. Now it follows from the principle of induction that all subsets of B with at least two elements have a lid in $\bar{B}$. The theorem now follows from a simple counting argument.

A Sylvester matroid of rank n for which equality is attained in (2) is isomorphic to the matroid of the example given in Section 2. A proof of this will be presented in [3].

4. Remarks

As was observed by Motzkin [2], Gallai's theorem holds for a class of geometries wider than the class of geometries over reals: Let F be a field. Then it is sufficient for F to be an ordered field in order that Gallai's theorem may hold for geometries over F. It would be of interest to investigate the classes of matroids among which there exists no Sylvester matroid.

References

1. H. Hadwiger, H. Debrunner, and V. Klee, *Combinatorial Geometry in the Plane*, Holt, New York, 1964.

2. T. MOTZKIN, The Lines and Planes Connecting the Points of a Finite Set, *Trans. Amer. Math. Soc.* **70** (1951), 451–464.
3. U. S. R. MURTY, Matroids with Sylvester Property (submitted for publication).
4. W. T. TUTTE, Lectures on Matroids, *J. Res. Nat. Bur. Standards Sect. B* **69** 1965, 1–47.
5. H. WHITNEY, On the Abstract Properties of Linear Dependence, *Amer. J. Math.* **57**, (1935), 509–533.

CYCLIC COLORATION OF PLANE GRAPHS*

Oystein Ore†

and

Michael D. Plummer‡

YALE UNIVERSITY, NEW HAVEN, CONNECTICUT

1. Angle Coloration and Cyclic Coloration

In this paper we shall consider two types of coloration for plane graphs which do not seem to have been examined previously. In the ordinary face coloration of a plane graph one assumes that the colors of the faces are such that *neighboring faces*, that is, faces with a common boundary edge, have different colors. Let us define two faces to be *contiguous* if they have at least one boundary vertex in common. We shall here investigate face colorations in which contiguous faces must have different colors. Such a coloration may be called an *angle coloration* since it is equivalent to requiring that at any vertex the angular face sections meeting there shall have distinct colors, provided the two angles do not belong to the same face.

We shall prefer to investigate this problem in the dual form. As one readily sees, this corresponds to a vertex coloration of a plane graph such that the vertices on any face boundary have different colors. Such a coloration we shall call a *cyclic* (*vertex*) *coloration*.

The terminology used in this paper is that of Ore [1, 2].

2. Notation

In the following, G shall be a finite plane graph (i.e., a graph with a given realization in the plane in which no two lines intersect except, perhaps, at

* This paper was partially supported by NSF Grant GP 6558.

† Deceased, August 1968.

‡ Present address: Department of Computer Science, City University of New York, New York, New York.

vertices) with vertex set V. The valence of a vertex $v \in V$ is the number of edges at v counting loops twice. The valence $\rho^*(F)$ of a face F is the number of different vertices on its boundary. It should be noted that this definition is different from the one used by Ore [2] in which the face valence was the number of boundary edges counting acyclic edges twice. For graphs without separating vertices, however, the two definitions coincide.

We shall examine the cyclic colorations of G. We denote by $\kappa(G)$ the *(cyclic) coloration number*, that is, the minimum number of colors in any cyclic coloration. Clearly $\kappa(G)$ cannot be less than the ordinary coloration number. From the definition one sees that

$$\kappa(G) \geq \max \rho^*(F) = \rho_0{}^*(G)$$

taken over the various faces F in G.

One has $\kappa(G) = 1$ only when G consists of a set of loops at a single vertex. It is not difficult to see that $\kappa(G) = 2$ if and only if G has two vertices v_1 and v_2 with families of loops at each and in addition a family of connecting edges (v_1, v_2). Thus we shall henceforth assume that $\kappa(G) \geq 3$.

3. Reductions

We shall discuss certain reductions which can be performed for the cyclic coloration problem. Suppose G is not connected so that there exists a decomposition

$$G = G_1 \cup G_2, \qquad G_1 \cap G_2 = \varnothing \tag{1}$$

(Of course G_1 and G_2 may themselves be disconnected.) Then G_2 will lie in one of the faces F_1 of G_1, and G_1 in a face F_2 of G_2. The face F in G defined by F_1 and F_2 has the valence

$$\rho^*(F) = \rho^*(F_1) + \rho^*(F_2). \tag{2}$$

To color G we begin by coloring the $\rho^*(F)$ vertices on the boundary of F in different colors. The $\rho^*(F_2)$ colors used on the boundary of F_2 can also be used on the interior vertices of G_1. Thus if $\kappa(G_1) - \rho^*(F_1) \leq \rho^*(F_2)$ we need no further colors. If $\kappa(G_1) - \rho^*(F_1) > \rho^*(F_2)$, an additional $\kappa(G_1) - \rho^*(F)$ colors are needed. The same argument applies to the coloration of G_2. We conclude:

$$\kappa(G) = \max(\kappa(G_1), \kappa(G_2), \rho^*(F)). \tag{3}$$

Exactly the same argument applies to the case where the graphs G_1 and G_2 in (1) have just one vertex in common. One finds the same formula (3) holds except that now (2) is replaced by

$$\rho^*(F) = \rho^*(F_1) + \rho^*(F_2) - \Delta \tag{4}$$

where $\Delta = 1$.

This reduces the coloration problem to the case where G is connected and inseparable. Also, one can suppose that

$$\rho^*(F) \geq 3 \tag{5}$$

for every face, since if a face should be bounded by two edges

$$E_1 = (a, b) = E_2 \tag{6}$$

one of them can be omitted without changing $\kappa(G)$. More generally, one can assume that G has no multiple edges (6). If two edges (6) should separate G so that there are vertices in both domains thus formed,

$$G = G_1 \cup G_2, \qquad G_1 \cap G_2 = \{E_1, E_2\} \tag{7}$$

then one readily sees that

$$\kappa(G) = \max(\kappa(G_1), \kappa(G_2)) \tag{8}$$

Also, separating triangles can be eliminated by a similar argument, since $\kappa(G) \geq 3$.

Next, suppose that G contains an edge $E = (a_1, a_2)$ where $\{a_1, a_2\}$ is a minimum separating set for G, i.e., G consists of two subgraphs G_1 and G_2 having only E in common. The previous argument applies and one obtains the same formula (3) with $\Delta = 2$ in (4).

Finally, let us assume that G has a separation by an arc

$$A = (a_1, a_2) \cdots (a_{k-1}, a_k). \tag{9}$$

To color G we first color the vertices on the boundary of the face F, requiring $\rho^*(F) = \rho^*(F_1) + \rho^*(F_2) - 2k + 2$ colors. We continue the coloration by giving the $k - 2$ inner vertices in A colors not used on F. Thus, in this special coloration we have so far used $\rho^*(F) + k - 2 = \rho^*(F_1) + \rho^*(F_2) - k$ colors. Among these, $\rho^*(F_2) - k$ colors have been applied to vertices not in G_1. These may be reused in the coloration of the interior vertices of G_1. They suffice to color them all if $\kappa(G_1) - \rho^*(F_1) \leq \rho^*(F_2) - k$ or $\kappa(G_1) \leq \rho^*(F) + k - 2$. If this inequality does not hold, altogether $\kappa(G_1)$ colors are needed to color G_1 and the remaining part of the boundary of G_2. The same argument applies to the coloration of the remaining part of G_2, so we arrive at the result:

Let G be separated into two components G_1 and G_2 by an arc (8), $k \geq 3$, connecting two vertices on the boundary of a face F. Then for $\kappa(G)$ we have the bound

$$\kappa(G) \leq \max(\kappa(G_1), \kappa(G_2), \rho^*(F) + k - 2). \tag{10}$$

4. Bounds for the Coloration Number

We shall now derive an upper bound for the cyclic coloration number for a plane graph G.

Let F_1 and F_2 be a pair of contiguous faces. We define the *contiguity* of this pair to be the number

$$\gamma(F_1, F_2) = \rho^*(F_1) + \rho^*(F_2) - \beta(F_1, F_2) \tag{11}$$

where the valences of F_1 and F_2 are defined as before while the deducted term $\beta(F_1, F_2)$ is the number of common boundary vertices $b_{1,2}$ for which

$$\rho(b_{1,2}) = 2. \tag{12}$$

It is clear from this definition that

$$\gamma(F_1, F_2) \geqq \max(\rho^*(F_1), \rho^*(F_2)). \tag{13}$$

One also verifies that the equality sign holds in (13) only when all boundary vertices for the two faces are common and satisfy (12). This can occur only when G is a single circuit with two faces F_1 and F_2.

The maximal contiguity in G we denote by

$$\gamma_0 = \gamma(G) = \max(F_1, F_2) \tag{14}$$

taken over all pairs of contiguous faces. We can then prove:

THEOREM 1. *The cyclic coloration number of a plane graph satisfies*

$$\kappa(G) \leqq \gamma_0. \tag{15}$$

One may notice that the equality sign holds in (15) for any graph consisting of a single circuit.

The proof of Theorem 1 shall be based upon induction with respect to the number $n = |V|$ of vertices in G. One readily verifies that it holds for the smallest values of n.

We first reduce the graph by means of the results in Section 3. Suppose first that G is disconnected and has a decomposition (1). The removal of G_1 or G_2 from G reduces the number of vertices, but no contiguity is increased as one sees. Hence the theorem holds for G_1 and G_2 so that

$$\kappa(G_1) \leqq \gamma(G_1), \qquad \kappa(G_2) \leqq \gamma(G_2),$$

so according to (3)

$$\kappa(G) \leqq \max(\gamma(G_1), \gamma(G_2), \rho^*(F)) \leqq \gamma(G).$$

Thus we may suppose that G is a connected graph.

Suppose next that G has a separation into two components G_1 and G_2 by a vertex s. If G_1 and G should have the same vertex set, the component G_2 must be a collection of loops at s. Since any such loop can be removed without changing $\kappa(G)$ or $\gamma(G)$ we may suppose that

$$|V(G_i)| < n, \qquad i = 1, 2.$$

Thus by the induction we have

$$\kappa(G_i) \leqq \gamma(G_i), \qquad i = 1, 2.$$

By the removal of G_1 and G_2 from G no contiguity can be increased and so we conclude from (3)

$$\kappa(G) \leqq \max(\gamma(G_1), \gamma(G_2), \rho^*(F)) \leqq \gamma(G).$$

So we may assume that our graph is connected and inseparable. The argument used in Section 3 serves to eliminate multiple edges. Our next step is to reduce the graph to the case where $\rho(v) \geqq 3$ for all vertices. Suppose $\rho(v) = 2$. For such a vertex let the two contiguous faces be F_1 and F_2 and let the edges be

$$E_1 = (v, v_1), \qquad E_2 = (v, v_2). \tag{16}$$

We replace E_1 and E_2 by a single edge $E' = (v_1, v_2)$. This produces a new graph G' with two new faces F_1' and F_2' for which we have

$$\rho^*(F_i') = \rho^*(F_i) - 1, \qquad \gamma(F_1', F_2') = \gamma(F_1, F_2) - 1$$

and no contiguity is increased. By the induction G' can be colored in $\gamma(G)$ colors. We reinsert the vertex v and give it a color different from those on the boundaries of F_1 and F_2. This is possible since the number of colors used on the boundaries of F_1' and F_2' is at most equal to $\gamma(G) - 1$.

In the final step, then, we may suppose that we have

$$\rho(v) \geq 3, \qquad \rho^*(F) \geq 3 \tag{17}$$

for all vertices and faces.

We now make use of the theory of Euler contributions as developed by Ore [2, Section 4.3]. For the vertex contribution of a vertex v, we have the expression

$$\phi(v) = 1 - \tfrac{1}{2}\rho(v) + \sum_i (1/\rho^*(F_i)) \tag{18}$$

where the sum is extended over all $\rho(v)$ faces F_i with a corner at v. The sum of all vertex contributions is

$$\sum_v \phi(v) = 2 \tag{19}$$

This shows that each planar graph must have vertices with

$$\phi(v) > 0. \tag{20}$$

According to our assumption (17) we have for the sum in (18), $\sum_i (1/\rho^*(F_i)) \leq \frac{1}{3}\rho(v)$, so for any vertex v satisfying (20), $\phi(v) \leq 1 - \frac{1}{6}\rho(v)$. As a consequence (20) can hold only for valences

$$\rho(v) = 3, 4, \text{ or } 5. \tag{21}$$

For each of these valences (21) there are a certain number of solution sets (x_i) where $x_i = \rho^*(F_i)$, $i = 1, 2, \ldots, \rho(v)$. These are the integers satisfying

$$\sum_i \frac{1}{x_i} > \tfrac{1}{2}\rho(v) - 1.$$

The examination of these inequalities for each of the values (20) leads to the solutions (x_i) given in Table I. In each solution the x_i are listed in increasing order. (We point out that the list given here is the dual of that by Ore [2, Section 4.3] for faces. The table has been repeated, since through an oversight, three entries were omitted there.)

TABLE I

$\rho(v) = 3$		A	$\gamma(F_1, F_2)$
$(3,\ 6, x)$,	$x = 6, \ldots$	$3 + x$	$6 + x$
$(3,\ 7, x)$,	$x = 7, \ldots, 41$	$4 + x$	$7 + x$
$(3,\ 8, x)$,	$x = 8, \ldots, 23$	$5 + x$	$8 + x$
$(3,\ 9, x)$,	$x = 9, \ldots, 17$	$6 + x$	$9 + x$
$(3, 10, x)$,	$x = 10, \ldots, 14$	$7 + x$	$10 + x$
$(3, 11, x)$,	$x = 11, 12, 13$	$8 + x$	$11 + x$
$(4,\ 4, x)$,	$x = 4, \ldots$	$2 + x$	$4 + x$
$(4,\ 5, x)$,	$x = 5, \ldots, 19$	$3 + x$	$5 + x$
$(4,\ 6, x)$	$x = 6, \ldots, 11$	$4 + x$	$6 + x$
$(4,\ 7, x)$,	$x = 7, 8, 9$	$5 + x$	$7 + x$
$(5,\ 5, x)$,	$x = 5, \ldots, 9$	$4 + x$	$5 + x$
$(5,\ 6, x)$,	$x = 6, 7$	$5 + x$	$6 + x$
$\rho(v) = 4$			
$(3, 3, 3, x)$,	$x = 3, \ldots$	$1 + x$	$3 + x$
$(3, 3, 4, x)$,	$x = 4, \ldots, 11$	$2 + x$	$4 + x$
$(3, 3, 5, x)$,	$x = 5, 6, 7$	$3 + x$	$5 + x$
$(3, 4, 4, x)$,	$x = 4, 5$	$3 + x$	$4 + x$
$\rho(v) = 5$			
$(3, 3, 3, 3, x)$,	$x = 3, 4, 5$	$2 + x$	$3 + x$

Each plane graph satisfying (17) must contain at least one vertex v_0 at which the face valences are those of one of the sets (x_i) listed above. We continue our reductions by selecting an edge $E_0 = (v_0, v_1)$ at v_0, taken such that E_0 is a boundary edge for a face which has the maximum valence at v_0. We contract G with respect to the edge E_0 to obtain a graph $G(v_0 = v_1)$. By our choice of E_0 one readily sees that the contiguity cannot increase by the contraction. It follows by induction that $G(v_0 = v_1)$ is colorable in $\gamma(G)$ colors.

We select any one of these colorations. It also defines a cyclic coloration of G except that v_0 and v_1 have the same color. However, if the total number of colors used on the vertices on the faces touching v_0 is less than $\gamma(G)$, one can assign a color to v_0 different from these to obtain a coloration of G in $\gamma(G)$ colors. The number of such colors cannot exceed the number of vertices $\neq v_0$ on these faces, hence it cannot exceed

$$A = \sum_i (x_i - 2) = \sum x_i - 2\rho(v_0).$$

This number is listed in the above table for the various cases. In the last column one finds the maximal contiguity $\gamma(F_1, F_2)$ for any pair of faces at v_0 and clearly $A < \gamma(F_1, F_2)$.

Since in each case one can perform a reduction of the types described, we conclude that Theorem 1 is valid.

A corollary of our theorem is:

THEOREM 2. *Let $\rho_0^* \geq \rho_1^*$ be the two largest face valences in a plane graph G. Then*

$$\kappa(G) \leq \rho_0^* + \rho_1^* \leq 2\rho_0^*.$$

REFERENCES

1. O. ORE, *Theory of Graphs*, Vol. 38, Colloquium Publ., Am. Math. Soc., Providence, Rhode Island, 1962.
2. O. ORE, *The Four-Color Problem*, Academic Press, New York, 1966.

PSEUDOSYMMETRY, CIRCUIT-SYMMETRY, AND PATH-SYMMETRY OF A DIGRAPH

J. M. S. Simões Pereira

UNIVERSITY OF COIMBRA
COIMBRA, PORTUGAL

1. Introduction

Let $G = (X, U)$ be a finite digraph whose set of vertices is X and whose set of arcs is U. Denoting by $d^+(x)$ and $d^-(x)$ the number of arcs incident from x and incident to x, respectively, we say *x is pseudosymmetric* if and only if $d^+(x) = d^-(x)$ and we say *G is pseudosymmetric* if and only if all of its vertices are pseudosymmetric. If x_i and x_j are the initial and terminal vertices of an arc $u \in U$, we identify u with the ordered pair (x_i, x_j). Then a path from x_1 to x_k is a sequence $P(x_1, x_k) = \{x_1, (x_1, x_2), x_2, \ldots, x_{k-1}, (x_{k-1}, x_k), x_k\}$ of vertices $x_i \in X$ and arcs $(x_i, x_{i+1}) \in U$. As in Gupta [1], no arc is allowed to appear more than once in the same path, and if $x_1 = x_k$ the path is called a *circuit*. Two paths (or circuits) are *arc-disjoint* (a.d.) if they have no arc in common. We denote by $w_G(x, y)$ the maximal number of (pairwise) a.d. paths from x to y and by $\sigma_G(x, y)$ the maximal number of (pairwise) a.d. circuits each passing through x and y. A digraph G is said to be *circuit-symmetric* if $\sigma_G(x, y) = w_G(x, y)$ for all $x, y \in X$, and *path-symmetric* if $w_G(x, y) = w_G(y, x)$ for all $x, y \in X$.

It is known (see Gupta [1] and Kotzig [2]) that pseudosymmetry implies circuit-symmetry and circuit-symmetry implies path-symmetry. Here a class of circuit-symmetric graphs which will be called *circularly symmetric* are shown to be pseudosymmetric. It remains an open question to find out whether every circuit-symmetric digraph is circularly symmetric or not. Moreover it will be proved that a graph with a nonpseudosymmetric vertex x such that $d^+(x)$ [or $d^-(x)$] < 2 may not be path-symmetric.

2. Circular Symmetry

Let us introduce the concept of circular symmetry as follows: Take a pair a, b of vertices of a circuit-symmetric graph G. Let $\sigma_G(a, b) = t$ and consider all possible systems of $2t$ a.d. paths, $P_1(a, b), \ldots, P_t(a, b)$ and $Q_1(b, a), \ldots, Q_t(b, a)$, which may be coupled to form t a.d. circuits passing through a and b. Take a circuit or set of a.d. circuits C in G and let r and s be the maximum number of P's and Q's, respectively, which form one of the above-mentioned systems and are a.d. from C. If we have always $r = s$ then the digraph G is said to be circularly symmetric.

We state the following:

Theorem 1. *Every finite circularly symmetric digraph* $G = (V, E)$ *is pseudosymmetric.*

Proof. Mathematical induction will be used.

Let G be circularly symmetric. By the definition of circuit-symmetry we have $\forall x, y \in V(G)$: $\sigma_G(x, y) = w_G(x, y)$; hence if $\sigma_G(x, y) \equiv 0$ then $w_G(x, y) \equiv 0$ and G is the completely disconnected graph ($E = \varnothing$) which is trivially pseudosymmetric.

Now suppose the theorem is true for digraphs G such that

$$\forall x, y \in V(G): \sigma_G(x, y) < k \qquad (k > 0 \text{ an integer})$$

and let $G = (V, E)$ be a digraph having at least a pair of vertices $x', y' \in V(G)$ such that $\sigma_G(x', y') = k$, k being the maximum value of σ for all pairs of vertices of G.

A system of $2k$ a.d. paths $P_1(x', y'), \ldots, P_k(x', y'), Q_1(y', x'), \ldots, Q_k(y', x')$ exists and the P's and Q's may be coupled to form k a.d. circuits passing through a, b. If we delete one of these circuits, say C, we do not affect the existence or nonexistence of pseudosymmetry. Neither do we affect the circular symmetry, as will be shown.

In fact for any pair of vertices x'', y'' of C (including x', y') the value of σ decreases by one. We can write

$$\forall x'', y'' \in C: \sigma_{G'}(x'', y'') = \sigma_G(x'', y'') - 1,$$

G' denoting the graph obtained from G by suppression of C. This is because it could never be $\sigma_{G'} > \sigma_G$ and if $\sigma_{G'}(x'', y'') = k$, then the corresponding system of k a.d. circuits in G' together with C would form a set of $k + 1$ a.d. circuits in G.

A similar reasoning shows that, for $x'', y'' \in C$,

$$w_{G'}(x'', y'') = w_G(x'', y'') - 1, \qquad w_{G'}(y'', x'') = w_G(y'', x'') - 1.$$

Hence

$$\sigma_{G'}(x'', y'') = w_{G'}(x'', y'') = w_{G'}(y'', x'').$$

Now let us take a pair of vertices a, b not both in C. By the hypothesis of circuit-symmetry:

$$\sigma_G(a, b) = w_G(a, b) = w_G(b, a) = t \qquad (0 \leq t \leq k).$$

Moreover (Fig. 1) since G is circularly symmetric, let $r(\leq t)$ be the value

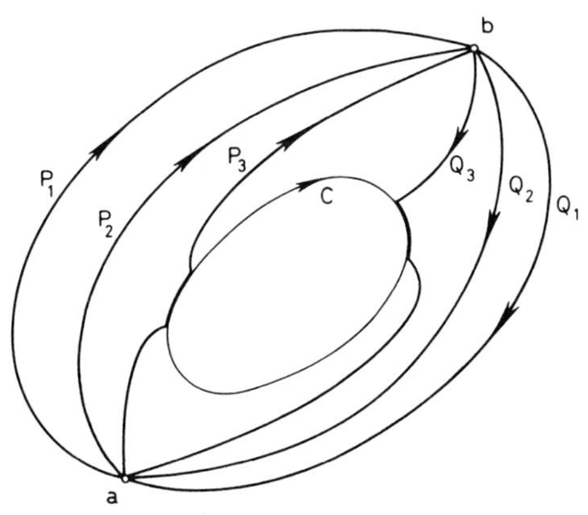

FIG. 1.

referred to in our definition of circular symmetry pertaining to C and the pair a, b. We have

$$\sigma_{G'}(a, b) = w_{G'}(a, b) = w_{G'}(b, a) = r.$$

Thus G' is circuit-symmetric as well as G and is also circularly symmetric: relatively to any circuit or set of a.d. circuits we still have, for a.d. paths connecting any pair of vertices, $r' = s'$ (r' and s' having a meaning similar to that of r and s). Failure to satisfy this condition in G' would obviously imply failure to satisfy it in G. (We emphasize the fact that, in the definition of circular symmetry, we cannot write "take a circuit" instead of "take a circuit or a set of a.d. circuits." See Fig. 2.)

As a consequence of our reasoning, the suppression of circuits does not affect the circular symmetry and so by successive deletion of circuits we obtain from G a new graph G^* such that

$$\forall x, y \in V(G^*): \sigma_{G^*}(x, y) < k.$$

By the induction hypothesis G^* is pseudosymmetric. Reinserting the deleted circuits the pseudosymmetry is not affected and we get G which is also pseudosymmetric.

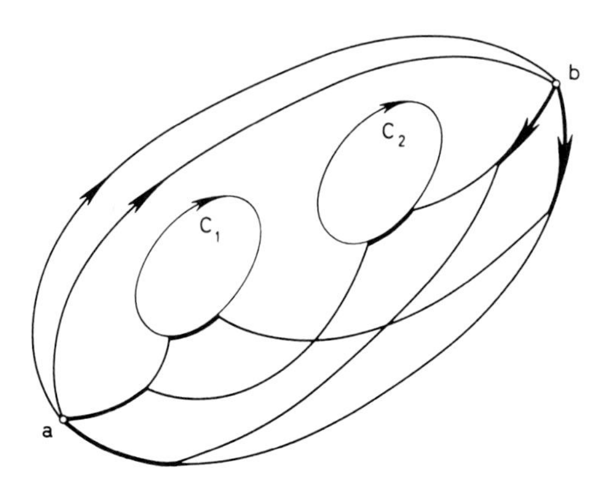

FIG. 2.

3. A SPECIAL CASE

Now the following theorem will be proved.

THEOREM 2. *No digraph having a nonpseudosymmetric vertex x with one of its degrees equal to* 1 *or* 0 *may be path-symmetric.*

PROOF. Let $G = (X, U)$ be nonpseudosymmetric. If $x \in X$ is a nonpseudosymmetric vertex such that $d^+(x) = 0$ [respectively; $d^-(x) = 0$] then $\exists y \in X$ such that $[y, x] \in U$ (respectively; $[x, y] \in U$); G is not path-symmetric since $w_G(y, x) \geq 1 \neq w_G(x, y) = 0$ [respectively; $w_G(x, y) \geq 1 \neq w_G(y, x) = 0$].

Now suppose $d^+(x) = 1 \neq d^-(x)$. If $d^-(x) = 0$ we have the preceding case. If $d^-(x) \geq 2$ (Fig. 3) at least two arcs exist incident to x, namely $[a_1, x]$ and $[a_2, x]$. As $d^+(x) = 1$, let $[x, b]$ be the arc incident from x. Denote by

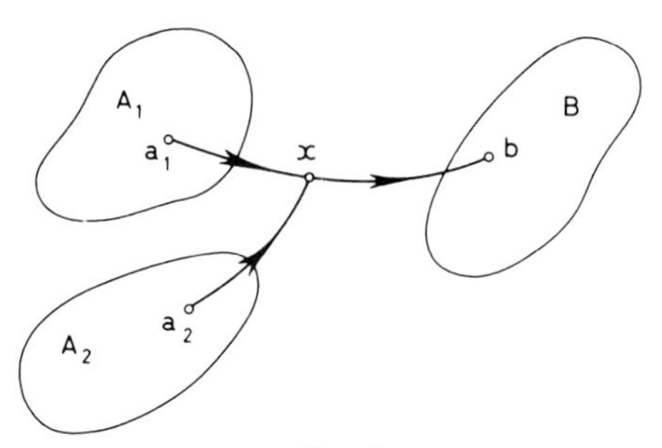

FIG. 3.

A_1 (respectively; A_2) the set of vertices from which a_1 (respectively; a_2) is reachable and by B the set of vertices reachable from b.

Suppose $A_1 \cap A_2 \neq \emptyset$. A vertex $x' \in A_1 \cap A_2$ exists from which a_1 and a_2 are reachable. Consider two paths $P_1(x', a_1)$ and $P_2(x', a_2)$. Set $u = x'$ if P_1 and P_2 are a.d.; if P_1 and P_2 are not a.d., then let u be the terminal vertex of the last arc they share. Clearly

$$w_G(u, x) \geq 2 \neq w_G(x, u) \leq 1.$$

Now suppose $A_1 \cap A_2 = \emptyset$. We have four subcases:

Subcase 1: $A_1 \cap B = A_2 \cap B = \emptyset$. We get

$$w_G(a_1, b) \geq 1 \neq w_G(b, a_1) = 0$$

and similarly

$$w_G(a_2, b) \geq 1 \neq w_G(b, a_2) = 0.$$

Subcase 2. $A_1 \cap B \neq \emptyset$, $A_2 \cap B = \emptyset$. Now

$$w_G(a_2, b) \geq 1 \neq w_G(b, a_2) = 0.$$

Subcase 3. $A_1 \cap B = \emptyset$, $A_2 \cap B \neq \emptyset$. Analogously

$$w_G(a_1, b) \geq 1 \neq w_G(b, a_1) = 0.$$

Subcase 4. $A_1 \cap B \neq \emptyset$, $A_2 \cap B \neq \emptyset$. This hypothesis is not compatible with $A_1 \cap A_2 = \emptyset$ since it implies $b \in A_1$ and $b \in A_2$, hence $b \in A_1 \cap A_2$.

The reasoning is the same for the hypothesis $d^-(x) = 1 \neq d^+(x)$. So the theorem is proved.

Acknowledgment

Thanks are due to Mr. J. A. Zimmer for pointing out to me a mistake in a previous version of this paper.

References

1. R. P. Gupta, Two Theorems on Pseudosymmetric Graphs, *SIAM J. Appl. Math.* **15** (1967), 168–171.
2. A. Kotzig, Beitrag zur Theorie der endlichen gerichteten Graphen, *Wiss. Z. Martin-Luther Univ. Halle-Wittenberg Math.-Natur. Reihe* **10** (1961–1962), 118–125 (cited in [1]).

ON THE BOXICITY AND CUBICITY OF A GRAPH

Fred S. Roberts

THE RAND CORPORATION
SANTA MONICA, CALIFORNIA

1. Introduction

Suppose A is a finite set of *points*, I is a reflexive, symmetric binary relation of *adjacency* on A, and G denotes the *graph* (A, I). All graphs in the following will be finite and have loops at each point as here. Several authors [1–3] have studied the notion of an *interval graph*, i.e., a graph G for which there is an assignment to each point x in A of an interval $N(x)$ on the real line such that for all $x, y, \in A$,

$$xIy \leftrightarrow N(x) \cap N(y) \neq \varnothing. \tag{1}$$

In the following, we consider the problem of finding an assignment satisfying (1) where the $N(x)$ are taken to be "boxes" in Euclidean n-space E^n, i.e., generalized rectangles with sides parallel to the coordinate axes.

Simultaneously, we consider a closely related problem. In [4], we studied the notion of *indifference graph*. This is a graph (A, I) for which there is a real-valued function f on A so that for all $x, y \in A$,

$$xIy \leftrightarrow d(f(x), f(y)) \leq 1, \tag{2}$$

where d is the usual metric on the real line, i.e., absolute value. A natural generalization is to ask for a representation (2) where f takes values in E^n and d is an appropriate metric on E^n. To maintain an analogy with the generalization of the interval graph problem, it is convenient to take the "product metric,"

$$d(\langle x_1, x_2, \ldots, x_n \rangle, \langle y_1, y_2, \ldots, y_n \rangle) = \max_i |x_i - y_i|,$$

because then the representation (2) corresponds exactly to a representation (1) with the $N(x)$ all closed[1] cubes of side-length 1. (Choice of this metric can also be motivated on economic and psychological grounds.)

These two representation problems will lead to two notions of dimension of a graph, the boxicity and cubicity,[2] and it will be our aim to establish a sharp upper bound for dimension in each case as a function of the number of points. We start in the next section with the second representation problem. The development for the first will be entirely analogous, and perhaps a little simpler.

2. The Cubicity

With this introduction, we now propose to study what graphs (A, I) can be *embedded into n-space* in the sense that there is a function $f: A \to E^n$ satisfying (2), where d is the product metric. It is somewhat easier to formulate the problem in terms of the coordinate functions $f_1, f_2, \ldots, f_n$ of f. Thus we ask: when do there exist real-valued functions $f_1, f_2, \ldots, f_n$ on A so that for all $x, y \in A$,

$$xIy \leftrightarrow (\forall i \leq n)[|f_i(x) - f_i(y)| \leq 1]. \tag{3}$$

It is convenient here to make one slight convention which is quite natural and makes the results a little neater to state. It seems fair to speak of a graph (A, I) as embeddable into 0-space in the above sense if and only if all points in A are adjacent, i.e., (A, I) is complete.

We begin by noting the not too surprising result that every graph with n points is embeddable into n-space in the sense of Eq. (3). To see this, simply list A as $a_1, a_2, \ldots, a_n$, and define for $i = 1, 2, \ldots, n$,

$$f_i(x) = \begin{cases} 0 & \text{if} \quad x = a_i \\ 1 & \text{if} \quad xIa_i,\ x \neq a_i \\ 2 & \text{if} \quad \sim xIa_i. \end{cases}$$

This observation permits us to define, motivated by the intersection interpretation, the *cubicity*[3] of a graph G, cub G, as the smallest n so that G is embeddable into n-space. Each graph has finite cubicity, and indeed a graph of n points has cubicity at most n.

We close this section with a few simple but basic remarks about embeddability and cubicity. In particular, it will be helpful to study the intersection of two graphs with the same point set. We note readily that if $n \geq 1$, a graph G

[1] Boxes are not necessarily closed, though it is not hard to show that if a representation (1) is attainable with boxes in E^n, it is attainable with closed boxes in E^n.

[2] See acknowledgments.

[3] More precisely, the unit cubicity.

is embeddable into n-space if and only if G is the intersection of n indifference graphs. For, each coordinate function gives a representation (2). It follows easily that if $G = H \cap K$, then cub $G \leq$ cub $H +$ cub K.

Following [4], it is convenient to introduce an equivalence relation E on the points of the graph $G = (A, I)$. This is defined by $xEy \leftrightarrow (\forall z)(xIz \leftrightarrow yIz)$. Note that since our graphs are reflexive, equivalence implies adjacency. Thus, two points are equivalent if and only if they have the same "closed neighborhoods." The relation E is significant because if aEb, then cub $G =$ cub $G - a$ $=$ cub $G - b$, where $G - x$ is the subgraph[4] generated by points different from x. (Map a and b onto the same point in n-space.) More generally, cub $G =$ cub G/E, where G/E is the graph obtained from G by cancelling out E.

3. The Cubicity of the Complete Partite Graphs

Having proved that every graph of n points can be embedded into n-space, we are interested in solving the following extremum problem: given n, what is the smallest k so that we can embed all graphs with n points into k-space? Put another way, what is the maximum cubicity $c(n)$ of all graphs with n points?

To study the function $c(n)$, it turns out to be extremely useful to know the cubicity of the so-called complete partite graphs, for the maximum cubicity is actually attained for each n in such a graph. We shall devote this section to calculating an explicit cubicity formula for the complete partites. A graph (A, I) will be called *complete partite* if A can be written as the disjoint union of nonempty classes so that no lines (except loops) within a class occur and all lines between points of different classes occur. We shall denote by $K(n_1, n_2, \ldots, n_p)$ the graph consisting of p such classes, containing $n_1, n_2, \ldots, n_p$ points, respectively ($n_i > 0$). Of particular interest are the graphs $K(1, n)$. We shall for simplicity denote $K(1, n)$ by $S(n)$, and call it a *star of n vertices*. The singleton point will be called the *center* of the star and the remaining n points the *vertices*. To calculate the cubicity of the graph $G = K(n_1, n_2, \ldots, n_p)$, we shall first calculate the cubicity of the star $S(n)$, for each n, and then express cub G in terms of the cubicities of the stars $S(n_i)$.

Theorem 1. *$S(n)$ is embeddable into k-space if and only if $n \leq 2^k$. Thus,* cub $S(n) = [\log_2(2n - 1)]$, *where $[x]$ is the greatest integer in x.*

[4] Subgraph will always mean "generated" subgraph, i.e., all adjacent lines (edges) are included.

PROOF: Suppose $n \leq 2^k$. If $k = 0$, then $n = 1$ and the graph is complete. If $k > 0$, embed $S(n)$ into k-space by sending the center into the origin $\langle 0, 0, \ldots, 0 \rangle$ and the vertices into points of the form $\langle \pm 1, \pm 1, \ldots, \pm 1 \rangle$.

Next, suppose $n > 2^k$. We show by induction on k that $S(n)$ cannot be embedded into k-space. If $k = 0$, then $n > 2^k$ implies $S(n)$ is not complete, and so not embeddable into k-space. If $k = 1$, then $n > 2^k$ implies that $S(n)$ contains $S(3) = K(1, 3)$ as a subgraph. But $S(3)$ is not embeddable into 1-space, as is easily verified directly or from the results of [4]. Thus, $S(n)$ is not embeddable into 1-space. We now assume the result for $k \geq 1$ and prove it for $k + 1$. Suppose $n > 2^{k+1}$ and suppose by way of contradiction that $f_1, f_2, \ldots, f_{k+1}$ is an embedding of $S(n)$ into $(k + 1)$-space. Let b denote the center of $S(n)$ and $a_1, a_2, \ldots, a_n$ its vertices, and let

$$A_1 = \{a_i : f_{k+1}(a_i) \geq f_{k+1}(b)\}$$

and

$$A_2 = \{a_i : f_{k+1}(a_i) \leq f_{k+1}(b)\}.$$

We may suppose that A_1 has $t > 2^k$ elements. But then if g_i is the restriction of f_i to $A_1 \cup \{b\}$, it follows that $g_1, g_2, \ldots, g_k$ is an embedding into k-space of the star $S(t)$ whose center is b and whose vertices are the elements of A_1. This contradicts the inductive assumption.

THEOREM 2. *Suppose* $G = K(n_1, n_2, \ldots, n_p)$. *Then,*

(a) *if* $p > 1$,

$$\text{cub } G = \sum_{j=1}^{p} \text{cub } S(n_j),$$

(b) *if* $p = 1$,

$$\text{cub } G = \begin{cases} 1 & \text{if } n_p > 1 \\ 0 & \text{if } n_p = 1. \end{cases}$$

PROOF: Part (b) is trivial. To prove (a), let $G = (A, I)$ and let $C_1, C_2, \ldots, C_p$ be the classes of $n_1, n_2, \ldots, n_p$ points, respectively. Define H_j to be the graph which is obtained from G by adding all lines (edges) within classes different from C_j. Note that $G = \bigcap_{j=1}^{p} H_j$. Also, cub $H_j =$ cub $S(n_j)$, since all points of $A - C_j$ are equivalent[5] in H_j. This gives one-half of the desired formula, namely an inequality in one direction:

$$\text{cub } G \leq \sum \text{cub } H_j = \sum \text{cub } S(n_j). \tag{4}$$

To get the inequality in the other direction, let $n =$ cub G and note first that G has cubicity 0 (i.e., G is complete) if and only if each n_j is 1, which is the case if and only if $\sum$ cub $S(n_j) = 0$. This proves the second inequality in

[5] Equivalence will always refer to the relation E defined previously.

the case $n = 0$. Next, suppose $n \geq 1$. We may write $G = G_1 \cap G_2 \cap \ldots \cap G_n$, where each $G_i = (A, I_i)$ is an indifference graph. If $n_j > 1$, let U_j be the collection of all I_i so that I_i is missing a line between two points of C_j, and define a graph $K_j = \bigcap \{G_i : I_i \in U_j\}$. Then, K_j is the same as the corresponding graph H_j described above. This follows easily from the observation that each I_i is in one and only one U_j. To verify the observation, note first that for each i, $I_i \supseteq I$. Thus, if I_i is in U_j and U_k, $j \neq k$, then I_i contains a square $K(2, 2)$ generated by points a, b in C_j, and points c, d in C_k. But it is easy to verify that indifference graphs cannot contain squares as subgraphs (cf. [4]). Conversely, if I_i is in none of the U_j, then since $I_i \supseteq I$, it is complete. It follows that $G = \bigcap_{r \neq i} G_r$, and so cub $G < n$.

Since each I_i is in one and only one U_j, it follows that $\sum_{n_j > 1} |U_j| = n$, where $|\cdot|$ denotes cardinality. Since cub $H_j =$ cub $K_j \leq |U_j|$ if $n_j > 1$, and cub $H_j =$ cub $S(1) = 0$ if $n_j = 1$, we have

$$\sum \text{cub } S(n_j) = \sum \text{cub } H_j = \sum_{n_j > 1} \text{cub } H_j \leq \sum_{n_j > 1} |U_j| = n = \text{cub } G.$$

Thus, $\sum$ cub $S(n_j) \leq$ cub G, and the theorem is proved.

It should be remarked that the same argument establishes a more general theorem, of which Theorem 2 is a special case. To state this, we need a notion which will be useful later. A point x in a graph (A, I) is a *focal point* if xIa, all $a \in A$. If G is a graph, we shall denote by G^f, its *focalization*, the graph obtained by adjoining a focal point. Thus, for example, if G consists of n points with no lines, G^f is $S(n)$. We have here the following theorem:

THEOREM 3. *Let $G = (A, I)$ be a graph. Suppose A can be written as the disjoint union of $A_1, A_2, \ldots, A_p$, where $p > 1$, and suppose that in G, all lines between points in different classes A_j occur. Then, if G_j denotes the subgraph generated by A_j, we have* cub $G = \sum$ cub G_j^f.

We are now in a position to calculate $d(n)$, the maximum cubicity of all complete partite graphs with n points, while leaving for the next section the proof that $c(n) = d(n)$. It is not too hard to see, using Theorems 1 and 2, that given n, we maximize the cubicity of $K(n_1, n_2 \ldots, n_p)$, with $n_1 + n_2 + \cdots + n_p = n$, if we take as many of the n_j as possible equal to 3, and the remaining one as 1 or 2 if necessary. Thus, for $n = 3k + i$, $0 \leq i \leq 2$, we have $d(n) =$ cub $K(3, 3, \ldots, 3, i)$, with k 3's. Using Theorems 1 and 2, we can show:

THEOREM 4. *$d(n) = [\frac{2}{3}n]$ if $n \neq 3$ and $d(3) = 1$.*

COROLLARY 4.1. If $n \geq 4$ and $n \neq 6$, $d(n) = d(n - 3) + 2$.

To close this section, we note a simple characterization of the complete partite graphs, which is useful in showing that $c(n) = d(n)$, and whose proof is left to the reader.

LEMMA 1. *A graph is complete partite if and only if it does not contain a subgraph of the form*

$$b \cdot \!\!\!-\!\!\!-\!\!\!-\!\!\!-\!\!\!-\!\!\!-\!\!\!-\!\!\!-\!\!\!-\!\!\!-\!\!\!\cdot\; c, \qquad \dot{a} \tag{5}$$

i.e., if and only if there are 3 distinct points a, b, c so that bIc but neither aIb nor aIc.

4. Calculation of the Maximum Cubicity

In this section, we present the proof of the following theorem.

THEOREM 5. $c(n) = d(n)$ for *all* n.

To begin with, we reduce the problem to the study of graphs which have focal points. If B is a set of points of the graph G, $G - B$ denotes the subgraph generated by points not in B.

LEMMA 2. *If a graph* $G = (A, I)$ *has a subgraph of the form* (5), *then* $\text{cub}\, G \leq \text{cub}(G - \{a, b, c\})^f + 2$.

PROOF: We note first that $G = H \cap K$, where H is the graph obtained from G by adjoining for each $x \in A$ (including a, b, c) the lines between x and a, b, c; and K is the graph obtained from G by adjoining all lines between points in $A - \{a, b, c\}$. It follows that $\text{cub}\, G \leq \text{cub}\, H + \text{cub}\, K$. Now, in H, the points a, b, and c are all equivalent and each is a focal point. Thus, H has the same cubicity as the graph $G - \{a, b, c\}$ with a focal point adjoined, i.e., $\text{cub}\, H = \text{cub}(G - \{a, b, c\})^f$. It is left to show that $\text{cub}\, K \leq 2$. To prove this, we write down an explicit embedding f_1, f_2 of K into 2-space. Let

$$f_1(a) = 0, \qquad f_1(b) = 2, \qquad f_1(c) = 3/2,$$
$$f_2(a) = 0 \qquad f_2(b) = 3/2, \qquad f_2(c) = 2;$$

and if $x \neq a, b, c$, define f_1 and f_2 according to Table 1. We leave it to the reader to check that f_1, f_2 actually do embed K into 2-space.

Now let $e(n)$ denote the maximum cubicity of all graphs with $n + 1$ points, including a focal point, or alternatively the maximum cubicity of the focalization of a graph with n points. Then, we have

LEMMA 3. *If* $n \neq 3$, $e(n) \leq d(n)$.

PROOF: The major step in the proof is to establish for $n \geq 4$ the recursion inequality:

$$e(n) \leq \max\{d(n), \operatorname{cub} S(n), e(n-3)+2\}. \tag{6}$$

To prove (6), suppose G has $n+1$ points, including a focal point x. If $G-x$ is complete partite, then $G-x = K(n_1, n_2, \ldots, n_p)$ and $G = K(1, n_1, n_2, \ldots, n_p)$. By Theorem 2, if $p > 1$, we have cub G = cub $G - x$, which is less than or equal to $d(n)$. If $p = 1$, then $n_p = n$, $G = S(n)$, and cub G = cub $S(n)$. To establish (6), it is now sufficient to assume that $G - x$ is not complete partite and to prove that cub $G \leq e(n-3)+2$. If $G - x$ is not complete partite, then $G - x$ has a subgraph of the form (5). Hence, (5) is a subgraph of G as well, and $x \neq a, b, c$. By Lemma 2, cub $G \leq \operatorname{cub}(G - \{a, b, c\})^f + 2$. But $(G - \{a, b, c\})^f$ consists of a focal point y added to $G - \{a, b, c\}$. Now, $x \in G - \{a, b, c\}$ and thus x and y are two distinct focal points of $(G - \{a, b, c\})^f$. It follows that x and y are equivalent here, and so $\operatorname{cub}(G - \{a, b, c\})^f = \operatorname{cub}(G - \{a, b, c, x\})^f$. But $G - \{a, b, c, x\}$ has $n - 3$ points, and so $\operatorname{cub}(G - \{a, b, c, x\})^f \leq e(n-3)$. Thus cub $G \leq \operatorname{cub}(G - \{a, b, c\})^f + 2 \leq e(n-3) + 2$. This establishes (6).

If $n \geq 4$, then as a final preliminary it is easy to verify that cub $S(n) \leq d(n)$, and so we get from (6) the simpler inequality

$$e(n) \leq \max\{d(n), e(n-3)+2\}. \tag{7}$$

The lemma is easily established for $n = 1, 2$; for $e(1) = d(1) = 0$, $e(2) = d(2) = 1$. Note next that $e(3) = 2$. For, cub $S(3) = 2$ and, as the reader can readily verify for himself, all 4-point graphs with focal points are embeddable into 2-space. The lemma for $n = 4, 5, 6$ now follows by a calculation using (7). Finally, for $n \geq 7$, by way of induction, $e(n-3) + 2 \leq d(n-3) + 2$, which by Corollary 4.1 is $d(n)$. Thus, $e(n) \leq \max\{d(n), e(n-3)+2\} = d(n)$.

TABLE I

If x is adjacent to	$f_1(x)$	$f_2(x)$
none of a, b, c	7/4	1/4
a only (among a, b, c)	1	1/4
b only	3/2	1/2
c only	3/4	5/4
a, b only	1	1/2
a, c only	3/4	1
b, c only	7/4	1
a, b, c	1	1

We are now ready to complete the proof of Theorem 5. Suppose first $n \neq 3, 6$, and suppose the graph G has n points. It is of course sufficient to prove cub $G \leq d(n)$. This is trivial if G is complete partite. Otherwise, G has a subgraph of the form (5) and so by Lemma 2, cub $G \leq \text{cub}(G - \{a, b, c\})^f + 2$. Clearly $n \geq 4$, since $n \neq 3$. Now $G - \{a, b, c\}$ has $n - 3$ points, so $\text{cub}(G - \{a, b, c\})^f \leq e(n-3)$. It follows that cub $G \leq e(n-3) + 2$, which by Lemma 3 and Corollary 4.1 is less than or equal to $d(n-3) + 2 = d(n)$, since $n \geq 4$ and $n \neq 6$. This completes the proof of Theorem 5 in the case $n \neq 3, 6$.

It is simple to verify that $c(3) = d(3) = 1$, thus settling the case $n = 3$. If $n = 6$, let x be an arbitrary point of G and note that $G = H \cap K$, where H is obtained from G by adjoining lines between x and all other points, and K is obtained from G by adjoining all lines between points of $A - \{x\}$. Thus, cub $G \leq$ cub $H +$ cub K. It is easy to show that cub $K \leq 1$, while cub $H \leq e(5) \leq d(5)$. Thus, cub $G \leq d(5) + 1 = d(6)$, and this completes the proof of Theorem 5.

5. The Boxicity

We turn now to the generalization of the notion of interval graph and ask whether a graph (A, I) is representable as the *intersection graph* [in the sense of Eq. (1)] of boxes in E^n. Many of the results of Sections 2–4 go over if we define the *boxicity* of a graph G, box G, as the smallest n such that G is representable as the intersection graph of boxes in E^n. As before, we take box $G = 0$ iff G is complete. We sketch the results here. Note first that by projecting into the coordinate axes, it is simple to prove that a graph G is representable as the intersection graph of boxes in E^n if and only if G is the intersection of n interval graphs. Thus, box $G \leq$ cub G, since each indifference graph is trivially an interval graph. Hence, each graph has finite boxicity. Also note that if $G = H \cap K$, then box $G \leq$ box $H +$ box K. Finally, if aEb, then box $G =$ box $G - a =$ box $G - b$.

Theorem 6.

$$\text{box } S(n) = \begin{cases} 1 & \text{if} \quad n > 1 \\ 0 & \text{if} \quad n = 1. \end{cases}$$

Proof: $S(n)$ is an interval graph.

Theorem 7. *box* $K(n_1, n_2, \ldots, n_p) = \sum \text{box } S(n_j) =$ *the number of* n_j *which are bigger than* 1.

PROOF: If $p > 1$, the proof of Theorem 2 applies almost verbatim. Otherwise, the result is trivial.

Note that if all n_j are greater than 1, the inequality "less than or equal to" in Theorem 7 also can be proved by taking p families of parallel $(p-1)$-dimensional hyperplanes in p-space. This naive representation turns out to be optimal. Suppose now $C(n)$, $D(n)$, and $E(n)$ are defined for boxicity analogously to $c(n)$, $d(n)$, and $e(n)$ for cubicity. Then we have

COROLLARY 7.1. *$D(n) = [n/2]$ for all n.*

COROLLARY 7.2. *$D(n) = D(n-2) + 1$ for all $n \geq 3$.*

LEMMA 4. *Suppose $G = (A, I)$ is a graph, a, $b \in A$, and $\sim aIb$. Then,* $\text{box } G \leq \text{box}(G - \{a, b\})^f + 1$.

PROOF: We write $G = H \cap K$, where H is obtained from G by adjoining for each $x \in A$ (including a, b) the lines between x and the points a and b; and K is obtained from G by adjoining all lines between points of $A - \{a, b\}$. We note $\text{box } G \leq \text{box } H + \text{box } K$. Moreover, in H, a and b are equivalent and each is a focal point, so $\text{box } H = \text{box}(G - \{a, b\})^f$. Finally, $\text{box } K \leq 1$, i.e., K is an interval graph. This is easy to see by the characterization of Lekkerkerker–Boland [3].

LEMMA 5. *$E(n) \leq D(n)$ for all n.*

PROOF: We first show that if $n \geq 3$, then $E(n) \leq E(n-2) + 1$. For, let G have $n + 1$ points and a focal point, x. If G is complete, then $\text{box } G = 0 \leq E(n-2) + 1$. Otherwise, there are a, $b \in G$ so that $\sim aIb$. Note that $a, b \neq x$ so $x \in G - \{a, b\}$. By the previous lemma, $\text{box } G \leq \text{box } (G - \{a, b\})^f + 1$. But the focal point added to $G - \{a, b\}$ in $(G - \{a, b\})^f$ is equivalent to x, so $\text{box}(G - \{a, b\})^f = \text{box}(G - \{a, b, x\})^f$. Now $G - \{a, b, x\}$ has $n - 2$ points, and therefore $\text{box}(G - \{a, b, x\})^f \leq E(n-2)$. Thus, $\text{box } G \leq E(n-2) + 1$, establishing the desired inequality.

The lemma follows by induction and Corollary 7.2. For, $E(1) = D(1) = 0$, $E(2) = D(2) = 1$, and $E(n) \leq E(n-2) + 1 \leq D(n-2) + 1 = D(n)$.

THEOREM 8. *$C(n) = D(n)$ for all n.*

PROOF: $C(1) = D(1) = 0$, $C(2) = D(2) = 1$. Suppose G has $n \geq 3$ points. If G is complete, then $\text{box } G = 0 \leq D(n)$. Otherwise, by Lemma 4, $\text{box } G \leq$

$E(n-2)+1$, which is less than or equal to $D(n-2)+1 = D(n)$ by Lemma 5 and Corollary 7.2. Thus, $C(n) \leq D(n)$.

Acknowledgments

The results appearing in this paper are contained in the author's doctoral dissertation in mathematics at Stanford University, written under the direction of Professor Dana Scott. The author would like to thank Professor Scott and the other members of his dissertation committee, Professors Halsey Royden and Patrick Suppes, for their helpful comments.

The research was supported in part by grants from the National Science Foundation and the National Institute of Health. This paper was written while the author was an N.I.H. "Postdoctoral Fellow" in the Department of Psychology at the University of Pennsylvania.

The author is indebted to Professor R. K. Guy for suggesting the terms boxicity and cubicity.

References

1. D. R. Fulkerson and O. A. Gross, Incidence Matrices and Interval Graphs, *Pacific J. Math.* **15** (1965), 835–855.
2. P. C. Gilmore and A. J. Hoffman, A Characterization of Comparability Graphs and of Interval Graphs, *Canad. J. Math.* **16** (1964), 539–548.
3. C. G. Lekkerkerker and J. C. Boland, Representation of a Finite Graph by a Set of Intervals on the Real Line, *Fund. Math.* **51** (1962), 45–64.
4. F. S. Roberts, Indifference Graphs, in *Proof Techniques in Graph Theory* (*Proc.* 2nd *Ann Arbor Graph Theory Conf.*, 1968) (F. Harary, ed.), Academic Press, New York, 1969.

ON THE NUMBER OF MUTUALLY DISJOINT TRIPLES IN STEINER SYSTEMS AND RELATED MAXIMAL PACKING AND MINIMAL COVERING SYSTEMS

*J. Schönheim**

DEPARTMENT OF MATHEMATICS
UNIVERSITY OF CALGARY
CALGARY, ALBERTA
AND
TEL AVIV UNIVERSITY
TEL AVIV, ISRAEL

1. INTRODUCTION

Let E be a set of n elements. A set of triples of E is called a *Steiner triple system*, a *maximal packing system*, or a *minimal covering system* (of E) if every pair of E is contained in *exactly* one, in *at most* one, or in *at least* one triple of the system, respectively.

It is very well known [1] that Steiner triple systems exist if and only if $n \equiv 1$ or 3 (mod 6) and the number of triples is $n(n-1)/6$.

It is also known [2, 3] that the number of triples in a maximal packing system is

$$\psi(n) = \begin{cases} [n/3[(n-1)/2]] & \text{if} \quad n \not\equiv 5 \pmod 6 \\ [n/3[(n-1)/2]] - 1 & \text{if} \quad n \equiv 5 \pmod 6 \end{cases} \tag{1}$$

and the number of triples in a minimal covering system is

$$\varphi(n) = \{n/3\{(n-1)/2\}\}, \tag{2}$$

where $[x]$ denotes the integer part of x and $\{x\}$ denotes the smallest integer y, $y \geq x$.

* This research was sponsored by the Canadian Mathematical Congress at the Summer Research Institute in Montreal.

By (1) and (2) a Steiner triple system may be considered as a maximal packing system and a minimal covering system as well.

P. Erdös [4] asked: what is the maximum number and what is the minimum number of mutually disjoint triples in a Steiner system?

We will give here an answer to the first part of this question, and more generally will show that, except for $n = 6$ and 7, there exist maximal packing systems and minimal covering systems containing the *a priori* maximum possible number, i.e., $[n/3]$ mutually disjoint triples (see the theorem of Section 6). This result is not trivial since not every extremal system contains $[n/3]$ mutually disjoint triples. An example for $n = 18$ will be given in Section 2. In spite of the difficulty of finding examples of extremal systems containing a smaller number of mutually disjoint triples than $[n/3]$, the proof of the announced result needs a surprisingly considerable effort.

2. Example of a Maximal Packing System of Eighteen Elements Which Does Not Contain Six Mutually Disjoint Triples

Let A, and A', be the sets $A = \{1, 2, \ldots, 9\}$ and $A' = \{1', 2', \ldots, 9'\}$, respectively. Let xyz be a member of a Steiner system of A. The triples xyz and $x'y'z$ contain every pair of $A \cup A'$ at most once. Their number is $9.8/6 + 9.8/2 = \psi(18)$ and therefore they form a maximal packing system. It does not contain six mutually disjoint triples. For, it contains at most four mutually disjoint triples of the second kind and then at most one more of the first kind; or, if it contains three mutually disjoint triples of the second kind, then it has at most two of the first kind; the other possibilities give less than five mutually disjoint triples.

3. The Minimal Covering Systems of Fort and Hedlund

In order to prove result (2) Fort and Hedlund [2] considered some supplementary properties of minimal covering systems and in particular proved the following two lemmas.

Lemma 1. *If a set E contains $n \equiv 3 \pmod 6$ elements then there exist Steiner triple systems of E containing $n/3$ mutually disjoint triples.*

Lemma 2. *If a set E contains $n \equiv 5 \pmod 6$ elements, then there exists a minimal covering system of E containing a minimal covering system R of a subset of E having five elements. By convenient notation of the elements of E,* $R = \{123, 124, 125, 345\}$.

Their method is recursive. We give some more details on it in a formulation which will be useful in the proof of the theorem of Section 6. For complete proofs see [2].

RECURSION 1 (from $6k + 2$ to $6k + 1$ and from $6k + 4$ to $6k + 3$). If S is a Steiner system of a set E of $6k + 1$ or $6k + 3$ elements and if S' is a minimal set of pairs of E containing every element of E at least once, then, x being a supplementary element, the system of triples $S^* = S \cup \{xyz\}_{yz \in S'}$ is a minimal covering system of $E \cup x$.

RECURSION 2 (from $6k + 6$ to $6k + 5$). If S is a minimal covering system of a set E having $n = 6k + 5$ elements, $S \supset R$ (R defined as in Lemma 2) and if S' is the set of pairs $S' = \{13, 23, 45, 67, \ldots, \overline{n-2}\,\overline{n-1}\}$, then, x being a supplementary element, the system of triples $S^* = (S - 123) \cup \{xyz\}_{yz \in S'}$ is a minimal covering system of $E \cup x$.

Further recursive theorems make use of combinatorial designs called *tricovers* which have been generalized more recently as *transversal systems* and *H-designs* [5, 6].

DEFINITION. Consider a partition of a set E having $3m$ elements, in three sets C_i $(i = 1, 2, 3)$ of m elements each, called columns. A *tricover* of E is a system of triples, containing a single element of each column, such that every pair of elements from different columns is contained in exactly one triple of the system.

LEMMA 3. *Tricovers exist for every m and the number of triples in a tricover is m^2.*

PROOF: Denote the elements of column C_i by $\{1_i, 2_i, \ldots, m_i\}$. The triples $x_1 y_2 z_3$

$$x + y + z \equiv 0 \pmod{m} \tag{3}$$

define a tricover of $C_1 \cup C_2 \cup C_3$ and their number is m^2.

LEMMA 4. *If $B_i \subset C_i$, $|C_i| = m$ $(i = 1, 2, 3)$, $|B_1| = |B_2| = |B_3| \leq m/2$, and T^* is a tricover of $B_1 \cup B_2 \cup B_3$, then there exists a tricover T of $C_1 \cup C_2 \cup C_3$ such that $T \supset T^*$.*

RECURSION 3 (from $18h + 1$ and $18h + 7$ to $6h + 1$ and $6h + 3$). If T is a tricover of a set E having $18h$ or $18h + 6$ elements, and partitioned as in the foregoing definition, and if x is a supplementary element and we denote by S_i

a Steiner system of $x \cup C_i$, then the system of triples $S = S_1 \cup S_2 \cup S_3 \cup T$ is a minimal covering system of $x \cup E$.

RECURSION 4 (from $18h + 5$ and $18h + 17$ to $6h + 3$ and $6h + 7$). If T is a tricover of a set E having $18h + 3$ or $18h + 15$ elements and partitioned as in the foregoing definition, and if we denote by S_i a Steiner system of $x \cup y \cup C_i$, x add y being two supplementary elements, then the system of triples $S = S_1 \cup S_2 \cup S_3 \cup T$ is a minimal covering system of $x \cup y \cup E$. Moreover $S \subset R$.

RECURSION 5 (from $18h + 13$ to $6h + 5$). Let $C_1 \cup C_2 \cup C_3$ be a partition of a set E having $18h + 12$ elements in three columns of $6h + 4$ elements each. Let S_i be a minimal covering system of $x \cup C_i$, x being a supplementary element. Moreover we suppose $S_i \supset R_i$. Denote the set of elements contained in the triples R_i and not equal to x by B_i. Let T^* be a tricover of $B_1 \cup B_2 \cup B_3$ and T a tricover of $C_1 \cup C_2 \cup C_3$ such that $T \supset T^*$ (T exists by Lemma 4). Let V be a minimal covering system of the 13 elements of $x \cup B_1 \cup B_2 \cup B_3$. The system of triples

$$S = (S_1 - R_1) \cup (S_2 - R_2) \cup (S_3 - R_3) \cup (T - T^*) \cup V \tag{4}$$

is a minimal covering system of $x \cup E$.

RECURSION 6 (from $18h + 11$ to $6h + 5$). Starting with a set E of $18h + 9$ elements and two supplementary elements x and y, a similar construction as in Recursion 5 shows that the system of triples as defined in (4) (but S_i being a minimal covering system of $x \cup y \cup C_i$, B_i being the set of elements contained in the triples R_i and different from x and y, and V being a minimal covering system of the 11 elements $x \cup y \cup B_1 \cup B_2 \cup B_3$) is a minimal covering system of $x \cup y \cup E$. Moreover $S \supset R$, two of the five elements of R being x and y.

4. THE MAXIMAL PACKING SYSTEMS

As mentioned in the introduction, the maximal packing systems for $n \equiv 1$ or 3 (mod 6) are the Steiner triple systems. Their recursive construction has been included in Section 3. The construction of maximal packing systems for the other values of n are given in [3]. The following three lemmas summarize these constructions and they will be used in the proof of the theorem of Section 6.

LEMMA 5. *If* $n \equiv 0$ *or* 2 (mod 6) *and if* S *is a maximal packing system for* $n+1$, *and if moreover we denote by* $S(x)$ *the set of triples in* S *containing a fixed element* x, *then* $S - S(x)$ *is a maximal packing system for* n.

LEMMA 6. *If* $n \equiv 5$ (mod 6) *and if* S *is a minimal covering system for* n *and* $S \supset R = \{123, 124, 125, 345\}$, *then* $S^* = S - \{124, 125\}$ *is a maximal packing system for* n.

LEMMA 7. *If* $n \equiv 4$ (mod 6) *and if* S^* *is a maximal packing system for* $n+1$ *as constructed in Lemma* 6, *and if* $S(1)$ *denotes the set of triples in* S^* *containing the element* 1 *then* $S^* - S(1)$ *is a maximal packing system for* n.

5. TRICOVERS WITH MAXIMUM POSSIBLE NUMBER OF MUTUALLY DISJOINT TRIPLES

LEMMA 8. *If* E *is a set of* $3m$ *elements, then there exist tricovers of* E *containing* m *mutually disjoint triples.*

PROOF: If m is odd, consider the tricover T of $3m$ elements defined in the proof of Lemma 3. The addition table (mod m) contains m different elements in its diagonal, and therefore the triples $\{x_1 x_2 (-2x)_3\}_{x=1}^{m}$ are members of T and are mutually disjoint. By a theorem of Hall [7] the foregoing construction will fail for groups of even order. If m is even, define a tricover of E using instead of condition (3) the condition $x \cdot y = z$, where $\cdot$ denotes an operation in a quasigroup. As proved in [8],[1] it is possible to construct quasigroups of every order m having m different elements in the diagonal of their operation table. Then the triples $\{x_1 x_2 (x_1 \cdot x_2)\}_{x=1}^{m}$ are members of T and are mutually disjoint.

6. THEOREM

If $n \neq 6, 7$, *then there exist maximal packing systems and minimal covering systems containing a set* D *of* $[n/3]$ *mutually disjoint triples.*

OUTLINE OF PROOF: The proof will be given for minimal covering systems first and then for maximal packing systems. We will begin constructing minimal covering systems containing the set D for small values of n, then start induction by recursion rules 1–6, using also Lemmas 1–4 and 8, and the

[1] Thanks are due to John Marica for calling my attention to this paper.

induction hypothesis on the existence of the set D. Moreover, if $n \equiv 5 \pmod 6$ it is useful to strengthen the assertion on D: namely, if the system contains the set of triples $R = \{123, 124, 125, 345\}$, then there are two ways of choosing the triples of D, one omitting the elements 1 and 2 ond the other having the pair 12 in one of the triples of D. The proof for maximal packing system is obtained using Lemmas 5–7.

7. Proof of the Theorem in Detail

Initial values for minimal covering systems and their set, D, are:

$n = 5$ {123, 124, 125, 345}; $D = 345$; $D' = 123$.
$n = 8$ {124, 235, 346, 457, 561, 672, 713, 812, 834, 856, 871};
$D =$ {124, 856}.
$n = 11$ {123, 124, 125, 345, 167, 189, 1 10 11, 268, 27 10, 29 11, 36 10, 379, 38 11, 46 11, 478, 49 10, 569, 57 11, 58 10}; $D =$ {379, 46 11, 58 10}.
$n = 13$ {123, 145, 16 13, 178, 19 12, 1 10 11, 24 10, 256, 279, 28 12, 2 11 13, 34 11, 357, 36 12, 38 13, 39 10, 467, 489, 4 12 13, 58 11, 59 13, 5 10 12, 68 10, 69 11, 7 10 13, 7 11 12}; $D =$ {19 12, 256, 34 11, 7 10 13}.

Induction procedure for minimal covering systems. In order to proceed by induction it is convenient to consider the remainder of $n \pmod{18}$.

Case 1 [$n \equiv 3, 9,$ or $15 \pmod{18}$]. The theorem is proved by Lemma 1.

Case 2 [$n \equiv 2, 4, 8, 10, 14,$ or $16 \pmod{18}$]. The theorem is a consequence of Recursion 1, the induction hypothesis, the fact that $D \subset S^*$, and the equality $[n/3] = [n - 1/3]$.

Case 3 [$n \equiv 1$ or $7 \pmod{18}$]. The theorem is a consequence of Recursion 3 and Lemma 8.

Case 4 [$n \equiv 5$ or $17 \pmod{18}$]. A minimal covering system containing the set D with omission of the elements x and y is obtained by Recursion 4 and Lemma 8. A set D' of $[n/3]$ mutually disjoint triples of the system with the pair xy occurring in one triple of D' is obtained by taking $D' = (D \cup xyz_1) - z_1z_2z_3$, where xyz_1 is the triple of S_1 which contains the pair xy, and $z_1z_2z_3$ is the triple of D containing z_1.

Case 5 [$n \equiv 13 \pmod{18}$]. Apply Recursion 5. By the induction hypothesis and Lemma 2 $S_i \supset D_i$, $|D_i| = 2h + 1$, $S_i \supset R_i = \{x2_i3_i; x2_i4_i, x2_i5_i;$

$3_i 4_i 5_i\}$, and the triples of D_i do not contain the elements x and 2_i. If $D_i \supset 3_i 4_i 5_i$ then $\bigcup_{i=1}^{3} (S_i - R_i)$ contains $6h$ mutually disjoint triples and together with four more of V, D contains $6h + 4$. If $D_i \not\supset 3_i 4_i 5_i$ then $\bigcup_{i=1}^{3} (S_i - R_i)$ contains $6h + 3$ mutually disjoint triples and, with adequate notation, V contains the triple $2_1 2_2 2_3$ and again D contains $6h + 4$ mutually disjoint triples.

CASE 6 [$n \equiv 11 \pmod{18}$]. Apply Recursion 6, denoting the supplementary elements by 1 and 2. If $h \neq 1 \pmod 3$ then by Case 4 [and if $h = 1 \pmod 3$ then by an argument using induction on $n = 11 \pmod{18}$], the covering S_i contains a set D_i of $2h + 1$ mutually disjoint triples and the pair 12 is contained in one of the triples of D_i. Therefore $\bigcup_{i=1}^{3} (S_i - R_i)$ contains $6h$ mutually disjoint triples; D is obtained by adding to them three more triples of V. Those triples may be chosen either not to contain the element 1 and 2 or to contain the pair 12.

CASE 7 [$n \equiv 0, 6$, or $12 \pmod{18}$]. $R = \{123, 124, 125, 345\}$, $D \supset 124$, and S' contain the pair whose elements are omitted in D.

Cases 1–7 exhaust all possibilities and the proof of minimal covering systems ends here.

Proof of the part of the theorem which concerns maximal packing systems. It is convenient to consider the remainder of $n \pmod 6$.

CASE 8 [$n \equiv 1$ or $3 \pmod 6$]. The maximal systems coincide with the minimal systems.

CASE 9 [$n \equiv 0 \pmod 6$]. The maximal packing system of Lemma 5 is also a minimal covering system and hence it contains a set D of $[n + 1/3]$ mutually disjoint triples. One of the $n + 1$ elements does not occur in those triples; if it is chosen for x then $D \subset S - S(x)$ and the theorem follows since $[n/3] = [n + 1/3]$.

CASE 10 [$n \equiv 2 \pmod 6$]. Consider the same construction as in Case 9. The theorem follows since the set $D' = (S - S(x)) \cap D$ contains $n + 1/3 - 1 = [n/3]$ mutually disjoint triples.

CASE 11 [$n \equiv 5 \pmod 6$]. Consider the construction in Lemma 6. The minimal covering system S contains a set D of $[n/3]$ mutually disjoint triples which do not contain the elements 1 and 2. Therefore $S^* \supset D$.

CASE 12 $[n \equiv 4 \pmod 6]$. If S^* is the system constructed in Case 11 and $S^* \supset D$ as before, then the system $S^* - S(x)$ of Lemma 7 with $x = 1$ contains the set D. The theorem follows since $[n + 1/3] = [n/3]$.

Minimal covering systems and maximal packing systems of six and seven elements do not contain two mutually disjoint triples.

The proof is herewith complete.

REFERENCES

1. M. REISS, Ueber eine Steinerische Combinatorische Aufgabe, *J. Reine Angew. Math.* **45** (1859), 326–344.
2. M. K. FORT, Jr. and G. A. HEDLUND, Minimal Coverings of Pairs by Triples, *Pacific J. Math.* **8** (1958), 709–719.
3. J. SCHÖNHEIM, On Maximal Systems of k-Tuples, *Studia Sci. Math. Hungar.* **1** (1966), 363–368.
4. P. ERDŐS, Oral communication, 1967.
5. H. HANANI, The Existence and Construction of Balanced Incomplete Block Designs, *Ann. Math. Statist.* **32** (1961), 361–386.
6. H. HANANI, Combinatorial Designs (to appear).
7. M. HALL, Jr., A Combinatorial Problem on Abelian Groups, *Proc. Amer. Math. Soc.* **3** (1952), 584–587.
8. R. H. BRUCK, Some Results in the Theory of Quasigroups, *Trans. Amer. Math. Soc.* **55** (1949), 19–52.

ON A PROBLEM OF ERDŐS

David P. Sumner

UNIVERSITY OF MASSACHUSETTS
AMHERST, MASSACHUSETTS

At the Symposium on the Theory of Graphs held in Smolenice in June of 1963 (see [5]), P. Erdős asked the following question: Let $l > 1$, and assume that G has ln vertices and every vertex has valency greater than or equal to $(l-1)n$. Is it then true that G contains n independent complete l-gons (i.e., no two of which have a common vertex)? By a complete *l-gon* is meant a set of l vertices each pair of which is joined by an edge. This problem does not seem to have been solved except in special cases.

The case $l = 2$ (and arbitrary n) follows from a theorem of Dirac [2] which states that any graph with n vertices in which each vertex is of valency greater than or equal to $n/2$ is Hamiltonian. The case $l = 3$ follows from a more general theorem due to Corrádi and Hajnal [1].

The problem was approached from the opposite viewpoint by Zelinka [4] who proved the conjecture was valid for $n = 1, 2, 3$, and arbitrary l.

More recently Grünbaum [3] proved the conjecture for $n = 4$. In his paper Grünbaum makes the following conjecture, which is slightly stronger than that of Erdős.

CONJECTURE. If a graph G has n vertices each of valence less than k, then it is possible to color the vertices of G with k colors in such a way as to assign each color to at least $[n/k]$ vertices and to at most $[n/k] + 1$ vertices.

Grünbaum proved this conjecture for $k = 4$.

In this paper we will give a new proof for the case $n = 3$ and some partial results for $n = 5$ and 6.

If a subset of a graph contains l vertices no two of which are joined by an edge it will be referred to as a TD l-set. We will denote the valence of a vertex x by $v(x)$. We will say a vertex x is separated from the set A in case for every

$a \in A$, x is not joined by an edge to a. If every point of the set A is joined by an edge to some vertex of the set B then we will say A is onto B.

THEOREM 1. *If G is a graph with $3l$ vertices ($l > 1$) and if for each $x \in G$, $v(x) \leq 2$, then G contains three independent* TD *l-sets.*

PROOF: The proof is by contradiction. Suppose that the theorem is false and let n be the smallest nonnegative integer such that for some $l > 1$ there exists a graph G with $3l$ vertices, n edges, and with $v(x) \leq 2$ for all $x \in G$; but G does not contain three independent TD l-sets. Clearly $n > 1$. Remove an edge of G thereby forming a new graph G^*. By our assumption, G^* contains three independent TD l-sets L_1, L_2, and L_3. As these sets cannot all be TD l-sets in G, the deleted edge must join two vertices a and b in the same L_i, say $a, b \in L_1$. But $v(b) \leq 2$, so b is separated from one of L_2 and L_3, say L_2. Thus L_2 is onto L_1, for if there were an $x \in L_2$ separated from L_1, then $(L_1 - \{b\}) \cup \{x\}$, $(L_2 - \{x\}) \cup \{b\}$, and L_3 would be three independent TD l-sets in G. Since there are at most $2l$ edges of G with an end point in L_1, there exists $c \in L_3$ separated from L_1. Thus it is clear that L_2 is onto L_3 [for if $x \in L_2$ were separated from L_3 then $(L_1 - \{b\}) \cup \{c\}$, $(L_2 - \{x\}) \cup \{b\}$, and $(L_3 - \{c\}) \cup \{x\}$ would be three independent TD l-sets in G]. But since b is separated from L_2 there must exist $d \in L_1$ such that d is joined by an edge to two vertices e, $f \in L_2$. But then, $(L_1 - \{b, d\}) \cup \{e, f\}$, $(L_2 - \{e, f\}) \cup \{b, d\}$, and L_3 are three independent TD l-sets in G. Thus we have a contradiction and our theorem is proved.

THEOREM 2. *Let G be a graph of $5l$ vertices ($l > 1$) such that for each $x \in G$, $v(x) \leq 4$. Then G contains four independent* TD *l-sets.*

PROOF: As before, the proof is by contradiction. We assume the theorem is false and let n be the smallest nonnegative integer such that there exists a graph G with n edges and $5l$ vertices for which the theorem fails. We remove an edge of G thus forming the graph G^*. Let L_1, L_2, L_3, and L_4 be four independent TD l-sets in G and let $K = G - \bigcup_{i=1}^{4} L_i$. As in the proof of Theorem 1, we may assume the deleted edge joins two vertices $a, b \in L_1$. It is clear that K is onto L_1. But K cannot be onto all of the L_i, for then K would be a TD l-set itself and G would contain K, L_2, L_3, and L_4 as four independent TD l-sets. Thus there exists $c \in K$ such that c is separated from some L_i, say L_2. From this it is easily seen that L_2 must be onto L_1. Since $a, b \in L_1$ are joined by an edge in G and $v(x) \leq 4$ for all $x \in G$, there must exist a vertex $d \in L_3 \cup L_4$ such that d is separated from L_1. We will suppose that $d \in L_3$. If there were a vertex $x \in K$ separated from L_3, then $(L_1 - \{b\}) \cup \{d\}$, $(L_3 - \{d\}) \cup \{x\}$, L_2, and L_4 would be four independent TD l-sets in G. Thus we have that K is onto L_3. If there were a vertex $y \in L_2$ separated from L_3, then $(L_1 - \{b\}) \cup \{d\}$, $(L_2 - \{y\}) \cup \{c\}$, $(L_3 - \{d\}) \cup \{y\}$, and L_4 would be four independent TD l-sets in G. Therefore L_2 is onto L_3. So if L_4 were onto L_1 and onto L_3 then the graph spanned by $L_2 \cup L_4 \cup K$ would contain $3l$ vertices each of valence less than or equal to 2 (in the generated graph). So $L_2 \cup L_4 \cup K$ would contain three independent TD l-sets which together with L_3 yield four independent TD l-sets in G. Thus there exists $e \in L_4$ such that e is separated from one of L_1 or L_3. In either event if there exists an $x \in K \cup L_2$ separated from L_4 we may easily obtain four independent TD l-sets in G. So we have that both of K and L_2 are onto L_4. But then the graph spanned by $K \cup L_2$ contains $2l$ vertices each of valence less than or equal to 1 (in $K \cup L_2$) so that $K \cup L_2$ contains two independent TD l-sets which together with L_3 and L_4 give four independent TD l-sets in G. This contradicts our original assumption, so that the theorem is proved.

THEOREM 3. *Let G be a graph of $5l$ vertices ($l > 1$) such that for each $x \in G$, $v(x) \leq 4$. Then G contains four independent* TD *l-sets and a fifth* TD *r-set independent of these, where $r \geq l/2$.*

PROOF: By Theorem 2 we know that G contains four independent TD l-sets. Let four such sets L_1, L_2, L_3, and L_4 be chosen in such a way that $K = G - \bigcup_{i=1}^{4} L_i$ contains a TD r-set with r as large as possible. Let A denote this TD r-set and let $B = K - A$. Since r is a maximum, we have that B is onto A. If $B = \varnothing$, then $r = l \geq l/2$. So we assume $B \neq \varnothing$. If $b \in B$ then since $v(b) \leq 4$, b is separated from some L_i, say L_1. Thus L_1 is onto A (if not, r would not be maximal). If there exists $c \in B$, c separated from some L_i, $i \neq 1$, say L_2, then L_2 must be onto A. In this case we have at least $3l - r$ edges with one end point in A and the other in $B \cup L_1 \cup L_2$. But there can be at most $4r$ edges with an end point in A, so $4r \geq 3l - r$ or $r \geq 3l/5 \geq l/2$. If no such c exists then B must be onto each of L_2, L_3, and L_4. But then B is a TD $(l - r)$-set. Thus by the maximality of r, $r \geq l - r$ or $r \geq l/2$. Thus in any event, $r \geq l/2$.

By considering the complementary graph we may restate Theorem 3 as:

THEOREM 4. *Let G be a graph of $5l$ vertices ($l > 1$) such that for $x \in G$, $v(x) \geq 5l - 5$. Then G contains four independent complete l-gons and a fifth complete r-gon independent of these where $r \geq l/2$.*

We can also prove in an exactly similar manner a similar result for $n = 6$. This can be stated more generally, however, as:

THEOREM 5. *If G is a graph with nl vertices ($n \geq 6, l > 1$) and if for each $x \in G$, $v(x) \leq 5$, then G contains $n - 1$ independent* TD *l-sets.*

A different kind of result is

THEOREM 6. *If G is a graph of nl vertices ($n \geq 5, l > 1$) each of valence less than or equal to $n - 1$ and if every* TD *$(l + 1)$-set in G contains a vertex of valence less than or equal to* 3, *then G contains n independent* TD *l-sets.*

We will prove Theorem 6 in the case $n = 5$. The general result then follows via a simple induction argument.

PROOF ($n = 5$): Clearly G must contain at least one TD l-set. Pick L to be a TD l-set with a maximum number of vertices of valence 4. Let $K = G - L$. Then K contains $4l$ vertices. If there exists $x \in K$ such that $v(x) = 4$ and x is separated from L, then $L \cup \{x\}$ is a TD $(l + 1)$-set so that there exists $a \in L \cup \{x\}$, $v(a) \leq 3$. But then $(L - \{a\}) \cup \{x\}$ is a TD l-set with more vertices of valence 4 than L. But this is a contradiction. Thus for each $x \in K$ either $v(x) \leq 3$ or $v(x) = 4$ and x is joined by an edge to some point of L. Therefore for each $x \in K$ the valence of x in the graph spanned by K is at most 3. Thus K contains four independent TD l-sets which together with L give five independent TD l-sets in G.

As an immediate consequence of Theorem 6 we have the following:

THEOREM 7. *If G is a graph of nl vertices ($n \geq 5, l > 1$) each of valence less than or equal to $n - 1$ and if there are at most l points of valence greater than or equal to* 4, *then G contains n independent* TD *l-sets.*

Of course we also have the theorems complementary to Theorems 5–7.

REFERENCES

1. K. CORRÁDI and A. HAJNAL, On the Maximal Number of Independent Circuits in a Graph, *Acta Math. Acad. Sci. Hungar.* **14** (1963), 423–439.
2. G. A. DIRAC, Some Theorems on Abstract Graphs, *Proc. London Math. Soc.* (3) **2** (1952), 69–81.
3. B. GRÜNBAUM, A Result on Graph-Coloring, *Michigan Math. J.* **15** (1968), 381–383.
4. B. ZELINKA, On the Number of Independent Complete Subgraphs, *Publ. Math. Debrecen* **13** (1966), 95–97.
5. M. FIELDER, ed., *Theory of Graphs and Its Applications* (*Proc. Symp. Smolenice*, 1963), Academic Press, New York, 1964 and Publ. House Czech. Acad. Sci., Prague, 1964.

SOME CLASSES OF HYPOCONNECTED VERTEX-TRANSITIVE GRAPHS

Mark E. Watkins

UNIVERSITY OF WATERLOO
WATERLOO, ONTARIO, CANADA

1. Introduction

Only finite simple graphs are considered in this paper, and the symbol G will always denote such a graph. If G is vertex-transitive, it is necessarily regular of valence ρ (G). Its vertex-connectivity is denoted by $\kappa(G)$.

It is well known that

$$\kappa(G) \leq \rho(G) \tag{1}$$

for any graph G. Examples abound of vertex-transitive graphs for which equality holds in (1); there are complete regular n-partite graphs, prisms, polygons, Moore graphs of diameter 2, n-cubes, and starred polygons (see [2]), to name just a few classes. Those vertex-transitive graphs for which inequality holds in (1), however, shall be termed *hypoconnected.*

In [4, Theorem 2] the author shows that if G is hypoconnected, then G is a spanning subgraph of a lexicographic product, although the converse is certainly false. It is the purpose of this paper to consider several classes of hypoconnected graphs. This paper thus complements the above-noted paper [4], which is of a more general nature.

It is safe to assume that $\kappa(G) \geq 1$, for otherwise G is the disjoint union of two or more copies of some vertex-transitive graph H with $\rho(H) = \rho(G)$; $\Gamma(G)$ will denote the automorphism group of G.

As a suitable reference for most of the terminology in this paper, W. T. Tutte [3] is recommended.

2. The Class H

If H_1 and H_2 are graphs, then following Sabidussi [1], we define their *lexicographic product*

$$G = H_1 \circ H_2 \tag{2}$$

to have vertex set

$$V(G) = V(H_1) \times V(H_2), \tag{3}$$

and

$$[(x_1, x_2), (y_1, y_2)] \in E(G) \tag{4}$$

if and only if either

$$[x_1, y_1] \in E(H_1), \qquad \text{or} \qquad x_1 = y_1 \quad \text{and} \quad [x_2, y_2] \in E(H_2).$$

The symbol **H** will denote the class of graphs of the form (2) where H_1 and H_2 are connected and vertex-transitive, H_1 is not complete, and $|V(H_2)| \geq 2$.

For such graphs G, clearly

$$\rho(G) = \rho(H_2) + \rho(H_1)\,|V(H_2)|\,. \tag{5}$$

THEOREM 1. *Let $G = H_1 \circ H_2$ where H_1 is not complete. Then*

$$\kappa(G) = \kappa(H_1)\,|V(H_2)|\,.$$

PROOF: Clearly $\kappa(G) = 0$ if and only if $\kappa(H_1) = 0$. Hence let C_1 be a minimum cut set of H_1. There are precisely $\kappa(H_1)|V(H_2)|$ vertices of G of the form

$$(x_1, x_2), \qquad x_1 \in C_1, \quad x_2 \in V(H_2),$$

and they form a cut set of G. Hence $\kappa(G) \leq \kappa(H_1)\,|V(H_2)|$.

Now let C be a cut set of G and pick vertices $\bar{x} = (x_1, x_2)$ and $\bar{y} = (y_1, y_2)$ lying in distinct components of $G[V(G) - C]$. It will suffice to show that G contains at least $\kappa(H_1)\,|V(H_2)|$ internally disjoint paths joining $\bar{x}$ and $\bar{y}$.

CASE 1 ($x_1 = y_1$). Let N be the set of vertices of H_1 adjacent to x_1. If $\rho_1(x_1)$ is the valence of x_1 in H_1, then there are $\rho_1(x_1)\,|V(H_2)| \geq \kappa(H_1)\,|V(H_2)|$ vertices of the form (u_1, u_2), $u_1 \in N$, $u_2 \in V(H_2)$, and all of these are adjacent to both $\bar{x}$ and $\bar{y}$.

CASE 2 ($x_1 \neq y_1$). By Whitney's theorem [5, Theorem 7, p. 160], there are $\kappa(H_1)$ internally disjoint paths A_i in H_1 joining x_1 and y_1. Proceeding along

A_i from x_1 we list the vertices in order $x_1, w_{i,1}, w_{i,2}, \ldots, w_{i,m_i}, y_1$. For each $i = 1, \ldots, \kappa(H_1)$ and for each $z \in V(H_2)$, we determine the path $A_{i,z}$ in G by naming its vertices in order

$$(x_1, x_2), (w_{i,1}, z), (w_{i,2}, z), \ldots, (w_{i,m_i}, z), (y_1, y_2). \tag{6}$$

Since $\bar{x}$ and $\bar{y}$ are not adjacent, there are $\kappa(H_1)\,|V(H_2)| \le |C|$ distinct arcs determined by (6).

Hence $\kappa(G) \ge \kappa(H_1)\,|V(H_2)|$, which proves the theorem.

It follows from (5) and Theorem 1 that if $G \in \mathbf{H}$ then G is hypoconnected.

It is not hard to show that every $G \in \mathbf{H}$ is vertex-transitive. For let (x_1, x_2), $(y_1, y_2) \in V(G)$. There exist $\phi_i \in \Gamma(H_i)$ such that

$$\phi_i(x_i) = y_i \qquad (i = 1, 2).$$

Define automorphisms $\Phi_1, \Phi_2 \in \Gamma(G)$ by

$$\Phi_1(u_1, u_2) = (\phi_1(u_1), u_2), \qquad \Phi_2(u_1, u_2) = (u_1, \phi_2(u_2)); \qquad (u_1, u_2) \in V(G).$$

Then $\Phi_2\,\Phi_1(x_1, x_2) = (y_1, y_2)$.

For an arbitrary connected graph G, we let $\mathbf{C}$ be the collection of cut sets of G having precisely $\kappa(G)$ vertices. If $C \in \mathbf{C}$, the components of $G[V - C]$ are the *parts of G with respect to* (w.r.t.) C. We let

$$p(G) = \min\{\min\{\,|V(P)| : P \text{ is a part w.r.t. } C\} : C \in \mathbf{C}\}.$$

If P is a part of G and $|V(P)| = p(G)$, then P is an *atomic part* of G. Corresponding to an atomic part P is a unique $C \in \mathbf{C}$ called the *minimum cut set* of P. If $c \in C$, then c is adjacent to a vertex of P and a vertex of $V(G) - (V(P) \cup C)$. Conversely, any vertex with this property belongs to C.

The set of vertices adjacent to a given vertex x is a minimum cut set if and only if $\{x\}$ is an atomic part. Thus

LEMMA 1. *If G is regular and connected,*

$$\kappa(G) < \rho(G) \Leftrightarrow p(G) \ge 2.$$

If $G = H_1 \circ H_2$, the graph G may also be thought of intuitively as being obtained from the graph H_1 by replacing each element of $V(H_1)$ by a copy of H_2 and each edge $[x, y]$ of H_1 by all the $|V(H_2)|^2$ possible edges joining the copies of H_2 which replace x and y. The theory presented in [4] tells us that when $G \in \mathbf{H}$, the atomic parts of G are precisely the $|V(H_1)|$ disjoint copies of H_2. This affords the following characterization of $\mathbf{H}$:

Let G be a hypoconnected vertex-transitive graph with the property that there exists an atomic part P of G and a vertex $a \in V(P)$ such that a is adjacent to every vertex in the minimum cut set C of P. Then $G \in \mathbf{H}$, and conversely.

The complement of a graph G is denoted by $\bar{G}$. It has been remarked [1, p. 693] that if $G = H_1 \circ H_2$, then $\bar{G} = \bar{H}_1 \circ \bar{H}_2$.

THEOREM 2. *Let $G = H_1 \circ H_2$ where H_1 and H_2 are connected, vertex-transitive, and not complete. Then*

(a) *$\bar{G} \in \mathbf{H}$ if and only if $\bar{H}_1$ is connected.*

(b) *$\bar{G}$ is the union of two or more disjoint isomorphic copies of a member of* $\mathbf{H}$ *if and only if $\bar{H}_1$ in not connected and either*

(i) *the components of $\bar{H}_1$ are not complete, or*

(ii) *$\bar{H}_2$ is not connected.*

PROOF: By Theorem 1, $\bar{G} \in \bar{H}$ only if $\bar{H}_1$ is connected.

Since H_i is vertex-transitive, so is $\bar{H}_i$ $(i = 1, 2)$. To prove the "if" parts of both conclusions (a) and (b), we consider $\bar{H}_1$ to be the union of disjoint isomorphic copies of some connected, vertex-transitive graph M. (Thus when $\bar{H}_1$ is connected, there is just one such copy. By hypothesis, $\bar{H}_1$ is not complete.) Then $\bar{G}$ will be the union of one of more disjoint copies of $M \circ \bar{H}_2$. It will suffice to demonstrate that $M \circ \bar{H}_2 \in \mathbf{H}$.

If $\bar{H}_2$ is connected, the result is immediate. Assume therefore that $\bar{H}_2$ is the union of disjoint isomorphic copies of some connected, vertex-transitive graph N. Let L be formed from $M \circ \bar{H}_2$ by identifying vertices of the form (z, x), $(z, y) \in V(M \circ \bar{H}_2)$, where $[x, y] \in E(\bar{H}_2)$, and by removing any loops formed by this identification. It is clear that L is connected, vertex-transitive, and not complete. By definition, $L \circ N \in \mathbf{H}$. But it is readily seen that $L \circ N$ is isomorphic to $M \circ \bar{H}_2$.

To complete the proof of (b), we may assume by Theorem 1, since $\bar{G}$ is not connected, that it is the union of at least two disjoint copies of $M \circ \bar{H}_2$. If it holds that both M is complete and $\bar{H}_2$ is connected, then $M \circ \bar{H}_2$ is readily seen (by techniques similar to those used in proving Theorem 1, for example) not to be hypoconnected. Thus $M \circ \bar{H}_2 \notin \mathbf{H}$, and the theorem is proved.

Suppose now that $H_0, H_1, \ldots, H_m$ are regular graphs and that H_0 is not complete. Let $v_i = |V(H_i)|$, and let

$$G = H_0 \circ (H_1 \circ (H_2 \circ (\cdots \circ H_m) \cdots)).$$

Repeated applications of (5) together with Theorem 1 give

$$\frac{\rho(G)}{\kappa(G)} = \frac{1}{\kappa(H_0)} \left[\rho(H_0) + \frac{1}{v_1} \rho(H_1) + \frac{1}{v_1 v_2} \rho(H_2) + \cdots + \frac{1}{v_1 v_2 \cdots v_m} \rho(H_m) \right]. \tag{7}$$

This quantity is maximized when H_0 is a circuit and $H_1, H_2, \ldots, H_m$ are all complete. The right-hand side of (7) then becomes

$$\frac{1}{2}\left[2+\frac{v_1-1}{v_1}+\frac{v_2-1}{v_1v_2}+\cdots+\frac{v_m-1}{v_1\cdots v_m}\right]$$
$$=\frac{1}{2}\left[2+\left(1-\frac{1}{v_1}\right)+\left(\frac{1}{v_1}-\frac{1}{v_1v_2}\right)+\cdots+\left(\frac{1}{v_1\cdots v_{m-1}}-\frac{1}{v_1\cdots v_m}\right)\right]$$
$$=\frac{3}{2}-\frac{1}{2v_1v_2\cdots v_m}.$$

If H_0 is not complete and $v_i \geq 2$ for some $i \geq 1$, then $G \in \mathbf{H}$.

The following theorem is one inference possible from these calculations.

THEOREM 3. *The least upper bound of the set* $\{\rho(G)/\kappa(G) : G \in \mathbf{H}\}$ *is* 3/2, *and this bound is never attained.*

(A generalization of Theorem 3 to the class of all vertex- transitive connected graphs is obtained in [4] by different methods.)

3. A Way to Generalize H

Not all hypoconnected vertex-transitive graphs are of the form $H_1 \circ H_2$. It follows from [4, Theorem 2] however that the others can be obtained from $G = H_1 \circ H_2$ by carefully deleting edges (4) for which $[x_1, y_1] \in E(H_1)$.

More precisely, suppose H_1 and H_2 are connected vertex-transitive graphs and suppose H_1 is not complete and $|V(H_2)| \geq 2$. Let k be an integer satisfying

$$0 \leq k \leq \rho(H_2)/\rho(H_1). \tag{8}$$

Let $\mathbf{U} = \{U_x : x \in V(H_2)\}$ be a sequence of k-subsets of $V(H_2)$, not necessarily distinct, but such that

$$\text{each } x \in V(H_2) \text{ is in precisely } k \text{ members of } \mathbf{U}, \tag{9}$$

and

$$\begin{aligned}&\text{for each } x, y \in V(H_2) \text{ there is an automorphism } \phi \in \Gamma(H_2)\\ &\text{such that } \phi(x) = y \text{ and } \phi(U_z) = U_{\phi(z)} \text{ for each } z \in V(H_2).\end{aligned} \tag{10}$$

If $\mathbf{U}$ satisfies, moreover, the condition

$$x \in U_y \Leftrightarrow y \in U_x \qquad \text{for all} \qquad x, y \in V(H_2),$$

we form the graph $G' = H_1 * H_2$ as follows:

$$V(G') = V(H_1) \times V(H_2),$$

and $[(x_1, x_2), (y_1, y_2)] \in E(G')$ if and only if

$$x_1 = y_1 \qquad \text{and} \qquad [x_2, y_2] \in E(H_2), \tag{11}$$

or

$$[x_1, y_1] \in E(H_1) \qquad \text{and} \qquad y_2 \notin U_{x_2}.$$

In this case $\rho(G') = \rho(H_1)(|V(H_2)| - k) + \rho(H_2)$ and $\kappa(G') \leq \rho(H_1)|V(H_2)|$. Since (8) is satisfied, G' is hypoconnected. The proof of vertex-transitivity is straightforward and is omitted.

Some special cases can be constructed as follows. For $V(H_1)$ we take ordered r-tuples of integers $(n_1, n_2, \ldots, n_r)$ where the jth coordinate is always read modulo m_j, and $m_j \geq 4$ $(j = 1, \ldots, r)$. We join $(n_1, \ldots, n_r)$ and $(n_1', \ldots, n_r')$ by an edge if and only if

$$n_j + 1 \equiv n_j' \pmod{m_j} \qquad \text{for some} \qquad j$$

and

$$n_i = n_i' \qquad \text{for} \qquad i = 1, \ldots, j-1, j+1, \ldots, r.$$

Thus H_1 is the Cartesian product $C_1 \times C_2 \times \cdots \times C_r$, where each C_i is a polygon of length m_i. Let H_2 and $\mathbf{U}$, satisfying (9) and (10), be as above. Let $V(G) = V(H_1) \times V(H_2)$ and let $[(x_1, x_2), (y_1, y_2)] \in E(G)$ if and only if (11) holds or if

$$y_1 - x_1 = (0, \ldots, 0, 1, 0, \ldots, 0) \qquad \text{and} \qquad y_2 \notin U_{x_2}.$$

Less specifically, the toroidal lattice H_1 in this example may be replaced by what reduces to a Klein bottle or a projective plane lattice when $r = 2$, yielding further examples of hypoconnected vertex-transitive graphs.

Acknowledgments

The author gratefully acknowledges some conversations with Professors C. St. J. A. Nash-Williams and Donald Miller which helped to realize this paper. Gratitude is also due to Professor Nash-Williams for his very conscientious reading of this manuscript prior to publication.

References

1. G. Sabidussi, The Composition of Graphs, *Duke Math J.* **26** (1959), 693–696.
2. J. Turner, Point-Symmetric Graphs with a Prime Number of Points, *J. Combinatorial Theory* **3** (1967), 136–145.
3. W. T. Tutte, *Connectivity in Graphs*, Univ. of Toronto Press, Toronto, 1966.
4. M. E. Watkins, Connectivity of Transitive Graphs, *J. Combinatorial Theory* (to appear).
5. H. Whitney, Congruent Graphs and the Connectivity of Graphs, *Amer. J. Math.* **54** (1932), 150–168.

MAXIMUM FAMILIES OF DISJOINT DIRECTED CUT SETS

D. H. Younger

DEPARTMENT OF COMBINATORICS AND OPTIMIZATION
UNIVERSITY OF WATERLOO
WATERLOO, ONTARIO

1. Introduction

We conjecture that for any finite directed graph G, a maximum family Δ of disjoint directed cut sets in G has cardinality equal to that of a minimum set of edges ϕ that meets every directed cut set of G. The equality is illustrated for the graph shown in Fig. 1, in which $\{\{\alpha_1, \alpha_4\}, \{\alpha_2, \alpha_5\}, \{\alpha_6, \alpha_7\}, \{\alpha_{10}, \alpha_{12}\},$

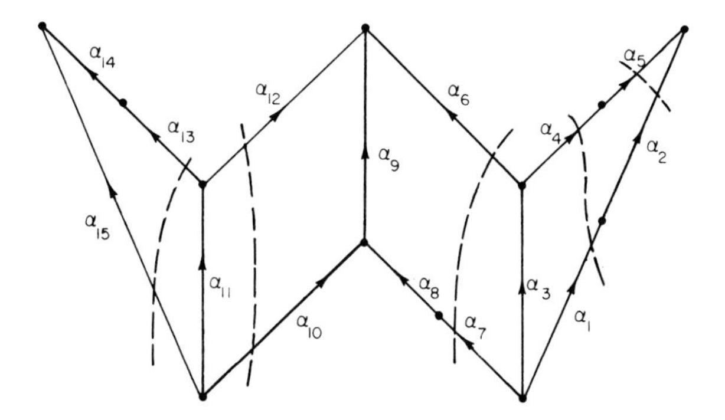

Fig. 1.

$\{\alpha_{13}, \alpha_{15}\}\}$ is a family Δ and $\{\alpha_1, \alpha_2, \alpha_6, \alpha_{12}, \alpha_{15}\}$ is a set ϕ. Since $|\phi| = |\Delta|$, Δ is indeed maximum and ϕ minimum.

In this paper we do not decide the conjecture but do establish that if the minimax equality holds for each proper contraction of G then it holds for G or there is in G a maximum disjoint directed cut set family Δ that consists solely of stars. The proof specifies a directed cut set in G of a type that can be found provided G has no maximum Δ consisting solely of stars. By contracting all edges on one side of this cut set, and then on the other, a pair of proper contractions of G is formed. Each contraction in this pair has a ϕ and a Δ that are equicardinal. The union of the ϕ's and the union of the Δ's are equicardinal and constitute a ϕ and a Δ for graph G.

2. Directed Cut Sets

Let G be a finite directed graph with vertex set vG and edge set γG. The *coboundary* δx of $x \subseteq vG$ is the set of edges each with one end in x, the other in $vG - x$; δ is a function that maps a subset of vG onto a subset of γG. We also apply the term coboundary and the symbol δ to a set of edges that is a value of the coboundary function. For a coboundary δ, an *argument* of δ is a minimum set x such that δ is the coboundary of x. If an argument of δ has cardinality 1, then δ is a *star*. A *cut set* is a minimal nonnull coboundary.

Graph G is *connected* if $\delta x \neq \varnothing$ for each nonnull proper subset x of vG. A *connected component* of G is a maximal nonnull connected subgraph of G.

In this section we record without proof basic properties relevant to the succeeding argument.

2.1. Let δx and δy be coboundaries in connected graph G; if $\delta x = \delta y$, then $x = y$ or $x = vG - y$.

2.2. A cut set contains edges from exactly one connected component of G.

For a directed edge α, one of its ends is specified the positive end $p\alpha$, the other the negative end $n\alpha$. If each edge of a coboundary δx has its negative end in x, or if each edge of δx has its positive end in x, then δx is a *directed coboundary*. A *directed cut set* is a cut set that is a directed coboundary.

For directed coboundary δ in connected graph G, $v^-\delta$ is the subset of vG whose coboundary is δ and which contains the negative end of each edge in δ. Let $v^+\delta = vG - v^-\delta$. By 2.1, $v^-\delta$ and $v^+\delta$ are uniquely defined.

2.3. A directed coboundary is expressible as the union of a family of disjoint directed cut sets.

2.4. If ζ and η are directed coboundaries, then $\delta(v^-\zeta \cap v^-\eta) = \delta_i$ and $\delta(v^-\zeta \cup v^-\eta) = \delta_u$ are directed coboundaries; moreover

$$\zeta \cap \eta = \delta_i \cap \delta_u, \qquad \zeta \cup \eta = \delta_i \cup \delta_u.$$

The operation of *contraction of an edge* consists of deletion of that edge from the graph and coalition of its ends. An *elementary contraction* of graph G

is a graph derived from G by contracting an edge; the graph derived by contraction of edge α is indicated by G_α. A *contraction* of G is a graph extracted from G by a sequence of edge contractions; if that *sequence is* nonempty, the extracted graph is called a *proper contraction* of G.

2.5. If $\delta \subseteq \gamma G$ and $\alpha \in \gamma G - \delta$, then δ is a directed cut set of G iff δ is a directed cut set of G_α.

3. The Minimax Relation

Let Δ be a family of directed cut sets of G that is pairwise disjoint. Let ϕ be a set of edges which meets every directed cut set of G.

3.1. For any ϕ and Δ, $|\phi| \geq |\Delta|$; consequently, if $|\phi| = |\Delta|$, then ϕ is minimum and Δ maximum.

Proof: Each cut set in Δ must meet ϕ. The cut sets in Δ are disjoint and so $|\phi| \geq |\Delta|$.

When equality holds, ϕ is a set of representatives of Δ.

3.2. Let G_α be an elementary contraction of G for which the minimax equality holds; if a maximum disjoint directed cut set family Δ_α for G_α is not maximum for G, then α is an element of a minimum set ϕ for G.

Proof: Let ϕ_α be a minimum edge set that meets each directed cut set of G_α. By 2.5, $\phi_\alpha \cup \{\alpha\}$ meets each directed cut set of G. Now $|\phi_\alpha| = |\Delta_\alpha|$ and so $|\phi_\alpha \cup \{\alpha\}| = |\Delta_\alpha| + 1 = |\Delta|$, where Δ is a maximum family for G. By 3.1, $\phi_\alpha \cup \{\alpha\}$ is a minimum set that meets every directed cut set of G.

Assume hereafter that G is connected; this simplifies notation but is not essential to the argument, in view of 2.2.

Two directed cut sets ζ, η *cross* if both $v^-\zeta - v^-\eta$ and $v^-\eta - v^-\zeta$ are nonnull. A family Δ of disjoint directed cut sets is *laminar* if no pair of its elements cross.

3.3. For any graph, there is a maximum family of disjoint directed cut sets that is laminar.

Proof: Let Δ be a maximum family that has a minimum number of pairs of cut sets which cross. Suppose, however, that there are ζ, η in Δ such that $v^-\zeta - v^-\eta \neq \varnothing$ and $v^-\eta - v^-\zeta \neq \varnothing$. By 2.4, $\delta_i = \delta(v^-\zeta \cap v^-\eta)$ and $\delta_u = \delta(v^-\zeta \cup v^-\eta)$ are nonnull directed coboundaries. Thus each is expressible as the union of a family of disjoint directed cut sets: $\delta_i = \bigcup \Delta_i$ and $\delta_u = \bigcup \Delta_u$. Since $\delta_i \cup \delta_u = \zeta \cup \eta$ and $\delta_i \cap \delta_u = \zeta \cap \eta$ by 2.4, thus $\Delta' = (\Delta - \{\zeta, \eta\}) \cup \Delta_i \cup \Delta_u$ is a disjoint directed cut set family for G. Since Δ is maximum for G and $|\Delta_i| \geq 1$, $|\Delta_u| \geq 1$, thus $\Delta_i = \{\delta_i\}$ and $\Delta_u = \{\delta_u\}$. So δ_i and δ_u are cut sets and Δ' is maximum for G. For any $\delta \in \Delta - \{\zeta, \eta\}$, if δ crosses one or both of δ_i, δ_u, then it crosses one or both of ζ, η. Thus the number of pairs of cut sets in Δ' that cross is less than the number which cross in Δ, a contradiction. Hence no pairs in Δ cross, i.e., Δ is laminar.

For Δ_0 a disjoint directed cut set family for graph G, and $\delta_0 \in \Delta_0$, let $\Delta_0{}^-$ be those $\delta \in \Delta_0$ for which either $v^-\delta$ or $v^+\delta$ is a subset of $v^-\delta_0$; let $\Delta_0{}^+$ be those for which either $v^-\delta$ or $v^+\delta$ is a subset of $v^+\delta_0$. Now $\Delta_0{}^-$ and $\Delta_0{}^+$ have one element in common, namely δ_0. Let G^- be the graph derived from G by contracting each edge having both ends in $v^+\delta_0$. Note that G^- is a proper contraction of G if $|v^+\delta_0| > 1$. Let G^+ be the graph derived from G by contracting each edge having both ends in $v^-\delta_0$.

3.4. If Δ_0 is a maximum cut set family for G that is laminar, and $\delta_0 \in \Delta_0$, then $\Delta_0{}^-$ and $\Delta_0{}^+$ are maximum cut set families for G^- and G^+, respectively; conversely, if Δ^- is a maximum cut set family for G^-, then $(\Delta_0 - \Delta_0{}^-) \cup \Delta^-$ is maximum for G, and similarly for a maximum family Δ^+ for G^+.

Proof: Every element in $\Delta_0{}^-$ is a cut set in contraction G^-, by 2.5. Thus $\Delta_0{}^-$ is a disjoint directed cut set family for G^-. If Δ^- is a maximum family for G^- then, since Δ_0 is laminar, $\Delta_1 = (\Delta_0 - \Delta_0{}^-) \cup \Delta^-$ is a disjoint family for G. Since $|\Delta_1| \leq |\Delta_0|$, thus $|\Delta^-| \leq |\Delta_0{}^-|$. Conversely, since $|\Delta^-| \geq |\Delta_0{}^-|$, thus $|\Delta_1| \geq |\Delta_0|$. So $\Delta_0{}^-$ and Δ_1 are maximum for G^- and G, respectively. This proof with superscript "+" in place of "−" establishes that $\Delta_0{}^+$ and $(\Delta_0 - \Delta_0{}^+) \cup \Delta^+$ are maximum for G^+ and G.

3.5. In any directed graph G there is a maximum laminar family Δ_0 that is composed solely of stars or else that contains an element δ_0 other than a star which either is in every maximum laminar family of G^- or in every maximum laminar family of G^+.

Proof: Let $\Delta\backslash$ be the collection of disjoint directed cut set families for G that are of maximum cardinality and laminar. Let $\Delta\backslash_0$ consist of those families $\Delta \in \Delta\backslash$ such that:

(1) The maximum cardinality of an argument of any cut set in Δ is minimum over $\Delta\backslash$. Call that cardinality m.

(2) Among $\Delta \in \Delta\backslash$ satisfying (1), the number of cut sets in Δ with arguments having cardinality m is minimum.

Let $\delta_0 \in \Delta_0 \in \Delta\backslash_0$ be such that the argument of δ_0 has cardinality m. We may take $|v^-\delta_0| = m$. Suppose that $m > 1$. By the conditions (1) and (2) imposed on Δ_0 and 3.4, a maximum laminar family for G^- must contain a cut set with argument in G of cardinality m. Since that argument must be a subset of $v^-\delta_0$ and since $|v^-\delta_0| = m$, that argument is $v^-\delta_0$. Thus a maximum laminar cut set family for G^- must contain δ_0.

3.6. If the minimax equality holds for each proper contraction of graph G, then it holds for G or there is in G a maximum family of disjoint directed cut sets that consists solely of stars.

PROOF: If G has no maximum cut set family consisting solely of stars, let Δ_0 be a maximum laminar cut set family of G containing a δ_0 other than a star that is an element of every maximum laminar cut set family of G^-; G^+ is a proper contraction of G and $\Delta_0{}^+$ is a maximum cut set family for G^+ by 3.4. Thus there is a ϕ^+ satisfying $|\phi^+| = |\Delta_0{}^+|$ that meets every directed cut set of G^+. For $\alpha \in \phi^+ \cap \delta_0$, let ϕ^- be a minimum such set for G^- that contains α; ϕ^- exists by 3.2, and satisfies $|\phi^-| = |\Delta_0{}^-|$ since G^- is a proper contraction of G. Thus $|\phi^- \cup \phi^+| = |\Delta_0{}^-| + |\Delta_0{}^+| = |\Delta_0|$.

We next show that $\phi^- \cup \phi^+$ meets every directed cut set of G. If ζ is a directed cut set of G, then $\delta_i = \delta(v^-\zeta \cap v^-\delta_0)$ and $\delta_u = \delta(v^-\zeta \cup v^-\delta_0)$ are directed coboundaries of G^- and G^+, respectively. For δ_i null, $v^-\zeta$ is disjoint from $v^-\delta_0$, and so is a subset of $v^+\delta_0$; thus ζ is in G^+ and meets ϕ^+. For δ_u null, ζ is in G^- and so meets ϕ^-. For δ_i and δ_u both nonnull, $\delta_i \cap \phi^- \supseteq \{\alpha^-\} \neq \varnothing$ and $\delta_u \cap \phi^+ \supseteq \{\alpha^+\} \neq \varnothing$ for edges α^-, α^+. If $\alpha^- = \alpha^+$ then, since $\delta_i \cap \delta_u \subseteq \zeta$ by 2.4, $\alpha^- \in \zeta$ and so ϕ^- meets ζ. If $\alpha^- \neq \alpha^+$, then

$$\{\alpha^-, \alpha^+\} \subseteq (\delta_u \cap \phi^-) \cup (\delta_i \cap \phi^+) \subseteq (\delta_u \cup \delta_i) \cap (\phi^- \cup \phi^+)$$
$$= (\zeta \cup \delta_0) \cap (\phi^- \cup \phi^+).$$

Since $|\delta_0 \cap (\phi^- \cup \phi^+)| = 1$ by virtue of $|\phi^-| = |\Delta_0{}^-|$ and $|\phi^+| = |\Delta_0{}^+|$, thus $\zeta \cap (\phi^- \cup \phi^+) \neq \varnothing$. Thus under the assumption that G has no maximum directed cut set family consisting solely of stars, the minimax equality holds for G.

ABSTRACTS

The No-Three-in-Line Problem*

RICHARD K. GUY AND PATRICK A. KELLY

An old puzzle (Dudeney, Rouse Ball) can be generalized to ask for sets of $2n$ points chosen from an $n \times n$ array of n^2 points of the unit lattice with no three of them in a straight line. Some evidence and a probabilistic argument were given in support of the following conjectures.

CONJECTURE 1. There is no solution which has the symmetry of a rectangle, without also having the full symmetry of the square (?).

CONJECTURE 2. If $n > 10$, there is no solution having the full symmetry of the square (?).

CONJECTURE 3. There is only a finite number of solutions to the no-three-in-line problem (?).

CONJECTURE 4. For large n, one may expect to be able to choose approximately $(2\pi^2/3)^{1/3}\, n$ points with no three in line, but no larger number (?).

A Packing Theory†

DAVID A. KLARNER

Let A be a set and let $\mathscr{A}$ be a collection of subsets of A. Conditions are given that must hold if a partition of A is a subset of $\mathscr{A}$. The main idea presented is a generalization of several methods that have been used to prove certain packing theorems.

* To be published in the *Canadian Mathematical Bulletin*.
† This paper will appear in the *Journal of Combinatorial Theory*.

Brooks' 2 × 1 Simple Perfect Rectangle

W. T. Tutte

A perfect rectangle is a dissection of a rectangle into a finite number, exceeding one, of unequal squares. The number of these squares is the *order* of the perfect rectangle. A perfect rectangle is *simple* if it contains no smaller perfect rectangle. Examples of simple perfect rectangles with equal sides (perfect squares) have long been known, but the problem of finding a simple perfect rectangle with sides in the ratio 2 : 1 has hitherto remained unsolved. Recently, however, R. L. Brooks has discovered a simple 2 × 1 perfect rectangle of order 1323.[1]

UNSOLVED PROBLEMS

Unsolved Problems

1. Is it true that every finite undirected graph is the representing graph of a family of convex sets of the two-dimensional space? of the three-dimensional space?

2. For a knot of crossing number n, what is the smallest genus of a torsion-free surface of R^3 upon which it is possible to place the knot in an alternating way?

C. BERGE

3. A *unitary* array of integers has the property that any square submatrix whose upper left corner falls on the boundary of the array has a determinant equal to 1.

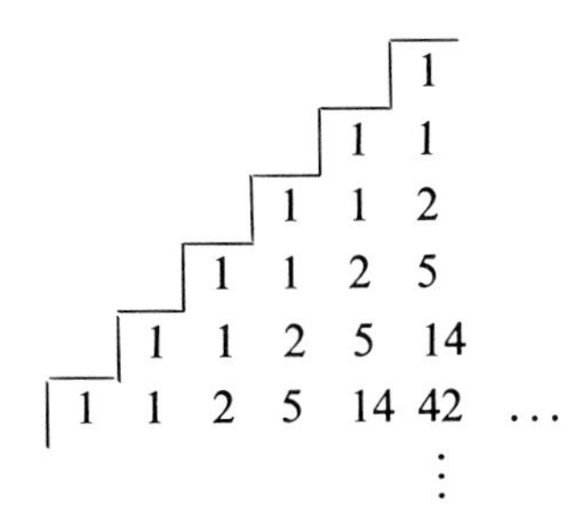

FIG. 1

For example, in Fig. 1,

$$\begin{vmatrix} 1 & 2 & 5 \\ 2 & 5 & 14 \\ 5 & 14 & 42 \end{vmatrix} = 1.$$

It is easily shown that there exists a unique unitary array having any given infinite monotonic boundary (i.e., the boundary goes only upward and rightward and reaches infinitely low and infinitely far to the right). It is also possible to show that the numbers are all positive and nondecreasing. The array corresponding to the staircase boundary is shown in Fig. 1. The values of the integers in any row of this array are the Catalan numbers,

$$\frac{1}{n+1}\binom{2n}{n}.$$

The array corresponding to the square array in Fig. 2 gives the binomial coefficients as in Pascal's triangle.

1	1	1	1	1	1
1	2	3	4	5	—
1	3	6	10	15	—
1	4	10	20	35	—

FIG. 2.

A *periodic quasilinear boundary* represents the best stair-step approximation to a straight line of rational slope. The boundary of Fig. 1 approximates a straight line of slope 1. Exact formulas are known for the values of the numbers in the unitary arrays generated by periodic quasilinear boundaries of slopes $1/n$ or n, but no such formulas are known (to me) for the values in the arrays with boundaries of slopes m/n where $1 < m < n$. The simplest such case is slope 2/3, as shown in Fig. 3.

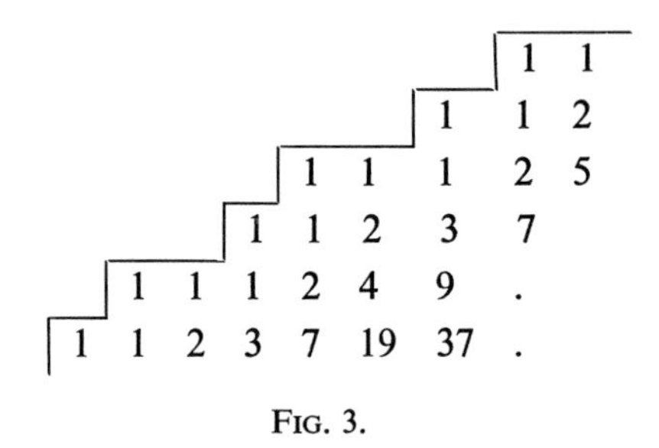

FIG. 3.

More partial results, including an easier way of finding the numbers in the array, are available from me upon request.

4.[1] Consider the two-person perfect-information game whose state is represented by a positive integer n. The player whose turn it is moves to the next state by adding *or* subtracting to n the greatest perfect square less than or equal to n, i.e.,

$$n \rightarrow n \pm [\sqrt{n}]^2.$$

A player wins if he manages to move to zero.

For example, if A moves from the initial state of 92, he can force a win as follows:

[1] Problem 4 is that of R. Epstein, Hughes Aircraft Co.

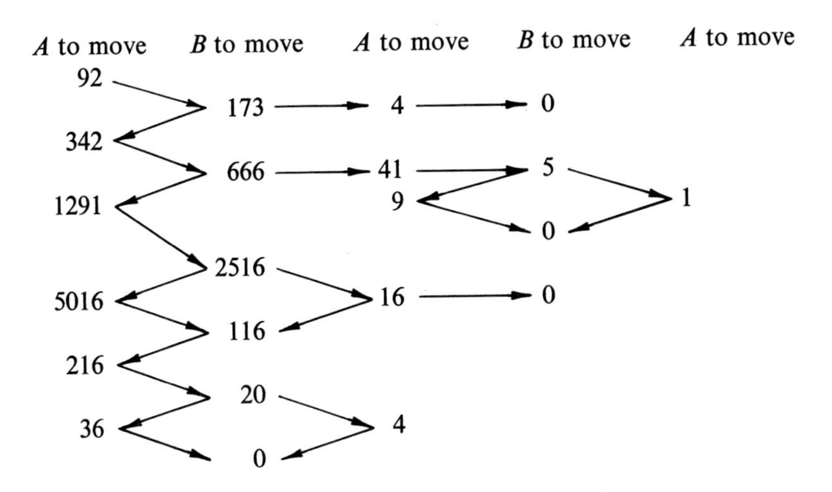

The problem is to exhibit an algorithm for deciding whether the first player should win, lose, or draw if the game is started from state n and both players play optimally. The answers are known for all $n \leq 2000$ (they are mostly draws, because each player usually has the option of forcing the other into the 2–3–2–3–$\cdots$ loop), but it is not easy to prove that any explicit algorithm will decide the value of the game for large n in a finite number of steps.

E. R. BERLEKAMP

5. CONJECTURE: Every planar, 3-connected, 3-valent, even (i.e., bipartite) graph admits a Hamiltonian circuit. (I heard the conjecture from David Barnette, though it seems that other people arrived at it independently.)

6. CONJECTURE: For every triangulation of each orientable surface it is possible to color the edges by three colors in such a fashion that the edges of each triangle have three different colors. [The conjecture cannot be generalized to nonorientable surfaces. If four colors are available, a coloring of the required type is trivially possible for all surfaces (orientable or not). Obviously, the truth of the conjecture for the sphere is equivalent to an affirmative solution of the 4-color problem.]

BRANKO GRÜNBAUM

7. A *cross-cut* of an ordered set P is a subset C of P such that

(i) C is a maximal antichain in P, and
(ii) if $x < x' \in C$, $y > y' \in C$, and $x = y$ then $x = z = y$ for some z in C.

Let X be any set and let $\mathscr{E}(X)$ be the set of all subsets of X, ordered by inclusion. If C is a cross-cut of $\mathscr{E}(X)$, let the dual C^* of C be the cross-cut which consists of all complements of members of C. Say that C is *trivial* if either C or C^* consists of all subsets of X having the same finite cardinality.

PROBLEM. Does $\mathscr{E}(X)$ have any nontrivial cross-cuts (for X infinite)?

[Using the generalized continuum hypothesis, it may be shown that all the sets in a cross-cut of $\mathscr{E}(X)$ have the same cardinality.]

D. A. HIGGS

8. Given a graph G and a positive integer k, let T be the set of all paths of G and let S be a subset of T. Let $\{H_\alpha\}$ be the set of graphs such that for each H_α: (a) every edge of H_α is a *path* of G; (b) every vertex of H_α is a vertex of G; and (c) for every path P in S there is a set of k or fewer edges in H_α whose union (as sets of edges of G) is the set of edges in P and which form a path in H_α from the first vertex of P to the last vertex of P.

Then:

A. Find $H \in \{H_\alpha\}$ which minimizes the sum of the total number of edges of G contained in edges of H (where each edge of G is counted as many times as it appears in different edges of H).

B. Find $H \in \{H_\alpha\}$ which minimizes the sum of the total number of edges of G contained in edges of H plus the total weighted number of vertices of H, with the vertices of H weighted by positive numbers dependent on their degrees in H.

Note that the path length limitation k is necessary in Problem 8A for the problem to be interesting. Furthermore, each H_α may include parallel edges, each tracing a different path in G. This problem has important applications in communications, and I believe it has some mathematical interest as well.

9. For what values of $n \geq 12$ is there a planar graph without loops or multiple edges which has 12 vertices of degree 5 and $n - 12$ vertices of degree 6?

It is easy to show that such a graph exists for $n = 12$ and that none exists for $n = 13$.

A. M. HOBBS

10. A *polytope* is the convex hull of a finite set of points in a vector space over an ordered field.[2] With each polytope P there is associated $F(P)$, the

[2] Not necessarily Archimedean.

lattice of all faces of P. Two polytopes P and Q are said to be *equivalent* provided that $F(P)$ and $F(Q)$ are isomorphic.

PROBLEM. Is every polytope equivalent to one in a *real* vector space?

VICTOR KLEE

11.[3] $P_1, P_2, \ldots, P_n$ are points of the unit square S and P_1 is at the bottom left-hand corner of S. Are there nonoverlapping rectangles $R_1, \ldots, R_n$ contained in S, with sides parallel to the sides of S so that P_i is at the bottom left-hand corner of R_i and the sum of the areas of the R_i exceeds $\frac{1}{2}$?

12. Let $f(m, l)$ denote the maximum number n such that there are sets $A_1, \ldots, A_n \subset \{1, 2, \ldots, m\}$, $|A_1| \leq l$, $A_i \not\subset A_j$ $(i \neq j)$, $A_i \cap A_j \cap A_k \neq \emptyset$, and $A_1 \cap A_2 \cap \ldots \cap A_n = \emptyset$.

It is conjectured [1] that if $l \leq \frac{1}{2}m$, $f(m, l) = \max(N_1, N_2)$ where

$$N_1 = 4\binom{m-4}{l-3} + \binom{m-4}{l-4}, \qquad N_2 = 2 + \binom{m-2}{l-2} - \binom{m-l-1}{l-2}.$$

This has been verified in the case $l = 4$ (see [2]) when $\max(N_1, N_2) = N_1 = 4m - 15$ and the same type of argument will probably work for other small values of l.

13. Let R_α denote the set of all 0, 1 sequences $(\varepsilon_\nu)_{\nu < \omega_\alpha}$ such that $\varepsilon_{\nu_0} \neq 0$ for some ν_0 and $\varepsilon_\nu = 0$ $(\delta \leq \nu < \omega_\alpha)$ for some $\delta < \omega_\alpha$. The order type of R_α ordered lexicographically is η_α. Erdös, Rado, and I have found a number of partition relations involving the η_α but we cannot prove or disprove the relation

$$(?) \qquad \eta_\omega \to (3, \eta_\omega)^2$$

In other words, if $G = (R_\omega, E)$ is any graph on R_ω without circuits of length 3, is there necessarily an independent set of vertices $S \subset R_\omega$ so that S also has order type η_ω (with the order relation induced by R)?

REFERENCES

1. A. J. W. HILTON and E. C. MILNER, On Some Intersection Theorems for Systems of Finite Sets, *Quart. J. Math. Oxford Ser.* (2) **18** (1967), 369–84.
2. R. K. GUY and E. C. MILNER, Graphs Defined by Coverings of a Set, *Acta Math. Acad. Sci. Hungar.* (1968).

E. C. MILNER

[3] Problem 11 is that of A. R. Freeman, Simon Frazer University.

14. Let D be a directed graph with n (≥ 3) vertices and no loops, and in which there do not exist two distinct edges λ, μ such that the tail (initial vertex) of λ is equal to the tail of μ and the head (terminal vertex) of λ is equal to the head of μ. If the in-degrees and out-degrees of the vertices of D are all greater than or equal to $n/2$, it is conjectured that D must have two edge-disjoint directed Hamiltonian circuits. (The existence of the one directed Hamiltonian circuit in these circumstances follows from a theorem of Ghouila–Houri.)

C. St. J. A. Nash-Williams

15. Conjecture: If $A_1, A_2, \ldots, A_n$ are different finite sets having empty intersection then there are are at least n different differences $A_i - A_j$ (a variation of an arithmetical problem of Graham).

J. Schönheim

16. Let G be a vertex-transitive graph with $|V(G)| = p_1 p_2$ where p_1 and p_2 are distinct primes. Does there exist an element $\phi_i \in \Gamma(G)$, the automorphism group of G, which is of order p_i $(i = 1, 2)$?

If so, we can probably generalize the result for when $|V(G)|$ is a product of n distinct primes. (By the way, the answer is in the affirmative if $n = 1$, as shown by James Turner [1].)

Reference

1. J. Turner, *J. Combinatorial Theory* **3** (1967).

Mark E. Watkins

17. Let G be a nonplanar graph.

(a) If G is drawn in the plane using straight lines for the edges, what is the *maximum* possible number of crossings of edges? Give an algorithm for drawing a given graph with the maximum number of crossings and determine all "nonisomorphic" drawings that achieve the maximum number. (*Definition:* In determining whether two drawings of a given graph are isomorphic, consider each crossing as a new vertex and determine whether the resulting plane graphs are isomorphic.)

(b) Answer the same questions as in (a), when straight lines are not required, with the restriction that a pair of arcs is permitted to cross only once.

18. Let G be a graph with be edges and n nodes. (If you wish, for simplicity, assume G is 2-connected.) Weight each of the b edges with an arbitrary positive weight w_i for $i = 1, \ldots, b$. Let D be the diagonal matrix

$$D = \text{diag}\,(w_1, w_2, \ldots, w_b).$$

Let C be a cut set matrix of G with $(n - 1)$ independent rows and *be* columns. Define the matrix T by the congruence transformation

$$T = CDC' \tag{1}$$

It is easy to see that T is real and symmetric. It can be shown also that it is a paramount matrix, that is, a matrix of which every principal minor is not less than the absolute value of any minor formed from the same rows.

(a) What are the necessary and sufficient conditions on a real matrix T for it to satisfy Eq. (1)? If T is of order less than or equal to 3, it is known that paramountcy is a necessary and sufficient condition, but for higher orders the conditions are not known in an explicit form. [An algorithm, however, is known for testing a given matrix to determine whether it is decomposable as in (1).]

(b) Now let P be a principal submatrix of the inverse of T. What are the necessary and sufficient conditions on P, if it is of order greater than 3?

(c) Answer the same questions when L, a circuit matrix of G, is substituted for C.

Note: These are basic problems in the synthesis of networks from their resistance or conductance matrices. Solving the problem when C is used as the transformation matrix does not solve the problem when L is substituted for C. It is known that there are decompositions when C is used but the same matrix is not decomposable when L is substituted for C. Thus considerations of planarity are crucial in the above questions.

LOUIS WEINBERG